Christopher Duston and Ma

Editors

I0045192

# Spacetime Physics
# 1907-2017

Selected peer-reviewed papers presented at the First Hermann Minkowski Meeting on the Foundations of Spacetime Physics, 15-18 May 2017, Albena, Bulgaria

MINKOWSKI
Institute Press

Christopher Duston
Merrimack College
Department of Physics
*dustonc@merrimack.edu*

Marc Holman
University of Western Ontario
*mholman8@uwo.ca*

Cover: (https://en.wikipedia.org/wiki/Hermann_Minkowski#/media/File:
De_Raum_zeit_Minkowski_Bild_(cropped).jpg)


© Minkowski Institute Press 2019


ISBN: 978-1-927763-48-3 (softcover)
ISBN: 978-1-927763-49-0 (ebook)


Minkowski Institute Press
Montreal, Quebec, Canada
http://minkowskiinstitute.org/mip/


For information on all Minkowski Institute Press publications visit our website
at http://minkowskiinstitute.org/mip/books/

# PREFACE

The current volume is a selection of peer-reviewed contributions to the *First Hermann Minkowski Meeting on the Foundations of Spacetime Physics*, which took place in Albena, Bulgaria, 15-18 May 2017, commemorating the 110th anniversary of two key lectures on the physics and mathematics of four-dimensional spacetime, delivered by Minkowski in 1907. Contributing papers have been classified into four main categories, represented by separate parts of the volume (this classification is not meant to be absolute, but should rather serve as a rough navigational tool).

Part I focuses on a number of well-known themes (and variations thereof) in the foundations of special and general relativity, such as "general covariance" and the nature of gravitation. The fact that many of these themes indeed have a long and well-recorded history in the relativity literature should however not be taken to suggest that they have been fully explored, or that they are of little relevance for the often "fashionable pursuits" of modern-day cutting-edge research in particle physics and cosmology – quite to the contrary. In the opening contribution, Vesselin Petkov presents a strong case, based on Minkowski's above-mentioned lectures and the subsequent research program founded on it, for "geometrizing gravity", i.e., for interpreting gravitation as a curved spacetime phenomenon. Although to many relativists such a conclusion will probably not come as a great surprise, as Petkov recalls, it was *not* the position Einstein himself adopted (at least at some stage in his thinking). This implication of Minkowski's program, i.e., that gravitation ought to be viewed as a manifestation of spacetime curvature, is at odds with the typical view in particle physics, according to which gravity is exhaustively described as a spin-two field (in spite of the fact that the very notion of such a field - in contrast to the case of spin-one, say - in general only makes sense when spacetime is flat). Einstein's struggles with the concepts of "general covariance" and coordinate independence in his attempts to formulate a relativistic theory of gravitation have led to a classic foundational debate. In this tradition, Dennis Dieks argues that when it comes specifically to time coordinates and clocks, modern commentaries have not always appreciated the details of the background of Einstein's early reasoning. As Dieks points out, this has led to false claims about early relativistic derivations by Einstein and others, concerning in particular the gravitational redshift of light (i.e., the first "classic test" of general relativity proposed by Einstein in 1908, although not unambiguously confirmed as such until 1960 in the Pound-Rebka experiment). Next, Tian Yu Cao discusses the various meanings that have been attached to the term "background independence", especially in relation to its potential role in a theory of quantum gravity. Just as with the closely related notion of general covariance, this term means rather different things to different investigators and Cao distinguishes three main interpretative ideas, all discussed in the contemporary literature as capturing the essence of what it means for a theory to be background independent. In relativity theory, the notion of proper time is restricted to a curve, representing the motion of a particle with zero spatial extension. In his contribution, Uri Ben-Ya'acov explains how to compare the prima facie different notions of proper time that arise for systems with finite spatial extension, in particular those with nonzero acceleration. Jerrold Franklin then presents a self-consistent approach to the kinematics of rigid bodies in special relativity, thereby resolving the historical debate about the existence of such physical objects between practitioners such as Pauli, Einstein, and Born. The contribution from Hou Y. Yau addresses a common misconception that one can derive the Schwarzschild metric without direct reference to the Einstein equations. He points out that the usual description of this approach relies on the "plausible analogy" of the proper time $\tau$ of a freely falling observer with the Newtonian absolute time, but really implies the inclusion of additional information beyond the Newtonian framework. He proposes an alternative to this analogy by the consideration of the fictitious motions around a thin shell in special relativity, which provides a clear relationship between matter and spacetime. Finally, Nadja Magalhaes investigates the concept of time from a physicist's perspective, emphasizing the importance of a definition in terms of a physical clock - e.g., a subatomic particle. As a result, she arrives at a new view of time that is crucially based on such an operational definition.

Part II of the volume deals with the question of whether, colloquially put, Einstein "was right" with his theory of general relativity. Since the very birth of the theory in 1915, there has been no shortage of proposals for alternative theories of gravity and this had led, over the years, to ever-more sophisticated experimental tests of general relativity against putative successor theories. But time and time again, general relativity has emerged as a victor from such tests, unlike many of its rivals (the detection of gravitational waves in the centennial year 2015, which essentially killed off a whole class of alternative theories almost immediately, forms a perfect illustration of this). On the other hand, the continued failure of physicists to directly observe or otherwise positively identify "dark matter" and/or "dark energy" continues to fuel the search for physically viable alternatives to general relativity. To this end, Myrzakulov Ratbay and his collaborators consider a very generic modification of general relativity, which includes contributions from the Ricci scalar, the torsion scalar, and a scalar field. This approach encompasses many of the popular modifications to general relativity. They find a structure which resembles the Monge-Ampère equations, and also present the simplest solutions for the scale factor within a standard cosmological setting. Yakov Itin's contribution explores Einstein's original idea for the unification of electromagnetism with general relativity, teleparallelism. He makes progress by finding that the variational approach on a teleparallel manifold must take into account the (anti-)commutivity with certain operators (the Hodge dual, the interior product, and the coderivative), and applying it to a pair of illustrative examples - the Maxwell Lagrangian and the translation Lagrangian.

In the final contribution on the topic of alternative gravity theories, Jerzy Kijowski approaches the issue from a very different perspective. Based on the dynamical equivalence between so-called $f(R)$ models and standard general relativity augmented with additional "matter fields" (a result established by himself and, independently, by others, several decades ago), he puts forward an extended equivalence that applies to actions with a more complicated dependence on (full) curvature. More precisely, all "alternatives" to general relativity described by such actions can be formulated equivalently as ordinary general relativity plus additional matter fields through use of an appropriate Legendre transformation.

Part III of this volume deals with the application of general relativity to cosmology and astrophysics, which is the primary physical arena in which the theory can be tested and new results uncovered. From compact objects to the entire Cosmos, there are a variety of astrophysical situations where one must take the non-linear effects of Einstein's equations into account to create models capable of explaining physical phenomena. The contributions in this section specifically work towards a better understanding of the source and detectability of astrophysical gravitational waves, as well as refining early Universe models to take into account the abundance of new data available from observational astronomy.

Reinoud Slagter's contribution deals specifically with the unexpected observation of the alignment of quasar polarization vectors on Mpc scales. His suggestion is that this behavior could be the result of a warped five dimensional spacetime model in which the spontaneous breaking of conformal symmetry creates transient super-massive cosmic strings. This results in alternative solutions to what are generally regarded as major conundrums, such as dark energy, the cosmological constant, and the hierarchy problem. Following the discovery of gravitational waves, one of the next frontiers will be the observation of relic gravitational waves from the early Universe.

In their work, Carlos Frajuca and his collaborators determine what effect the self-interaction of the gravitational field will have on the detectability of the background waves, and find a frequency-dependent damping effect, relevant for some LISA detection scenarios. In the final paper of this section, Aizhan Myrzakul and Ratbay Myrzakulov explore the Hojman symmetry as an alternative fundamental symmetry, in the context of a cosmological model. Using an equation of state corresponding to a Chaplygin gas, they determine the equations of motion and conserved quantities for such a model, which corresponds to a de Sitter spacetime.

The final section of the volume focuses on the quantum nature of space-time. It has been known for some time that quantum field theory, the theoretical underpinning of the standard model of particle physics, and general relativity, are essentially incompatible. The simplest argument for this is that general relativity is not perturbatively renormalizable - that is, the basic tool used in quantum field theory to make predictions at particular energy scales fails for the gravitational field. It is a currently open question whether general relativity may still be renormalizable at a *non-perturbative* level - as some theorists have contemplated, even though others have pointed out that the only more-or-less well-understood renormalization framework is perturbative (and even there, possible inconsistency of course still lurks for theories that are not completely asymptotically free, such as the particle physics standard model). But there are many additional structural issues as well, such as the prima facie impossibility of applying the quantum superposition principle to different spacetime geometries, the role of background independence, the precise meaning of nonlocality, the different roles of geometry, the origin of causality and the classical limit (or measurement problem), and the interpretation of quantum theory more generally. The nature of quantum geometry is one of the great challenges for theoretical physics in the 21st century. While often characterized as "String Theory vs Loop Quantum Gravity", it may be more appropriately characterized as "geometry vs fields" - or, at an even deeper level, as "quantizing gravity vs. gravitizing quantum (field) theory". In accordance with the historical impact of the original 1907 Minkowski presentations, most of the contributions to this volume are geometric in nature, and illustrate a frontier of knowledge that results from explorations of the essential idea that *gravity is geometry.*

In Christopher Duston's contribution, he explores the idea that a reparametrization of the gravitational degrees of freedom that respects the underlying geometry and topology can be used in the classification of spacetime models. Specifically, he uses 1- and 2-knots to characterize spacetime structures, and shows they can be used for explicit calculations, such as the semiclassical entropy.

Garnet Ord makes the interesting case that the relativistic description of physical processes in terms of worldlines is at odds with the microscopic description of such processes offered by quantum theory. As a remedy, Ord proposes to replace the notion of a worldline with something that he refers to as a "world signal", which is essentially a binary clock. Paul de Vegvar addresses the origin of causality in a background independent way in his contribution. By positing non-local behavior above the Planck scale described by deformed Hopf algebras, he is able to generate causality at longer scales using an approach found in solid-state physics. This leads to a new description of dark matter, as well as resolving the so-called horizon problem without inflation.

In the contribution from Ivan Gutièrrez-Sagredo and his collaborators, quantum properties are encoded in the geometry by deforming the classical symmetry groups of spacetime. After a review of this process in the maximally symmetric Lorentzian spacetimes of constant curvature, they present explicit examples in 1+1 and 2+1 dimensions, as well as discuss the 3+1 case.

One of the major developments in quantum gravity over the past 20 years is the conjectured holographic duality between geometry in the bulk and field theory on the boundary. In Masoud Ghezelbash's contribution, he reviews recent developments in a particular example of this holography, the extremal Kerr-Sen black hole, and focuses on the agreement between macroscopic and microscopic entropy through this correspondence. Mohammed Sanduk discusses a cogwheel model for unifying quantum theory and relativity, based on two perpendicular rolling circles (of not necessarily the same radius) - a system which mechanical engineers will recognize as the mathematical idealization of a "bevel gear". The model is argued to provide a concrete instantiation of certain ideas going back to de Broglie about the existence of a common origin to both relativity and quantum physics. In the final essay of the current collection, Vesselin Gueorguiev considers the role of diffeomorphism invariance in both classical and quantum systems. As is well known, many physical processes, such as the worldline of a particle or a dynamical field configuration in spacetime, can be described in mathematically different ways, i.e., through the use of different parametrizations, but are also characterized by *invariants*, e.g. a particle's proper time or a physical action, which do not depend on the chosen parametrization for the particular physical process in question. Gueorguiev reviews some of the conceptual and technical difficulties associated with reparametrization independent systems and in particular focusses on the role of time in such systems.

It is our hope that this volume will serve to communicate many diverse aspects of the foundations of spacetime physics – specifically those which follow from Minkowski's original vision – to the community at large.

THE EDITORS
*March 2019*

# CONTRIBUTORS

*The list follows the order of chapters.*

VESSELIN PETKOV
Minkowski Institute
Montreal, Canada
vpetkov@minkowskiinstitute.org


DENNIS DIEKS
History and Philosophy of Science
Utrecht University
d.dieks@uu.nl


TIAN YU CAO
Boston University
tycao@bu.edu


URI BEN-YA'ACOV
School of Engineering
Kinneret Academic College on the Sea of Galilee
D.N. Emek Ha'Yarden 15132, Israel
uriby@kinneret.ac.il


JERROLD FRANKLIN
Department of Physics
Temple University
Philadelphia, PA 19122
Jerry.F@TEMPLE.EDU


HOU YAU
FDNL Research
Daly City, California, USA
hyau@fdnresearch.us

NADJA MAGALHAES
Federal University of Sao Paulo
Diadema, SP 09913-030, Brazil
nadjasm@gmail.com


MYRZAKULOV RATBAY, YERZHANOV KOBLANDY,
BAUYRZHAN GULNUR, MEIRBEKOV BEKDAULET
Eurasian International Center for Theoretical Physics and Department of General Theoretical Physics
Eurasian National University
Astana 010008, Kazakhstan
yerzhanovkk@gmail.com


YAKOV ITIN
The Hebrew University of Jerusalem
Jerusalem College of Technology
Jerusalem, Israel
itin@math.huji.ac.il


JERZY KIJOWSKI
Center for Theoretical Physics, Polish Academy of Sciences,
Al. Lotników 32/46; 02-668 Warszawa, Poland
kijowski@cft.edu.pl


REINOUD SLAGTER
Asfyon, Astronomisch Fysisch Onderzoek Nederland
1405EP Bussum
*and*
Department of Physics
University of Amsterdam
The Netherlands
info@asfyon.com


CARLOS FRAJUCA, FABIO DA SILVA BORTOLI, FRANCISCO Y. NAKAMOTO,
GIVANILDO A. SANTOS
Department of Mechanics
Sao Paulo Federal Institute
Sao Paulo, SP 01109-010, Brazil
frajuca@gmail.com


CHISTOPHER DUSTON
Merrimack College
Department of Physics
dustonc@merrimack.edu

GARNET ORD
Department of Mathematics
Ryerson University
Toronto, Ont. M5K25B
Canada
gord@ryerson.ca


PAUL DE VEGVAR
SWK Research
1438 Chuckanut Crest Dr.
Bellingham, WA 98229 USA
Paul.deVegvar@post.harvard.edu


IVAN GUTIEREZ-SALGREDO, ANGEL BALLESTEROS, GIULIA GUBITOSI, FRAN-
CISCO J. HERRANZ
Departamento de Física
Universidad de Burgos
E-09001 Burgos, Spain
igsagredo@ubu.es, angelb@ubu.es, ggubitosi@ubu.es, fjherranz@ubu.es


A. M. GHEZELBASH
Department of Physics and Engineering Physics
University of Saskatchewan
Saskatoon, Saskatchewan S7N 5E2, Canada
masoud.ghezelbash@usask.ca


MOHAMMED SANDUK
Chemical and Process Engineering Department
University of Surrey
Guildford GU2 7XH, UK
m.sanduk@surrey.ac.uk


V. G. GUEORGUIEV
Ronin Institute
Montclair, NJ, USA
*and* Institute for Advanced Physical Studies
Sofia, Bulgaria
Vesselin@MailAPS.org

# CONTENTS

## II   Alternatives to General Relativity       105

## 12 Propagation of Gravitational Waves through the Stochastic Background of Gravitational Waves

**Carlos Frajuca, Fabio da Silva Bortoli, Francisco Y. Nakamoto, Givanildo Alves dos Santos**     **181**

## IV    Quantum Aspects of Space and Time      191

## 13 Modeling Spacetime as a Branched Covering Space over 2-Knots

**Christopher Duston**     **193**

## 14 Can Minkowski Spacetime Resolve Quantum Superposition?

**G. N. Ord**     **223**

## 15 Who asked for causality?

**Paul G.N. de Vegvar**     **241**

## 16 Quantum groups, non-commutative Lorentzian space-times and curved momentum spaces

I. Gutierrez-Sagredo, A. Ballesteros, G. Gubitosi, F.J. Herranz

## 17 Some developments in the holography of black holes and conformal field theories

A. M. Ghezelbash

# Part I

# Classic Themes in the Foundations of Special and General Relativity

C. Duston, M. Holman (Eds), *Spacetime Physics 1907 - 2017. Selected peer-reviewed papers presented at the First Hermann Minkowski Meeting on the Foundations of Spacetime Physics, 15-18 May 2017, Albena, Bulgaria* (Minkowski Institute Press, Montreal 2019). ISBN 978-1-927763-48-3 (softcover), ISBN 978-1-927763-49-0 (ebook).

# 1 Minkowski's program of geometrizing physics and general relativity

Vesselin Petkov

**Abstract**  In order to examine whether Einstein's general relativity is a development of Hermann Minkowski's program of geometrizing physics I will discuss three main issues. First, I will summarize how in his groundbreaking lecture "Space and Time" Minkowski decoded the profound physical message hidden in the failed experiments to detect absolute motion, which led him to initiate the first program of geometrizing physics, and how he started to implement it. Second, I will show that Einstein did not follow entirely the internal logic of his general relativity, according to which gravity is a manifestation of the non-Euclidean geometry of spacetime, because, contrary to common opinion, he himself did not believe that general relativity geometrized gravity. Third, employing Minkowski's original idea to examine the internal logic of the mathematical formalism of classical mechanics (that led him to the discovery of spacetime physics) to the mathematical formalism of general relativity suggests that general relativity *itself* is indeed a further development of Minkowski's program of geometrizing physics because gravitational phenomena seem to be fully described in general relativity as nothing more than manifestations of the non-Euclidean geometry of spacetime without the need to introduce the notion of gravitational interaction.

*Gravitation as a separate agency becomes unnecessary*
Arthur S. Eddington [1]

*An electromagnetic field is a "thing;" gravitational field is not, Einstein's theory having shown that it is nothing more than the manifestation of the metric*
Arthur S. Eddington [2]

C. Duston, M. Holman (Eds), *Spacetime Physics 1907 - 2017. Selected peer-reviewed papers presented at the First Hermann Minkowski Meeting on the Foundations of Spacetime Physics, 15-18 May 2017, Albena, Bulgaria* (Minkowski Institute Press, Montreal 2019). ISBN 978-1-927763-48-3 (softcover), ISBN 978-1-927763-49-0 (ebook).

## 1.1 How Minkowski decoded the profound message hidden in the failed experiments to detect absolute motion

In the 17th century Galileo realized, by analyzing thought experiments and doing actual experiments, that one cannot detect absolute motion with constant velocity (or absolute rest – the special case of zero velocity). We call this unexplained (at the time) experimental fact Galileo's principle of relativity: *absolute motion with constant velocity cannot be discovered through mechanical experiments*.

In 1862 James Clerk Maxwell described all electromagnetic phenomena effectively by four equations (which we now call Maxwell's equations) which predicted that light is in fact an electromagnetic wave. But, as any wave is a disturbance in a medium, the very fact that light exists in empty space implied that what we had been calling "space" is some kind of a medium – called a luminiferous aether. Maxwell's discovery that light waves are electromagnetic waves, which require the luminiferous aether for their very existence, made the experimental fact, captured in Galileo's principle of relativity, even more inexplicable: the luminiferous aether seemed to fit all features of Newton's absolute space and therefore it appeared to be a deep puzzle why absolute motion with constant velocity with respect to the absolute space (the luminiferous aether) cannot be detected; what makes the puzzle even deeper is that accelerated motion is always detected because an accelerated particle offers resistance to its acceleration (it is this resistance that makes accelerated motion experimentally detectable). In 1887 Michelson and Morley performed an experiment with light beams intended to detect the absolute motion of the Earth with respect to the luminiferous aether. It turned out that, like mechanical experiments, electromagnetic experiments involving light cannot detect absolute motion either.

To explain the Michelson-Morley experiment, Lorentz [3] introduced (as an abstract auxiliary mathematical quantity) the time $t'$, calling it the *local* time of a moving observer, in addition to the true time $t$ of a stationary observer (at rest with respect to the luminiferous aether). In his 1905 paper Einstein simply postulated that $t$ and $t'$ should be treated equally as physical times. That meant that time is not absolute, but is relative to an observer, that is, *observers in relative motion have different times*. The next step Einstein mad? was to generalize Galileo's principle of relativity – absolute motion with constant velocity cannot be discovered through electromagnetic and mechanical experiments – and used it as a basis of his special relativity calling the generalized principle simply the principle of relativity.[1] Its modern version,

---

[1] Einstein merely *postulated* that absolute motion did not exist and declared that only relative motion is observed in Nature. He wrote that the experimental evidence implied that the notion of aether was not needed to explain physical phenomena, but did not explain how light, as electromagnetic wave, would exist if the aether did not exist (being a wave – a disturbance in a medium – light required the aether for its very existence). As in 1905 Einstein also published his "Nobel" paper, in which he argued that light was not only emitted

which unfortunately hides the deep puzzle mentioned above, is: all physical phenomena (and laws) are the same in all inertial reference frames. In fact, Einstein used a second postulate – the constancy of the velocity of light: "in empty space light is always propagated with a definite velocity $V$ which is independent of the state of motion of the emitting body" [6]. However, it is not an independent postulate since it is a consequence of the principle of relativity – as Maxwell's equations predict that light propagates with constant velocity and as, by the principle of relativity, Maxwell's equations are valid in all inertial frames, it does follow from the principle of relativity that light propagates with constant velocity in all inertial frames.

In 1908 Hermann Minkowski (Einstein's mathematics professor) resolved the deep puzzle that all experiments failed to detect absolute motion with constant velocity; that resolution was totally unexpected – the explanation of the failure of all those experiments was that. . . the world is four-dimensional with time as the fourth dimension [8].

On September 21, 1908 Minkowski began his famous and groundbreaking lecture "Space and Time" by announcing the revolutionary views of space and time, which he deduced from experimental physics by successfully decoding the profound physical message hidden in the failed experiments to discover absolute motion [9, p.111]:

> The views of space and time which I want to present to you arose from the domain of experimental physics, and therein lies their strength. Their tendency is radical. From now onwards space by itself and time by itself will recede completely to become mere shadows and only a type of union of the two will still stand independently on its own.

In his 1908 lecture and in his 1907 lecture "The Relativity Principle" [10] Minkowski repeatedly stressed that it was an *experimental* fact that absolute motion and absolute rest cannot be discovered (which implied that these notions do not represent anything in Nature):

> All efforts directed towards this goal, especially a famous interference experiment of Michelson had, however, a negative result [9, p. 116]

> In light of Michelson's experiment, it has been shown that, as Einstein so succinctly expresses this, the concept of an absolute state of rest entails no properties that correspond to phenomena [10].

---

(as Planck suggested) but also absorbed as quanta, he probably thought that light existed rather as particles (quanta) and therefore did not need the aether for its existence. However, such a justification applies only to the particle aspect of light, whereas its wave aspect still needed explanation if the aether did not exist. Einstein never addressed that difficulty explicitly, but he did indicate that he revised his 1905 view of the aether pointing out it was not expelled from physics: "the aether became, as it were, four dimensional" [4]. Eddington was of the same opinion and suggested that the aether should be "called by Minkowski's term *world*" [5].

Minkowski had apparently felt that the enormous experimental evidence captured in Galileo's principle of relativity and the failed experiments involving light beams to detect the Earth's motion with respect to the aether contained some hidden information about the physical world that needed to be decoded. That is why he had not been satisfied with the principle of relativity which merely *postulated* that absolute motion with constant velocity cannot be detected. To decode the suspected hidden information, Minkowski employed an original idea[2] – to explore the internal logic of the mathematical formalism of classical mechanics. As a mathematician he first examined the fact that "The equations of Newtonian mechanics show a twofold invariance" [9, p. 111]. As each of the two invariances represents a certain group of transformations for the differential equations of mechanics Minkowski noticed that the second group (representing invariance with respect to uniform translations, i.e. Galileo's principle of relativity) leads to the conclusion that the "time axis can then be given a completely arbitrary direction in the upper half of the world $t > 0$." This strange implication made Minkowski ask the crucial question that led him to the new views of space and time: "What has now the requirement of orthogonality in space to do with this complete freedom of choice of the direction of the time axis upwards?" [9, p. 111].

In answering this question Minkowski first showed *why* the time $t$ of a stationary observer and the introduced by Lorentz *local time* $t'$ of a moving observer (whose $x'$ axis is along the $x$ axis of the stationary observer), should be treated equally (which Einstein simply *postulated* in his 1905 paper) [9, p. 114]:

> One can call $t'$ time, but then must necessarily, in connection with this, define space by the manifold of three parameters $x'$, $y$, $z$ in which the laws of physics would then have exactly the same expressions by means of $x'$, $y$, $z$, $t'$ as by means of $x$, $y$, $z$, $t$. Hereafter we would then have in the world no more *the* space, but an infinite number of spaces analogously as there is an infinite number of planes in three-dimensional space. Three-dimensional geometry becomes a chapter in four-dimensional physics. You see why I said at the beginning that space and time will recede completely to become mere shadows and only a world in itself will exist.

The profound implication of "the requirement of orthogonality in space" is evident in the beginning of this quote – as $t$ and $t'$ are *different* times, it necessarily follows that two different spaces must be associated with these times since each space is orthogonal to each time axis. Minkowski easily saw the obvious for a mathematician fact that different time axes imply different spaces and remarked that "the concept of space was shaken neither by Einstein nor by Lorentz" [9, p. 117]. Then, as the quote demonstrates, Minkowski had

---

[2]The idea of exploring the internal logic of a postulate or a view can be traced back at least to the Eleatic school of thought. However, it was Galileo who most systematically employed this idea to examine critically Aristotle's view of motion and to obtain two important results – the principle of inertia and what we now call Galileo's principle of relativity.

immediately realized that many spaces and times imply that the world is four-dimensional with *all* moments of time forming the fourth dimension (Poincaré published before Minkowski his observation that the Lorentz transformations can be regarded as rotations in a four-dimensional space with time as the fourth axis but, unlike Minkowski, he did not believe that that mathematical space represented anything in the world; see the Introduction of [9], particularly pages 19-23, and the references therein).

Minkowski excitedly and confidently announced the new views of space and time since he seemed to have clearly recognized that their strength comes from the fact that they "arose from the domain of experimental physics" – the arguments that many times imply many spaces as well, which in turn implies that the world is four-dimensional, are deduced unambiguously from the *experiments* captured in the principle of relativity (i.e. the impossibility to discover absolute uniform motion and absolute rest). Minkowski did stress that it is the experiment which revealed that the world is four-dimensional, but he did not emphasize that all experimental facts demonstrating the impossibility to detect absolute uniform motion would be *impossible* if the world were *not* four-dimensional – perhaps it looked obvious to him, as a mathematician, that the common perception that the world is three-dimensional *contradicts* those experimental facts: as there is just one space in a three-dimensional world, all experiments captured in the principle of relativity (*whose explanation requires many spaces*) would be impossible if the world were three-dimensional.[3]

After Minkowski had successfully decoded the profound message hidden in the failed experiments to detect absolute uniform motion and absolute rest – that the world is four-dimensional – he had certainly realized that the physics of this four-dimensional world was in fact the geometry of this world (or in Minkowski's terminology "world-geometry" [11]) since all particles which *appear* to move in space are in reality a forever given web of the particles' worldlines

---

[3]The most direct way to evaluate Minkowski's confidence in the strength of the new views of space and time and his insistence that they were deduced from experimental physics is to assume, for the sake of the argument, that spacetime is nothing more than "an abstract four-dimensional mathematical continuum" [12] and that the physical world is three-dimensional. Then there would exist a *single* space (since a three-dimensional world presupposes the existence of one space), which as such would be absolute (the same for all observers). As a space constitutes a class of simultaneous events (the space points at a given moment), a single (absolute) space implies absolute simultaneity and therefore absolute time as well. Hence a three-dimensional world allows *only* absolute space and absolute time in contradiction with the experimental evidence that uniform motion with respect to an absolute space cannot be discovered as encapsulated in the principle of relativity (which implies that such an absolute space does not exist in Nature). Minkowski's realization that the world must be four-dimensional in order that absolute motion and rest do not exist explains naturally his dissatisfaction with the principle of relativity, which postulates, but does not explain the non-existence of absolute motion and rest [9, p. 117]:

> I think the word *relativity postulate* used for the requirement of invariance under the group $G_c$ is very feeble. Since the meaning of the postulate is that through the phenomena only the four-dimensional world in space and time is given, but the projection in space and in time can still be made with a certain freedom, I want to give this affirmation rather the name *the postulate of the absolute world.*

in spacetime. Then Minkowski outlined the first program of geometrization of physics [9, p. 112]:

> The whole world presents itself as resolved into such worldlines, and I want to say in advance, that in my understanding the laws of physics can find their most complete expression as interrelations between these worldlines.

Then Minkowski started to implement his program by explaining that inertial motion is represented by a timelike *straight* worldline, after which he pointed out that [9, p. 115]:

> With appropriate setting of space and time the substance existing at any worldpoint can always be regarded as being at rest.

In this way he explained not only why the times of inertial observers are equivalent (their times can be chosen along their timelike worldlines and all straight timelike worldlines in spacetime are equivalent) but also explained the physical meaning of all experimental facts demonstrating that absolute motion and absolute rest cannot be detected, that is, he explained the physical meaning of the principle of relativity: all physical phenomena look in the same way to two observers $A$ and $B$ in uniform relative motion (so they cannot tell who is moving as the experimental evidence proved) *because* $A$ and $B$ have different times (as Lorentz formally proposed, Einstein postulated and Minkowski explained) and different spaces (as Minkowski first noticed) – each observer performs experiments in his own space and time and for this reason the physical phenomena look in the same way to $A$ and $B$ (e.g. the speed of light is the same for them since each observer measures it in his own space by using his own time). This *explanation* of the profound meaning of the principle of relativity, extracted from experimental physics, makes the non-existence of absolute motion and absolute rest quite evident – absolute motion and absolute rest do not exist since they are defined with respect to an absolute (single) space, but such a single space does not exist in the world; all observers in relative motion have their own spaces and times, which is *impossible* in a three-dimensional world.

Minkowski also explained the other deep puzzle – why accelerated motion is absolute: he explained that accelerated motion is represented by a *curved* or, more precisely, *deformed* worldline and noticed that "Especially the concept of *acceleration* acquires a sharply prominent character" [9, p. 117].

As Minkowski knew that a particle moving by inertia offers no resistance to its motion with constant velocity (which explains why inertial motion cannot be detected experimentally as Galileo first demonstrated), whereas the accelerated motion of a particle can be discovered experimentally since the particle *resists* its acceleration, he might have very probably linked the sharp physical distinction between inertial (non-resistant) and accelerated (resistant) motion with the sharp geometrical distinction between inertial and accelerated motion represented by straight and deformed (curved) worldlines, respectively.

The realization that an accelerated particle (which resists its acceleration) is a deformed worldtube in spacetime would have allowed Minkowski (had he lived longer) to notice two virtually obvious implications of this spacetime fact [13]:

- The acceleration of a particle is absolute not because it accelerates with respect to some absolute space, but because its worldtube is deformed, which is an absolute geometrical and physical fact.

- The resistance a particle offers to its acceleration (i.e. its inertia) originates from a four-dimensional stress in its deformed worldtube. That is, the inertial force with which the particle resists its acceleration turns out to be a static restoring force arising in the deformed worldtube of the accelerated particle. I guess Minkowski might have been particularly thrilled by this implication of his program to regard physics as spacetime geometry because inertia happens to be another manifestation of the fact that reality is a four-dimensional world.

## 1.2 Einstein did not follow entirely the internal logic of his general relativity

Most physicists seem to believe that it was Einstein's general relativity which first geometrized physical phenomena. This is an unfortunate historical injustice on two counts:

- many physicists (including relativists) are probably not fully aware that it was Minkowski who first introduced the program to geometrize *all* physics – to regard the four-dimensional physics as spacetime geometry – and who started to employ this program to the physics of flat spacetime;

- contrary to common belief, Einstein himself did not believe that general relativity geometrized gravitation: "I do not agree with the idea that the general theory of relativity is geometrizing Physics or the gravitational field" [14].

Einstein seemed to have regarded the mathematical formalism of general relativity rather as pure mathematics and believed that gravitation was a physical interaction involving exchange of gravitational energy and momentum.

However, the internal logic of the mathematical formalism of general relativity seems to imply that it fully explains gravitational phenomena as mere effects of the non-Euclidean geometry of spacetime without assuming that there exists gravitational interaction – as in general relativity, a particle, whose worldline is geodesic, is a free particle moving by inertia, the motion of bodies, for example, falling toward the Earth's surface and of planets orbiting the Sun (whose worldlines are geodesic) is inertial, i.e., interaction-free, because the very essence of inertial motion is motion which does not involve any interaction (and any exchange of energy momentum) whatsoever.

Despite this fact, according to the currently accepted understanding of general relativity, which was initiated and greatly influenced by Einstein himself, gravitation is a physical interaction involving exchange of gravitational energy and momentum. However, there is no proper tensorial expression of gravitational energy and momentum;[4] for a 100 years no one has managed to find such an expression.

Einstein made a gigantic step by linking gravitation and the geometry of spacetime but even he was unable to accept fully what his general relativity was saying – (i) gravitational phenomena are fully explained as effects of the non-Euclidean geometry of spacetime without the need to assume that there exists gravitational interaction and (ii) the mathematical formalism stubbornly refuses to yield a proper tensorial expression of gravitational energy and momentum.[5] That is why, it appears Einstein did not follow entirely the internal logic of the mathematical formalism of general relativity and smuggled the concept of gravitational interaction into its framework.

Since Einstein it has been taken for granted that gravity is a physical interaction and gravitational energy and momentum do exist, but it is admitted that there is some annoying difficulty to represent them in a proper mathematical form. Indeed, the prevailing view among relativists is that there exists indirect astrophysical evidence for the existence of gravitational energy – coming from the interpretation of the decrease of the orbital period of the binary pulsar system PSR 1913+16 discovered by Hulse and Taylor in 1974 [17] (and other such systems discovered after that), which is believed to be caused by the loss of energy due to gravitational waves emitted by the system (which carry away gravitational energy).

This interpretation that gravitational waves carry gravitational energy should be carefully scrutinized (especially after the recent detection of gravitational waves) by taking into account the above arguments against the existence of gravitational energy and momentum and especially the fact that there does not exist a rigorous (analytic, proper general-relativistic) solution for the two body problem in general relativity. I think the present interpretation of the decrease of the orbital period of binary systems contradicts general relativity, particularly the geodesic principle (a particle, whose worldline is geodesic, moves by inertia, that is, moves non-resistantly) and the experimental evidence which confirmed it (the experimental fact that falling particles do not resist their fall), because by the geodesic principle the neutron stars, whose worldlines had been regarded as exact geodesics (since the stars had been modelled dynamically as a pair of orbiting *point* masses), *move by inertia without losing energy since the very essence of inertial motion is motion without any loss of energy.* For this reason no energy can be carried away by the gravitational waves emitted by the binary pulsar system. Let me stress it: the geodesic principle and the assertion that bodies, whose worldlines are geodesic, emit gravitational energy (carried away by gravitational waves), cannot be both correct.

---

[4]Only tensorial expressions represent real physical quantities; in general relativity gravitational energy and momentum are represented by a pseudo-tensor, which is coordinate-dependent and therefore does not represent anything in the physical world.

[5]The fact that "in relativity there is no such thing as the force of gravity" [16] implies that there is no gravitational energy either since such energy is defined as the work done by gravitational forces. Whether or not gravitational energy is regarded as local does not affect the very definition of energy.

In fact, it is the very assumption that the binary system emits gravitational waves which contradicts general relativity in the first place, because motion by inertia does not generate gravitational waves in general relativity. The inspiralling neutron stars in the binary system were modelled (by Hulse and Taylor) as *point* masses and therefore their worldlines are exact geodesics, which means that the stars move by inertia and no emission of gravitational radiation is involved; if the stars were modelled as extended bodies, then and only then they would be subject to tidal effects and energy would be involved, but that energy would be negligibly small (see next paragraph) and would not be gravitational (see the explanation of the origin and nature of energy in the sticky bead argument below). So, the assertion that the inspiralling neutron stars in the binary system PSR 1913+16 generate gravitational waves is incorrect because it contradicts general relativity.

Gravitational waves are emitted only when the stars' timelike worldlines are not geodesic,[6] that is, when the stars are subject to an absolute (curved-spacetime) acceleration (associated with the absolute feature that a worldline is not geodesic), not a relative (apparent) acceleration between the stars caused by the geodesic deviation of their worldlines. For example, in general relativity the stars are subject to an absolute acceleration when they collide (because their worldlines are no longer geodesic); therefore gravitational waves – carrying no gravitational energy-momentum – are emitted only when the stars of a binary system collide and merge into one,[7] that is, "Inspiral gravitational waves are generated during the end-of-life stage of binary systems where the two objects merge into one" [18].

Indeed, when the stars follow their orbits in the binary system, they do not emit gravitational waves since they move by inertia according to general relativity (their worldlines are geodesic and no absolute acceleration is involved); even if the stars were modelled as extended bodies, the worldlines of the stars' constituents would not be geodesic (but slightly deviated from the geodesic shape) which will cause tidal friction in the stars, but the gravitational waves generated by the very small absolute accelerations of the stars' constituents will be negligibly weak compared to the gravitational waves be-

---

[6]The original prediction of gravitational wave emission, obtained by Einstein (*Berlin. Sitzungsberichte*, 1916, p. 688; 1918, p. 154), correctly identified the source of such waves – a mass whose curved-spacetime acceleration is different from zero (i.e., whose worldline in not geodesic); for example, a spinning rod (what Einstein considered), or any rotating material bound together by cohesive force. None of the particles of such rotating material (except the centre of rotation) are geodesic worldlines in spacetime and, naturally, such particles will emit gravitational waves. This is not the case with double stars; as the stars are modelled as point masses, their worldliness are exact geodesics (which means that the stars are regarded as moving by inertia) and no gravitational waves are emitted. If the stars are regarded as extended bodies their worldtubes will be close to geodesic and their motion will not be entirely non-resistant, because of the tidal friction within the stars (caused by the fact that the worldlines of the stars' constituents are not congruent due to geodesic deviation).

[7]That gravitational waves are emitted only when two members of a binary system collide is explicitly stated also in the LIGO Press Release regarding the third detection of gravitational waves "LIGO Catches its Third Gravitational Wave!" [19]: "As was the case with the first two detections, the waves were generated when two black holes collided to form a larger black hole."

lieved to be emitted from the spiraling stars of the binary system (that be-lief arises from using not the correct general-relativistic notion of acceleration $(a^\mu = d^2x^\mu/d\tau^2 + \Gamma^\mu_{\alpha\beta}(dx^\alpha/d\tau)(dx^\beta/d\tau))$, but the Newtonian one).

The famous sticky bead argument has been regarded as a decisive argument in the debate on whether or not gravitational waves transmit gravitational energy because it has been perceived to demonstrate that gravitational waves do carry gravitational energy which was converted through friction into heat energy [20]:

> The thought experiment was first described by Feynman (under the pseudonym "Mr. Smith") in 1957, at a conference at Chapel Hill, North Carolina. His insight was that a passing gravitational wave should, in principle, cause a bead which is free to slide along a stick to move back and forth, when the stick is held transversely to the wave's direction of propagation. The wave generates tidal forces about the midpoint of the stick. These produce alternating, longitudinal tensile and compressive stresses in the material of the stick; but the bead, being free to slide, moves along the stick in response to the tidal forces. If contact between the bead and stick is 'sticky,' then heating of both parts will occur due to friction. This heating, said Feynman, showed that the wave did indeed impart energy to the bead and rod system, so it must indeed transport energy.

However, a careful examination of this argument reveals that kinetic, not gravitational, energy is converted into heat because a gravitational wave changes the shape of the geodesic worldline of the bead (from a geodesic in spacetime before the arrival of the gravitational wave to a new geodesic in the addition-ally distorted by the wave spacetime) and the stick prevents the bead from following its changed geodesic worldline, i.e., prevents the bead from moving by inertia; as a result the bead resists and exerts an *inertial* force on the stick (exactly like when a particle moving by inertia away from gravitating masses is prevented from its inertial motion, it exerts an inertial force on the obstacle and the kinetic energy of the particle is converted into heat).

It appears more adequate if one talks about *inertial*, not kinetic, energy, because what is converted into heat (as in the sticky bead argument) is the energy corresponding to the work done by the inertial force (and it turns out that that energy, originating from the inertial force, is equal to the kinetic energy [21]; see Appendix). The need to talk about the adequate inertial, not kinetic, energy is clearly seen in the explanation of the sticky bead argument above – initially (before the arrival of the gravitational wave) the bead is at rest and *does not possess any kinetic energy*; when the gravitational wave arrives, the bead starts to move but by inertia (non-resistantly) since the shape of its geodesic worldline is changed by the wave into another geodesic worldline (which means that the bead goes from one inertial state – rest – into another inertial state, i.e., without any transfer of energy from the gravitational wave; transferring energy to the bead would occur if and only if the gravitational wave changed the state of the bead from inertial to non-inertial), and when the stick tries to prevent the bead from moving by inertia, the bead resists and exerts an inertial force on the stick (that is why, what converts into heat through friction is inertial energy).

Finally, it is a fact in the rigorous structure of general relativity that gravitational waves do not carry gravitational energy,[8] which, however, had been inexplicably ignored despite that Eddington explained it clearly in his comprehensive treatise on the mathematical foundations of general relativity *The Mathematical Theory of Relativity* [2, p. 260]: "The gravitational waves constitute a genuine disturbance of space-time, but their energy, represented by the pseudo-tensor $t_\mu^\nu$, is regarded as an analytical fiction" (it cannot be regarded as an energy of any kind for the well-known reason that "It is not a tensor-density and it can be made to vanish at any point by suitably choosing the coordinates; we do not associate it with any absolute feature of world-structure," *ibid*, p. 136).

## 1.3 Employing Minkowski's idea of exploring the internal logic of mathematical formalism

If Minkowski had lived longer but did not discover what Einstein called general relativity, I believe he would have certainly reformulated it like he reformulated[9] Einstein's special relativity. Minkowski would have certainly ex-

---

[8] An immediate and misleading reaction "A wave that carries no energy?!" should be resisted, because it is from the old times of three-dimensional thinking – assuming that a wave really *travels* in the external world. There is no such thing as a propagating wave in spacetime – what is there is a spacetime region whose "wavelike" geometry is interpreted in three-dimensional language as a wave which propagates in space (exactly like a timelike worldline is interpreted in three-dimensional language as a particle which moves in space); also, keep in mind that there is no such thing as space in the external world, because spacetime is not divided into a space and a time.

[9] A recollection of Max Born shows that Minkowski was already discussing his work on decoding the physical message hidden in all experimental facts captured in the principle of relativity and on the physical meaning of the Lorentz transformations as rotations in a four-dimensional space ("I remember that Minkowski occasionally alluded to the fact that he was engaged with the Lorentz transformations, and that he was on the track of new interrelationships" [23]) at the seminar in the summer of 1905. Note that at that time neither Poincaré's paper "Sur la dynamique de l'électron" [24] nor Einstein's 1905 paper were published; Minkowski asked Einstein to send him the 1905 paper hardly on October 9, 1907 [25]. So, Minkowski's insight had occurred sufficiently long before his December 1907 lecture "The Fundamental Equations for Electromagnetic Processes in Moving Bodies" when he presented the fully developed mathematical formalism of the four-dimensional physics of spacetime (or the *World* as Minkowski called it) introduced by him, because such a revolutionary four-dimensional formalism (published in 1908 [26]) could not have been created in just several months. It appears Minkowski needed two years – from 1905 to 1907 – to develop the mathematics of spacetime. It is precisely the complexity of this novel mathematical apparatus specifically developed to describe spacetime which indicates that Minkowski had developed his own ideas at which he arrived independently of Poincaré and Einstein. A second recollection by Born conveys what Minkowski had told him in 1908 after his lecture "Space and Time:" "He told me later that it came to him as a great shock when Einstein published his paper in which the equivalence of the different local times of observers moving relative to each other were pronounced; for he had reached the same conclusions independently but did not publish them because he wished first to work out the mathematical structure in all its

14

plored the internal logic of its mathematical formalism, exactly like he explored the internal logic of the mathematical formalism of classical mechanics (which helped him decode the profound physical message hidden in the experimental facts demonstrating that absolute motion cannot be detected and which led him to the discovery of spacetime physics) and would have very likely demonstrated that the only rigorous interpretation of the mathematical formalism of general relativity is that gravitation is *nothing more* than a manifestation of the non-Euclidean geometry of spacetime and would have concluded that general relativity is a triumph of his program to present physics as spacetime geometry.

As a mathematician (who would not allow anything external, like gravitational energy and momentum, to be smuggled into the theory) Minkowski might have concluded that all gravitational phenomena are completely explained in general relativity as manifestations of the non-Euclidean geometry of spacetime without the need to assume that gravitational interaction is causing the gravitational phenomena[10] – a particle, whose worldline is geodesic, is a free particle moving by inertia (the geodesic principle); therefore the motion of bodies falling toward the Earth's surface and of planets orbiting the Sun (whose worldtubes are geodesic) is inertial, i.e., interaction-free, because the very essence of inertial motion is motion which does not involve any interaction (and any exchange of energy and momentum) whatsoever.

Minkowski would have also easily explained the force acting on a particle on the Earth's surface, i.e. the particle's weight. The worldtube of a particle falling toward the ground is geodesic, which, in ordinary language, means that the particle moves by inertia (non-resistantly) in full agreement with the experimental evidence that falling particles do not resist their fall (which proves that no gravitational force causes the particles' fall[11]). When the particle lands on the ground it is prevented from moving by inertia and it resists the change of its inertial motion by exerting an inertial force on the ground. Like in flat spacetime the inertial force originates from the deformed worldtube of the particle which is at rest on the ground. So the weight of the particle that has been traditionally called gravitational force turns out to be inertial force, which naturally explains the observed equivalence of inertial and gravitational forces. While the particle is on the ground its worldtube is deformed (due to

splendour. He never made a priority claim and always gave Einstein his full share in the great discovery" [27]. Born's recollections only confirm what Minkowski's papers demonstrate – that he arrived independently on what Einstein called special relativity and on the notion of spacetime, but Einstein and Poincaré published their results first while Minkowski was developing the four-dimensional formalism of spacetime "in all its splendour," which was reported in 1907 and published in 1908 as a 59-page treatise, whose depth and length alone indicate that Minkowski developed his own ideas.

[10]The failures so far to create a theory of quantum gravity may have a simple but unexpected explanation – it might turn out that gravitation is not a physical interaction and therefore there is nothing to quantize.

[11]It should be emphasized that the experimental fact that falling particles do not resist their fall proves that no gravitational force is acting on the particles – a gravitational force would be required to accelerate particles downwards *if and only if* the particles resisted their acceleration, because only then a gravitational force would be needed to *overcome* that resistance.

the curvature of spacetime), which means that the particle is being constantly subjected to a curved-spacetime acceleration (keep in mind that acceleration means *deformed* worldtube!); the particle resists its acceleration through the inertial force and the measure of the resistance the particle offers to its acceleration is its inertial mass, which traditionally has been called (passive) gravitational mass. This fact naturally explains the equivalence between a particle's inertial and gravitational masses, which turned out to be the same thing. In this way, Minkowski would have explained Einstein's equivalence postulate exactly like he explained Einstein's relativity postulate.

Minkowski would have dismissed the attempts to forcefully insert in the formalism of general relativity the pseudo-tensor $t^{\nu}_{\mu}$ and regard it as representing gravitational energy and momentum. Minkowski would have pointed out that such attempts have no chances of success for two reasons:

- the mathematical formalism of general relativity fully describes gravitational phenomena as mere effects of the non-Euclidean geometry of spacetime

- the mathematical formalism of general relativity stubbornly refuses to yield a proper tensorial expressions for gravitational energy and momentum.

# Appendix: Kinetic Energy is Actually Inertial Energy

As the energy involved in gravitational phenomena is *inertial*, not gravitational, it will be helpful to emphasize what appears to be virtually obvious – that what has been traditionally called kinetic energy is in fact inertial energy because it is related to the work done by inertial forces.

But first let me address a long-standing confusion on the status of inertial forces – physicists usually call them fictitious forces, whereas engineers regard them as real forces. In fact both are correct. To see why, let us imagine an Einstein lift (with transparent walls) moving with constant velocity (by inertia) far away from any masses in the cosmos [22]. Imagine also that a metal ball is floating in the middle of the elevator. At the moment the lift starts to accelerate an observer $A$ in it sees that the ball starts to fall (accelerate) toward the lift's floor. $A$ would say that a fictitious inertial force is causing the fall of the ball in the lift. That force is indeed fictitious because $A$ knows perfectly well that the inertial state of motion of the ball was not changed (which is confirmed by an outside inertial observer $B$). However, when the ball hits the lift's floor, or, more precisely, when the lift's floor hits the ball, the ball's inertial motion is disturbed and it resists its acceleration through a *real* inertial force, which (as engineers know well) is quite real, because it does work when it deforms the lift's floor at the spot where the collision occurred.

In the above example the deformation on lift's floor (resulting from the collision of the ball and the floor) is caused by the real inertial force with which

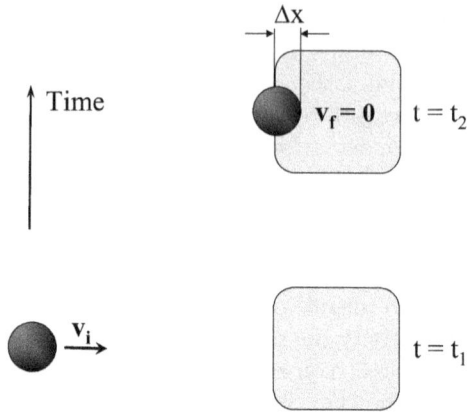

**Figure 1.1:** A massive plastic block is deformed when hit by a ball moving by inertia. Traditionally, it is stated that the ball's kinetic energy converts into a deformation energy. However, a deep physical explanation demonstrates that the ball's energy is inertial energy since the deformation is caused by the work done by the real inertial force with which the ball resists its deceleration

the ball resists its acceleration. Therefore the work done by the ball's inertial force, which is equal to the ball's inertial energy, converts into a deformation energy and ultimately heat. So far inertial energy has been called kinetic energy. But such a name does not reveal the true nature of the ball's energy responsible for the deformation on the lift's floor – the ball's inertia, i.e. its resistance to the change in its inertial state.

The qualitative argument that kinetic energy is actually inertial energy has a straightforward quantitative counterpart. That inertial energy – the work done by inertial forces – is equal to kinetic energy is easily demonstrated by an example depicted in the figure above. At moment $t = t_1$ a ball travels at constant "initial" velocity $v_i$ towards a huge block of some plastic material; we can imagine that the block is mounted on the steep slope of a mountain. Immediately after that the ball hits the block, deforms it and is decelerated. At moment $t = t_2$ the block stops the ball, that is, the ball's final velocity at $t_2$ is $v_f = 0$ (the block's mass is effectively equal to the Earth's mass, which ensures that $v_f = 0$). According to the standard explanation it is the ball's kinetic energy $E_k = (1/2)mv_i^2$ which transforms into a deformation energy. But a proper physical explanation demonstrates that the energy of the ball, which is transformed into deformation energy, is its inertial energy $E_i$, because the ball resists its deceleration $a$ and it is the work $W = F\Delta x$ (equal to $E_i$) done by the *inertial* force $F = ma$ that is responsible for the deformation of the plastic material.

Using the relation between $v_i$, $v_f$, $a$ and the distance $\Delta x$ in the case of deceleration

$$v_f^2 = v_i^2 - 2a\Delta x$$

and taking into account that $v_f = 0$ we find

$$a = \frac{v_i^2}{2\Delta x}.$$

Then for the ball's inertial energy $E_i$ we have

$$E_i = W = F\Delta x = ma\Delta x = \frac{1}{2}mv_i^2.$$

Therefore the inertial energy of the ball is indeed equal to what has been descriptively (lacking physical depth) called kinetic energy.

The same result is obtained when we consider a falling ball hitting a plastic block on the Earth's surface. In this case $v_i$ will be the instantaneous velocity of the ball at the moment it hits the block and obviously again $v_f = 0$.

# References

[1] A.S. Eddington, The Relativity of Time, *Nature* **106**, 802-804 (17 February 1921); reprinted in: A. S. Eddington, *The Theory of Relativity and its Influence on Scientific Thought: Selected Works on the Implications of Relativity* (Minkowski Institute Press, Montreal 2015) pp. 27-30, p. 30

[2] A.S. Eddington, *The Mathematical Theory of Relativity* (Minkowski Institute Press, Montreal 2016) p. 233.

[3] H. A. Lorentz, *The Theory of Electrons and Its Applications to the Phenomena of Light and Radiant Heat*, 2nd ed. (Dover, Mineola, New York 2003); see also his comment on p. 321

[4] A. Einstein, "Über den Aether", Verhandlungen der Schweizerischen Naturforschenden Gesellschaft 105:2 (1924) 85-93. See English translation "Concerning the Aether" (http://www.askingwhy.org/blog/speeches-of-einstein-and-schrodinger/concerning-the-aether-einstein/)

[5] A. S. Eddington, "Space" or "Aether" Nature 107, 201 (April 14, 1921). Reprinted in A. S. Eddington, *Space, Time and Gravitation: An Outline of the General Relativity Theory* (Minkowski Institute Press, Montreal 2017)

[6] A. Einstein: On the Electrodynamics of Moving Bodies. In: [7] pp. 37–65

[7] H.A. Lorentz, A. Einstein, H. Minkowski, and H. Weyl: *The Principle of Relativity: A Collection of Original Memoirs on the Special and General Theory of Relativity* (Dover, New York 1952)

[8] H. Minkowski, Space and Time, new translation in [9].

[9] H. Minkowski, *Space and Time: Minkowski's Papers on Relativity*, edited by V. Petkov (Minkowski Institute Press, Montreal 2012).

[10] H. Minkowski, The Relativity Principle, in [9] pp. 42–43

[11] A. Sommerfeld, To Albert Einstein's Seventieth Birthday. In: *Albert Einstein: Philosopher-Scientist*. P. A. Schilpp, ed., 3rd ed. (Open Court, Illinois 1969) pp. 99- 105, p. 102

[12] N. D. Mermin, What's Bad About This Habit? *Physics Today* **62** (5) 2009, p. 8.

[13] V. Petkov, *Relativity and the Nature of Spacetime*, 2nd ed. (Springer, Heidelberg 2009) Chapter 5

[14] A letter from Einstein to Lincoln Barnett from June 19, 1948; quoted in [15].

[15] D. Lehmkuhl, Why Einstein did not believe that General Relativity geometrizes gravity. *Studies in History and Philosophy of Physics*, Volume 46, May 2014, pp. 316-326.

[16] J. L. Synge, *Relativity: the general theory*. (Nord-Holand, Amsterdam 1960) p. 109

[17] R.A. Hulse, J.H. Taylor, Discovery of a pulsar in a binary system, *Astrophys. J.* **195** (1975) L51–L53

[18] Introduction to LIGO and Gravitational Waves: Inspiral Gravitational Waves http://www.ligo.org/science/GW-Inspiral.php

[19] LIGO Catches its Third Gravitational Wave! https://www.ligo.caltech.edu/news/ligo20170601

[20] Sticky bead argument, https://en.wikipedia.org/wiki/Sticky_bead_argument

[21] V. Petkov, On Inertial Forces, Inertial Energy and the Origin of Inertia, published as Appendix B in [22]

[22] V. Petkov, *Inertia and Gravitation: From Aristotle's Natural Motion to Geodesic Worldlines in Curved Spacetime* (Minkowski Institute Press, Montreal 2012).

[23] Quoted from T. Damour, What is missing from Minkowski's "Raum und Zeit" lecture, *Annalen der Physik* **17** No. 9-10 (2008), pp. 619-630, p. 626.

[24] H. Poincaré, Sur la dynamique de l'électron, *Rendiconti del Circolo matematico Rendiconti del Circolo di Palermo* **21** (1906) pp. 129-176

[25] Postcard: Minkowski to Einstein, October 9, 1907, in: M.J. Klein, A. J. Kox, and R. Schulmann (eds) *The Collected Papers of Albert Einstein*, Volume 5: The Swiss Years: Correspondence, 1902-1914 (Princeton University Press, Princeton 1995), p. 62.

[26] H. Minkowski, Die Grundgleichungen für die elektromagnetischen Vorgänge in bewegten Körpern, Nachrichten der K. Gesellschaft der Wissenschaften zu Göttingen. Mathematisch-physikalische Klasse (1908) S. 53-111; reprinted in H. Minkowski, *Zwei Abhandlungen über die Grundgleichungen der Elektrodynamik*, mit einem Einführungswort von Otto Blumenthal (Teubner, Leipzig 1910) S. 5-57, and in *Gesammelte Abhandlungen von Hermann Minkowski*, ed. by D. Hilbert, 2 vols. (Teubner, Leipzig 1911), vol. 2, pp. 352-404.

[27] M. Born, *My Life: Recollections of a Nobel Laureate* (Scribner, New York 1978) p. 131.

# 2    TIME COORDINATES AND CLOCKS: EINSTEIN'S STRUGGLE

DENNIS DIEKS

**Abstract**    In his Autobiographical Notes, Einstein mentioned that on his road to the final theory of general relativity it was a major difficulty to accustom himself to the idea that coordinates need not possess an immediate physical meaning in terms of lengths and times. This appears strange: that coordinates are conventional markers of events seems an obvious fact, already familiar from pre-relativistic physics. In this paper we explore the background of Einstein's difficulties, going from his 1905 paper on special relativity, through his 1907 and 1911 papers on the consequences of the equivalence principle, to the 1916 review paper on the general theory. As we shall argue, Einstein's problems were intimately connected to his early methodology, in which clarity achieved by concrete physical pictures played an essential role; and to the related fact that on his route to the general theory he focused on special situations that were easily accessible to physical intuition. The details of this background of Einstein's early reasoning have not always been sufficiently appreciated in modern commentaries. As we shall see, this has led to erroneous judgments about the status and validity of some of the early relativistic derivations by Einstein and others, in particular concerning the gravitational redshift.

## 2.1   Introduction

In his Autobiographical Notes [16, pp. 66–67], Einstein remarks that during his work on the general theory of relativity an important obstacle for progress was his own reluctance to abandon the idea that coordinates should possess an immediate physical meaning in terms of measurable distances and times. At first sight, this strikes one as strange: that coordinates are conventional and can be chosen arbitrarily as long as they uniquely identify physical events seems an obvious fact already familiar from pre-relativistic physics. In this paper we explore the background of Einstein's difficulties, going from his 1905

C. Duston, M. Holman (Eds), *Spacetime Physics 1907 - 2017. Selected peer-reviewed papers presented at the First Hermann Minkowski Meeting on the Foundations of Spacetime Physics, 15-18 May 2017, Albena, Bulgaria* (Minkowski Institute Press, Montreal 2019). ISBN 978-1-927763-48-3 (softcover), ISBN 978-1-927763-49-0 (ebook).

paper on special relativity [6], through his 1907 and 1911 papers [7, 9] on the consequences of the principle of equivalence, to the definitive 1916 review paper on the general theory [10] and some later developments.

An essential ingredient of Einstein's 1905 paper was the analysis of physical time, in particular the notion of distant simultaneity. The purpose of this analysis was to become as clear and concrete as possible about the role of simultaneity in concrete physical situations, in order to show that what classical physics assumed about time and simultaneity was partly unfounded. For this reason Einstein discussed an almost everyday synchronization procedure, via light signals exchanged between (ideal) macroscopic clocks. Clearly, simultaneity specified this way does not consist in the equality of arbitrarily chosen time coordinates, but is a concrete physical notion relating to the indications of clocks.

We encounter the same strategy of arguing on the basis of concrete physical situations that are not too different from familiar cases in Einstein's 1907 and 1911 papers. Here Einstein discusses the principle of equivalence, and derives the gravitational redshift as one of its consequences. In his 1911 article Einstein focuses on the comparison of similarly constructed clocks placed at different positions in a homogeneous gravitational field. As Einstein argues on the basis of symmetry considerations, from the point of view of local observers, near to the clocks, such clocks must all behave identically and tick at the same rate. However, Einstein continues, in addition to these local times we should also introduce a *global* time, in order to compare processes at different positions. This alternative time is also introduced via clocks and signals that connect them, via procedures suggested by the physical characteristics of the global situation. It then turns out that the original ("local") clocks and the new "global" ones do not agree in their indications.

This introduction of a global time is not a matter of introducing a conventionally chosen coordinate. Rather than presenting his global time as an arbitrary marker, Einstein explains that it is a physical quantity that is measured by actual clocks, synchronized via a physically justified procedure. The same line of thought occurs in the beginning of the 1916 general relativity paper, in which Einstein discusses temporal relations on a rotating disk. Here he emphasizes that clocks at the periphery of the disk *really* run slower than a clock at the center, again via the introduction of a physically significant global time that is measured by a set of suitable clocks. The notion that there exists a "real" global time which is represented by the indications of a privileged set of clocks apparently had a strong hold on Einstein, even at the time when he was presenting an overview of his general theory of relativity.

It is clear why Einstein felt that the introduction of a global notion of time was necessary: we need to compare and connect local situations in order to be able to make predictions concerning processes that are spatially extended. A prime example is the propagation of light: how long does it take for a light signal to go from $A$ to $B$, and how do the frequencies at emission and reception compare? Another and related example is the gravitational deflection of light that grazes the Sun (the best-remembered subject of Einstein's 1911 paper). Einstein's theoretical derivations of these effects, in 1907, 1911 and 1916, make essential use of the differences between the indications of local clocks in a gravitational field and clocks that measure global time. As we shall argue, not all aspects of these derivations have always been represented correctly in modern commentaries, which has led to a number of unjustified criticisms of Einstein's (and others') early arguments, in particular concerning the gravitational redshift. Consideration of this episode in the history of relativity theory thus opens up both a perspective on Einstein's early methodological thinking and on the status of a number of early relativistic derivations.

Present-day relativists are well aware that there is no pre-determined metrical structure in general relativity; that the theory is "background-independent". This background independence has the consequence that the preferred global time coordinates that were assumed to exist by Einstein in his early work, cannot be expected to exist in general. The transition from Einstein's global time as a privileged coordinate, measurable by physical clocks, to the modern concept of arbitrarily chosen temporal coordinates is closely connected to the transition from special relativity, with its *a priori* symmetries, to general relativity with its lack of a pre-given metrical structure. The account given in this paper is intended to contribute to understanding why Einstein only gradually became comfortable with the idea that space-time coordinates cannot always be given a direct physical and intuitive meaning.

## 2.2   Einstein's 1905 paper

In the Introduction of his 1905 paper Einstein famously declared [12, p. 38]: "The theory to be developed is based—like all electrodynamics—on the kinematics of the rigid body, since the assertions of any such theory have to do with the relationships between rigid bodies (systems of coordinates), clocks, and electromagnetic processes." On the same page he continues (section 1 of the paper): "If we wish to describe the *motion* of a material point, we give the values of its coordinates as a function of the time. Now we must bear carefully in mind that a mathematical description of this kind has no physical meaning unless we are quite clear as to what we understand by 'time'."

The necessary clarification of the meaning of time is given by Einstein in two steps: first we need to avail ourselves of clocks, calibrated in such a way that they locally indicate the familiar time of mechanics (so that "the equations of Newtonian mechanics hold good to the first approximation" [12, p. 38]). Now, if we have two such clocks, one at position A, and another of exactly the same construction at B, we have defined an "A time" and a "B time", but not yet a common time for A and B. In order to establish this common, global time Einstein in his second step introduces the celebrated special relativistic synchronization rule [6, p. 894], [12, p. 40]: "Let a ray of light start at the 'A time' $t_A$ from A towards B, let it at the 'B time' $t_B$ be reflected at B in the direction of A, and arrive again at A at the 'A time' $t'_A$. By definition, the two clocks synchronize if $t_B - t_A = t'_A - t_B$."

This procedure settles the physical content of synchronicity of stationary clocks at different places. We can now imagine an inertial system filled with such clocks, all synchronized with one standard clock (for example, the clock at the origin of the coordinate system). The times thus assigned plus the spatial coordinates provided by measuring rods give a concrete physical meaning to the spatiotemporal description of processes.[1]

In his 1905 paper Einstein thus introduces a local and a global time, both measurable by clocks (in macroscopic laboratory contexts). In fact, *one* set of clocks suffices: synchronization of the local clocks makes them into indicators of global time as well. This reflects a peculiarity of special relativistic inertial frames, which is due to the symmetry of Minkowski spacetime. As we shall see in the next two sections, *accelerated* frames in Minkowski spacetime, and *a fortiori* frames in general relativistic spacetimes, in general do not admit such a simple connection between local and global time.

## 2.3   The 1907 paper: the principle of equivalence

In 1907 Einstein wrote an extensive review of his recently proposed relativity theory: "On the relativity principle and the conclusions drawn from it" [7]. After having reviewed known territory, in the final part of the paper Einstein poses the question whether the relativity principle might be extended so as to apply also to *accelerated* frames of reference. In order to answer this question Einstein considers two reference frames: an inertial system $\Sigma_2$, which is located in a homogeneous gravitational field with a free-fall acceleration of $-\gamma$ in the direction of the $X$-axis, and another frame $\Sigma_1$ that is uniformly accelerated in empty space with an acceleration $\gamma$ along the $X$-axis. Now, Einstein states, as far as we know the physical laws in $\Sigma_1$ do not differ from those in $\Sigma_2$: all bodies are equally accelerated in a gravitational field, so that we see exactly the same motions in $\Sigma_2$ as relative to the accelerated frame $\Sigma_1$. Einstein concludes that in our present state of knowledge we have no reason to assume that the systems $\Sigma_1$ and $\Sigma_2$ differ in any respect. In other words, we may suppose that there exists a physical equivalence between an inertial system with a gravitational field and an accelerated frame of reference. This is the first appearance in the literature of the celebrated principle of equivalence.[2]

---

[1]Einstein's 1905 explanation of the physical meaning of space and time coordinates is sometimes interpreted as being part and parcel of an operationalist philosophy of science, according to which the meaning of concepts is nothing but what is *defined* (in the mathematical-logical sense) by a procedure with measuring instruments. However, the 1905 paper is obviously meant to make the introduction of new physics, with new concepts, acceptable, and is not intended to defend some particular position in the philosophy of science—operationalism is not the issue at stake here. Accordingly, the synchronization procedure proposed by Einstein should not be seen as a strict *definition* of simultaneity in the logical sense, but rather as a concrete implementation of the simultaneity relation that is close to everyday experience and physical intuition (cf. [1]). To mention just one relevant argument, special relativistic physics, and relativistic space and time, will clearly still be well-defined in situations in which there can exist no clocks or measuring rods. However, this question and the related one whether the value of the one-way velocity of light is purely conventional are tangential to the theme of this article and will not be discussed here.

[2]Einstein´s text may suggest that the principle of equivalence *follows* from the empirical fact that all bodies are equally accelerated in a homogeneous gravitational field. That is not the case, of course; the *complete* equivalence posited by Einstein constitutes a hypothesis

As Einstein emphasizes, the equivalence principle has great heuristic value, because it enables us to make predictions about what will happen in a gravitational field: we can replace a constant and homogeneous gravitational field by a uniformly accelerated frame of reference, which is accessible to theoretical treatment via the special theory of relativity.

In order to elaborate this latter point, Einstein invites us to consider the space-time relations between an inertial system $S$ and a system $\Sigma$ that is uniformly accelerated along the $X$-axis of $S$, with acceleration $\gamma$. Both systems are supposed to be equipped with measuring rods and clocks of exactly the same construction. At $S$-time $t = 0$ the frames $S$ and $\Sigma$ are assumed to coincide and to be instantaneously at rest with respect to each other. At that point every clock in $\Sigma$ is adjusted so as to indicate the same time as the corresponding clock in $S$. The time indicated by the individual clocks in $\Sigma$, after this initial synchronization, is called the "local time" $\sigma$ of $\Sigma$. As Einstein observes, in terms of this local time the description of physical processes will be locally everywhere the same in system $\Sigma$, i.e. independent of spatial position (indeed, this is clear when we realize that all points in $\Sigma$ move in exactly the same way against the homogeneous background of Minkowski spacetime).

However, Einstein continues, we cannot simply consider $\sigma$ as giving us "the time" of system $\Sigma$. This is because the accelerating clocks of $\Sigma$, indicating $\sigma$, will not remain synchronous with respect to each other: after their initial synchronization at $t = 0$ two clocks that show the same value of $\sigma$ will no longer satisfy the simultaneity criterion (as introduced in the 1905 paper). This can easily be seen in the following way: since all clocks in $\Sigma$ execute exactly the same accelerated motion, with respect to $S$, they will remain synchronized as judged from $S$ (in which they were synchronized at $t = 0$)—indeed, at any instant of $S$, all $\Sigma$-clocks will have ticked away the same amount of time since t=0, as judged from $S$. But this synchronicity with respect to $S$ entails that the $\Sigma$-clocks can no longer be synchronized as viewed from an inertial system $S'$ that is instantaneously at rest relative to $\Sigma$ at any later time: at any moment of $S$-time later than t=0, $\Sigma$ and therefore also $S'$, will have a non-zero velocity with respect to $S$ so that the simultaneity relations in $S$ and $S'$ differ, as we know from special relativity.

In addition to the local time $\sigma$ Einstein therefore defines a "global time": the global time $\tau$ of an event in $\Sigma$ is the time indicated by a clock at the origin of $\Sigma$ at the instant that is simultaneous with the event in question, according to the simultaneity of the instantaneously comoving inertial system $S'$. As we have just seen, it follows that the local time $\sigma$ and this global time $\tau$ in $\Sigma$ are different. The global time $\tau$ coincides with the time of the instantaneously comoving frame $S'$ (if the clocks at their common origin are set to agree).

The quantitative relation between $\sigma$ and $\tau$ can be determined on the basis of the special theory of relativity. Two events that take place at positions $x_1$ and $x_2$ and times $t_1$ and $t_2$ of $S$, respectively, are simultaneous with respect to $S'$ if $t_1 - x_1 v/c^2 = t_2 - x_2 v/c^2$, in which $v$ is the speed of $S'$ with respect to $S$. In the case of a small time difference with $t = 0$, and a small velocity

---

that goes far beyond the direct empirical evidence of equality of accelerated motions.

$v$, we have in first approximation $x_2 - x_1 = x_2' - x_1' = \xi_2 - \xi_1,$[3] with $\xi$ the coordinate in system $\Sigma$ along the common $X$-axis. Moreover, we have in this approximation $t_1 = \sigma_1$, $t_2 = \sigma_2$, and $v = \gamma\tau$.[4]

It follows that $\sigma_2 - \sigma_1 = (\xi_2 - \xi_1)(\gamma\tau)/c^2$. When we take the origin of the coordinate system as the place of the first event, so that $\sigma_1 = \tau$ and $\xi_1 = 0$, we obtain:

$$\sigma = \tau(1 + \frac{\gamma\xi}{c^2}). \tag{2.1}$$

Now, according to the equivalence principle, this equation should also hold in a system in which there is a homogeneous gravitational field. In this case we can replace $\gamma\xi$ by the gravitational potential $\Phi$, so that we obtain

$$\sigma = \tau(1 + \frac{\Phi}{c^2}). \tag{2.2}$$

So, summarizing, there are two time systems in $\Sigma$: the local time $\sigma$ and the global time $\tau$. Both are indicated by sets of clocks: the local time by the clocks that were initially synchronized with the clocks in $S$ and then move along with $\Sigma$ without further interference; and the successive sets of synchronized clocks in the instantaneously comoving frames $S'$, with the understanding that the clock at the origin of the comoving frame is set to show the same time as the $\sigma$-clock at the origin of $\Sigma$.

When we are interested in local processes and local measurements, it is natural to use the $\sigma$-clocks, which are everywhere in the same physical state and run at the same local rate. However, when we describe extended processes we need a notion of simultaneity to formulate the pertinent physical laws (for example those governing the propagation of light signals), and in this case the $\tau$-clocks will be appropriate because equality of $\tau$ values signifies simultaneity in the standard sense.

It follows from Eq. (2.1) that clocks in a gravitational field, at a position with gravitational potential $\Phi$, indicate time values that differ from the time of the clock at the origin (where $\Phi = 0$) by a factor $(1 + \Phi/c^2)$. It is important to note, as Einstein stresses, that this difference has an immediate experimental significance: a stationary observer in $\Sigma$ who looks at the two clocks will observe that they tick at different rates, given by exactly this factor. This is because the time intervals needed by the light to reach the observer are independent of $\tau$ [8, p. 458]—there is no time-dependence in the factor by which $\tau$ differs from $\sigma$, according to formula (2.1). This constitutes the background for the statement that clocks at a position with a higher gravitational potential *really* run faster: the global time system $\tau$ makes it possible to compare the local rates in a way that is directly empirically verifiable.

An example of the influence of gravity on the rate of clocks is provided by the behavior of atoms and molecules that emit spectral lines: these atoms and

---

[3]As Einstein points out, only speeds and not accelerations have a systematic effect on lengths.

[4]In his 1908 brief correction and addition to the 1907 paper [8] Einstein points out that in a rigorous treatment the constant acceleration $\gamma$ should be defined with respect to the instantaneously comoving system $S'$, so that it is not constant with respect to $S$.

molecules can be considered to be clocks with a highly stable frequency. From the just-given argument it can be concluded that spectral light coming from the surface of the Sun will arrive on Earth with a frequency that is slightly shifted to the red end of the spectrum.

Another prediction is that electromagnetic radiation will be bent in a gravitational field [7, p. 461]. The derivation of this prediction in the 1908 paper is rather cumbersome, though—Einstein comes back to the topic in his 1911 paper, to which we turn now.

## 2.4   The 1911 paper on gravity, time and light

In the Introduction of his 1911 paper Einstein states to return to a number of questions already treated in his 1907 paper, partly because his earlier discussion no longer satisfies him, but primarily because he now realizes that one of his earlier predictions can be tested experimentally. In fact, star light grazing the Sun will be bent by the Sun's gravitational field, and this will lead to a detectable apparent shift in the positions of fixed stars that appear near to the Sun in the sky. The declared objective of the paper is to explain this and other gravitational phenomena by very elementary considerations, so as to make the basic assumptions and arguments of the theory easily understandable.

Einstein again bases his considerations on the principle of equivalence, now making it more explicit than before that this principle represents a fundamental new hypothesis about the nature of gravitation and is meant to extend to all possible physical phenomena. Then he introduces a simple thought experiment in which the equivalence principle is used to derive several results concerning gravity; we shall only discuss the parts pertinent to the behavior of clocks in a gravitational field.

Let there be two bodies $A$ and $B$,[5] in a system of reference $K$ in which there exists a homogeneous gravitational field. Both systems, assumed to be infinitely small, are positioned on the $z$-axis of $K$; the acceleration due to gravity is $-\gamma$ (directed downward along the $z$-axis) and system $B$ is located higher in the field, at a distance $h$ from $A$, so that the gravitational potential at $B$ is greater than at $A$, $\Phi(B) - \Phi(A) = \gamma h$. The system $B$ emits a quantity of electromagnetic radiation in the direction of $A$, and we are interested in the influence of gravity on the properties and propagation of this radiation.

By virtue of the principle of equivalence we can replace system $K$ with a system $K'$ that possesses a constant acceleration $\gamma$ in the positive $z$-direction and in which there is no gravitational field. In order to have a situation that is equivalent to the original one we have to assume that $A$ and $B$ are located at fixed positions on the $z'$-axis, with the constant mutual distance $h$. Finally, let $K_0$ be an inertial system that at the moment of the emission of the radiation is instantaneously at rest with respect to $K'$.

When we describe the process of the emission, propagation and reception of the radiation from system $K_0$, we have that $B$ has no velocity relative to

---

[5]We change the notation from Einstein's own in order to avoid confusion with symbols used in the previous section.

$K_0$ when the radiation is emitted, that the radiation then takes a time $h/c$ to arrive at $A$ (in first approximation), and that $A$ possesses the approximate speed $(h\gamma)/c = v$ when the radiation arrives. Now, Einstein notes, if the radiation had the frequency $\nu_2$ when it was emitted by $B$, as measured by a standard clock positioned at $B$, the radiation received in $A$ will have a different frequency $\nu_1$ as measured by a clock of the same construction comoving with $A$. Indeed, when the signal arrives at $A$, $A$ (and its clock) will possess a speed $v$ relative to $K_0$, so that there will be a change in measured frequency on account of the Doppler effect. The relation between $\nu_1$ and $\nu_2$ is given by the Doppler formula

$$\nu_1 = \nu_2(1 + \frac{\gamma h}{c^2}). \tag{2.3}$$

According to the equivalence principle this result also holds for the system $K$ in which there is a gravitational field. That means that light emitted at a higher value of te gravitational potential will arrive at positions with a lower potential with a higher frequency (as measured by a local clock). Rewriting Eq. (2.3) for this case, we find

$$\nu_1 = \nu_2(1 + \frac{\Phi}{c^2}), \tag{2.4}$$

where $\Phi$ is the gravitational potential at $B$ and the value of the potential at $A$ has been set to 0. This is the same gravitational redshift formula as derived in the 1907 paper, Eq. (2.2).

The 1911 derivation of the redshift formula may seem very different from that in the 1907 paper: in 1907 Einstein emphasized the necessity of a global time, whereas the 1911 derivation appears to involve only local times, measured by clocks of the same kind positioned at different positions in the gravitational field. However, that impression would be deceptive. First, the application of the Doppler formula presupposes that there is a global time system by means of which we can compare the frequencies at $A$ and $B$. This global time is provided by inertial system $K_0$ with its standard simultaneity. As we have seen in section 2.3, however, the clock that is stationary at $A$ will not agree with this $K_0$ time, because during the transmission process it obtains a velocity relative to $K_0$. Conversely, if we want to describe the process from the viewpoint of a stationary observer at $A$, with the clock at $A$ as his fiducial clock, we should introduce a global time via the standard synchronization procedure from $A$. This is in essence the same procedure as in section 2.3 and leads to the same global time that was introduced there. Of course, compared to this global time the local clock in $B$ will be out of step, and formula (2.1) applies.

Einstein discusses the situation as follows [9, pp. 905–906; pp. 105–106 in the English translation]:

> On superficial consideration equation (2.4) seems to assert an ab-
> surdity. If there is constant transmission of light from $B$ to $A$, how
> can any other number of periods per second arrive at $A$ than is
> emitted from $B$? But the answer is simple. We cannot regard $\nu_2$
> or respectively $\nu_1$ simply as frequencies (as the number of periods

per second) since we have not yet determined a time in system $K$. What $\nu_2$ denotes is the number of periods per second with reference to the time-unit of the clock $U$ at $B$, while $\nu_1$ denotes the number of periods per second with reference to the identical clock at $A$. Nothing compels us to assume that the clocks $U$ in different gravitation potentials must be regarded as going at the same rate. On the contrary, we must certainly define the time in $K$ in such a way that the number of wave crests and troughs between $B$ and $A$ is independent of the absolute value of time: for the process under observation is by nature a stationary one. ... Therefore the two clocks at $A$ and $B$ do not both give the "time" correctly. If we measure time at $A$ with the clock $U$, then we must measure time at $B$ with a clock which goes $1 + \Phi/c^2$ times more slowly than the clock $U$ when compared with $U$ at one at the same place. For when measured by such a clock, the frequency of the light-ray which is considered above is at its emission from $B$ given by $\nu_2(1 + \Phi/c^2)$, and is therefore, by (2.4), equal to the frequency $\nu_1$ of the same light-ray on its arrival at $A$.

Einstein here explicitly introduces a global time that differs from the local time; the global time corresponds to the time $\tau$ defined in the 1907 article, while the clocks $U$ correspond to the local time $\sigma$. As in the 1905 and 1907 articles, both the local and global times are assumed to be directly measured by sets of clocks. In the 1907 article this material implementation of global time was realized by standard clocks in the instantaneously comoving inertial systems like $S'$, whereas in the 1911 paper the clocks indicating global time are introduced directly, via the rule that they be constructed thus that they tick $1 + \Phi/c^2$ times more slowly than local clocks at the same location.[6]

The 1911 paper ends with a calculation of the bending of light in a gravitational field, and is most famous for this prediction. This calculation is based on the observation that the velocity of light, measured in global time, will not be constant but will vary with the gravitational potential according to

$$c = c_0(1 + \frac{\Phi}{c^2}),$$ (2.5)

where $c_0$ is the value at the origin (where $\Phi = 0$). Huygens's principle tells us that as a consequence a ray of light will be deflected in the direction in which the gravitational potential diminishes. For a ray grazing the Sun, Einstein finds a deflection of $0.83''$, and comments [12, p. 108] that "it would be a most

---

[6] In the Collected Papers of Einstein, Volume 3, the editors comment that Einstein's train of thought in the 1911 paper is quite different from the one in 1907 [13, p. 497]. In particular, they claim that the slow clocks of the 1911 paper played no role in the 1907 article, and refer to [15, pp. 198–199] for a further analysis of the significance of these clocks. In the indicated passage Pais suggests that clocks produced at $A$ will automatically run slower by the factor $1 + \Phi/c^2$ when transported to $B$. This seems a misunderstanding: clocks transported to other positions will tick at the rate of *local* clocks, whereas the slower clocks were introduced by Einstein to measure *global* time. In fact, the slow clocks *did* occur in the 1907 paper, although in the guise of standard clocks in instantaneously comoving frames.

desirable thing if astronomers would take up the question here raised."[7]

## 2.5    Einstein's 1916 review of general relativity

In 1916 Einstein published a self-contained and comprehensive overview of his just-finished general theory of relativity [10], in the first part of which ("Fundamental Considerations on the Postulate of Relativity") he pays ample attention to the conceptual foundations of the new theory. Here he also investigates the consequences of gravity for the notions of space and time, on the basis of the equivalence principle. In particular, Einstein gives the example of a frame $K'$ that rotates with respect to an inertial frame $K$ [pp. 115–116][12]. Concerning time in the rotating frame he writes (after having discussed the failure of Euclidean geometry)[8]:

> Neither can we introduce a time in $K'$ that meets the physical requirements if this time is to be indicated by clocks of identical construction at rest relatively to $K'$. To see this, let us imagine two such identical clocks, placed one at the origin of the coordinates and the other at the circumference of the circle and both considered from the "stationary" frame $K$. By a familiar result of the special theory of relativity, the clock at the circumference—judged from $K$—goes more slowly than the other, because the former is in motion and the other at rest. An observer at the common origin of coordinates, capable of seeing the clock at the circumference by means of light, would therefore see it lagging behind the clock beside him. As he will not make up his mind to let the velocity of light along the path in question depend explicitly on the time, he will interpret his observations as showing that the clock at the circumference "really" goes more slowly than the clock at the origin. So he will be obliged to define time in such a way that the rate of a clock depends upon where the clock may be.

The structure of the argument here is the same as the one in the 1907 and 1911 articles: stationary clocks in an accelerated system, and therefore also in a system in which there is a gravitational field, will indicate a local time—but these local times do not combine into one physically reasonable global time. In the situations considered (uniform linear acceleration and uniform rotation) we should require of a physically reasonable global time that in its terms physical laws, in particular the law governing the propagation of light, should not depend explicitly on time. The latter requirement has the consequence that differences between the rates of clocks at different locations become "objectified": an observer who receives light signals from the clocks will be able to directly see these differences between local rates. In this way he will be able

---

[7]The value found by Einstein in 1911 reflects the influence of gravity on time, but does not take into account the influence of gravity on spatial geometry. The full general theory of relativity predicts a value that is twice the value predicted by the 1911 theory.

[8]Translation following [12], but with corrections.

to verify that distant clocks "really" tick slower or faster than his own clock. In all the given examples, both the local and the global time are supposed to be indicated by sets of clocks. In the earlier examples the global time corresponded to the indications of standard clocks in instantaneously comoving frames, in the rotating disc case global time is given by the standard clocks in the inertial frame relative to which the rotation takes place.

Of course, in the formal part of the 1916 paper things become much more abstract and general. In particular, the restriction to homogeneous gravitational fields (or special cases like fields corresponding to uniform rotation) is dropped so that, generally speaking, there will be no physically privileged global frames that provide a natural arena for the definition of clocks showing a global time. Nevertheless, it is not difficult to recognize traces of the treatment of the earlier cases, even in this part of the paper. Einstein's discussion of the gravitational redshift at the end of the 1916 review is a case in point.

In the last section of his review paper, §22, entitled "Behaviour of Rods and Clocks in the Static Gravitational Field. Bending of Light-rays. Motion of the Perihelion of a Planetary orbit", Einstein applies his new and general theory to a number of crucial cases [12, pp. 160–164]. The influence of gravity on clocks and the gravitational redshift are now dealt with very quickly. For a unit clock that is at rest in a static gravitational field we have for a clock period $ds = 1$ and $dx_1 = dx_2 = dx_3 = 0$. Therefore, $g_{44}dx_4{}^2 = 1$, so that $dx_4 = 1/\sqrt{g_{44}}$. If there is a point mass with mass $M$ at the origin of coordinates it follows from the Einstein field equations that in first approximation $g_{44} = 1 - \kappa M/4\pi r$, with $\kappa$ the gravitational coupling constant appearing in the field equations ($\kappa = 8\pi G/c^2$, with $G$ Newton's constant) and $r$ the radial spatial distance from the point mass. Therefore,

$$dx_4 \approx 1 + \frac{\kappa M}{8\pi r}. \tag{2.6}$$

Einstein immediately concludes [12, p. 162]:

> Thus the clock goes more slowly if set up in the neighbourhood of ponderable masses. From this it follows that the spectral lines of light reaching us from the surface of large stars must appear displaced towards the red end of the spectrum.[9]

Einstein's reasoning here is basically the same as in his earlier discussions of the gravitational redshift, as we shall discuss in a moment—and only this historical context of the earlier derivations makes the meaning and correctness of Einstein's derivation completely clear. Without this context misunderstandings may easily arise, as shown below.

---

[9] Einstein subsequently shows that a light-ray grazing the Sun will be deflected by 1.7″, twice the magnitude of the 1911 prediction, and that the orbits of the planets undergo a slow rotation, which in the case of Mercury will be 43″ per century.

## 2.6 Appraisal of the early redshift derivations

Einstein's 1916 derivation of the gravitational redshift soon became the standard one—it can still be found in general relativity textbooks. An important role in making it widely known and popular was played by the work of Eddington. Eddington was the first to make the general theory of relativity known in the English-speaking world, and his seminal publications (first of all [4] and [5]) were widely read. Of the two just-mentioned titles, especially the less technical *Space, Time and Gravitation* was very influential. In this book Eddington discusses the comparison of frequencies emitted by atoms of the same kind but located at different positions, e.g. on the Sun and on Earth, respectively. Eddington explains the situation as follows [5, pp. 128–129] (italics in the original):[10]

> Consider an atom momentarily at rest at some point in the solar system... If $ds$ corresponds to one vibration ... we have $ds^2 = g_{44}dt^2$. The *time* of vibration $dt$ is thus $1/\sqrt{g_{44}}$ times the *interval* of vibration $ds$.
>
> Accordingly, if we have two similar atoms at rest at different points in the system, the interval of vibration will be the same for both; but the time of vibration will be proportional to the inverse square-root of $g_{44}$, which differs for the two atoms. Since
>
> $$g_{44} = 1 - \frac{2M}{r},$$
> $$\frac{1}{\sqrt{g_{44}}} = 1 + \frac{M}{r},$$
>
> very approximately.
>
> Take an atom at the surface of the sun, and a similar atom in a terrestrial laboratory. For the first, $1 + M/r = 1.00000212$, and for the second $1 + M/r$ is practically 1. The time of vibration of the solar atom is thus longer in the ratio 1.00000212, and it might be possible to test this by spectroscopic examination.
>
> There is one important point to consider. The spectroscopic examination must take place in the terrestrial laboratory; and we have to test the period of the solar atom by the period of the waves emanating from it when they reach the earth. Will they carry the period to us unchanged? Clearly they must. The first and second pulse have to travel the same distance $r$, and they travel with the same velocity $dr/dt$; for the velocity of light in the mesh-system used is $1 - 2M/r$, and though this velocity depends on $r$, it does not depend on $t$. Hence the difference $dt$ at one end of the waves is the same as that at the other end.

---

[10]For the sake of consistency of notation we use $g_{44}$ where Eddington wrote $\gamma$. For the comparison with Einstein's formulas it is important to note that Eddington uses units in which $\kappa/8\pi = 1$.

Eddington's account faithfully represents Einstein's 1907, 1911 and 1916 derivations. First, the atoms at different locations function as *local* clocks—in modern terms, they measure *proper time*, and this justifies taking $ds = 1$ as the interval for a unit period, at all locations. But second, in addition to this local time there is a global time, the "*time*" as Eddington writes just as Einstein did in 1907 and 1911. Einstein underlined the physical importance of this global time by associating it with the indications of sets of actual clocks. Eddington does not delve into this, but he does make it clear that the global time must be used to make time comparisons between different places. This comparison is very simple: the interval $ds$ of one vibration on the Sun corresponds to a lapse $dt$ of global time at that position. Now, what is the corresponding global time lapse on Earth between the received light signals that were emitted at the beginning and the end of the atom's vibration at the sun, respectively? Because the velocity of light does not depend on $t$ (although it does depend on position), the period taken by the signals to go from the Sun to the Earth will remain the same over time, and this means that the time interval $dt$ will be transmitted unchanged.

As we have seen in the previous sections, the latter is exactly Einstein's argument of 1907, 1911 and 1916: "physical requirements" tell us that global time should be introduced in such a way that physical laws, in particular the law governing the propagation of light, will not depend on time. The immediate consequence is that global time intervals at different positions can be directly "seen": we do not need to make calculations about the propagation of light signals because we know that intervals of $dt$ will be transferred without change by light (or other signals). Therefore, if we want to judge at a spatial position $A$ what is the duration of a physical processes at another spatial position $B$, we can simply measure the global time interval $dt_B$ taken by the process at $B$: we know that the $B$-process will be seen at $A$ as taking up precisely the same amount of global time. This time interval can be directly compared to the time interval $dt_A$ associated with a similar process that takes place at $A$—or equivalently, both $dt_A$ and $dt_B$ can be translated into proper time intervals at $A$, which can then be compared; global and local (proper) time intervals at one position differ only by a constant factor. The frequencies of spectral lines emitted at different places can in this way be compared directly.

In an influential article from 1980, John Earman and Clark Glymour [3] criticized the early derivations of the gravitational redshift by Einstein, Eddington, and other authors who followed in their footsteps. They characterize Einstein's 1907 derivation as cumbersome, obscure and lacking clarity concerning the meanings of "time" and "local time" [3, p.178], without offering a detailed discussion of the article. However, in their subsequent discussion of the 1911 paper they provide a short description of the thought experiment in which radiation is emitted from $B$ to $A$ in a homogeneous gravitational field. As we have seen in some detail in section 2.4, and as Earman and Glymour mention, Einstein concluded that clocks at different positions run at different rates (measured in global time), and that this implies that the velocity of light is position-dependent (as measured in global time). Earman and Glymour do not analyze Einstein's reasoning on this point, but comment [3, pp.181–182]:

All of the heuristic derivations of the red shift can be faulted on various technical grounds. But to raise such objections is to miss the purpose of heuristic arguments, which is not to provide logically seamless proofs but rather to give a feel for the underlying physical mechanisms. It is precisely here that most of the heuristic red shift derivations fail—they are not good heuristics. For they are set in Newtonian or special relativistic space-time; but the red shift strongly suggests that gravitation cannot be adequately treated in a flat space-time. Einstein's resort to the notions of a variable speed of light and variable clock rates in a gravitational field can be seen as an acknowledgment, albeit unconscious, of this point; but as we will now see, these notions served to obscure the role of curvature of space-time as the light ray moves from source to receiver.

It is true that the 1907 and 1911 papers only use the (somewhat vague) principle of equivalence and can be considered faulty from the point of view of the completed general theory of relativity. In particular, there is no principled discussion of inhomogeneous fields in the 1907 and 1911 papers (although Einstein mentions in several places that he conjectures that his results for homogeneous fields will also apply to inhomogeneous ones). But it seems unjustified to condemn Einstein's early work on this ground for being based on *bad heuristics*. Moreover, and more importantly, the reproach that Einstein's treatment does not take into account the process of the transmission of light from source to receiver, and that the early derivations of Einstein, Eddington and their followers are for this reason fallacious [3, p. 176] is simply incorrect— this should already be evident from our explanation of Eddington's derivation, but we shall discuss some more details below.

After reporting on Einstein's 1916 derivation leading to Eq. (2.6) and quoting Einstein's immediate conclusion ("The clock goes more slowly if set up in the neighbourhood of ponderable masses. From this it follows that the spectral lines of light reaching us from the surface of large stars must appear displaced towards the red end of the spectrum."), Earman and Glymour continue [3, pp. 182–183]:

> To the modern eye, Einstein's derivation is no derivation at all, for the formula (2.6) expresses only a co-ordinate effect, and ... Einstein provided no deduction from the theory to explain what happens to a light ray or photon as it passes through the gravitational field on its way from the Sun to the Earth. Unfortunately, Einstein's 'derivation' was dressed up by he expositors of the general theory, and it quickly became codified in the literature as the official derivation.

They then explain Eddington's role in the dissemination of Einstein's error, and criticize the red shift derivation Eddington gave in his *Report on the Relativity Theory of Gravitation* [4]: Earman and Glymour assert that the derivation confuses coordinate time with physical time and does not enter into the essential question of how the radiation emitted at the Sun is received on Earth.

However, Eddington's derivation in [4] is essentially identical to his derivation in [5], which we have reproduced above. As should be clear from that discussion, Eddington *did* take account of the role of the propagating light signal and remarked expressly on the fact that in terms of the global time $t$ the velocity of light did nor depend on time—this justified the essential point in the proof, namely that the time interval $dt$ is transferred unchanged from Sun to Earth.

It is surprising, then, to find that Earman and Glymour present their own (admittedly flawless!) derivation of the redshift formula by arguing at length that in a static gravitational field coordinates can be chosen in such a way that the coordinate time interval is transmitted without change. Via a rather roundabout use of this premise they finally arrive at a formula that is equivalent with Einstein's [3, pp. 184–185].

Apparently, the underlying reason of the confusion is that Earman and Glymour have looked at Einstein's (and Eddington's) formulas with all too modern eyes. In modern expositions of general relativity coordinates are considered to be purely conventional markers; in particular, the time coordinate $x_4$ does not need to have any direct physical interpretation in terms of clock indications. Formulas like (2.1) then indeed express nothing but a coordinate effect, which does not have to possess a physical significance. But Einstein did not originally approach the subject from that direction. It is true that Einstein in his definitive work on general relativity [10] took a step towards the modern notion, via his insistence that frames of reference in arbitrary motion should be equivalent and that this implies that arbitrary coordinate systems can be used [10, p. 776],[11] but the ideas from his earlier work, in which he deemed it important to give the time coordinate always a clear concrete physical meaning, still lingered on. From this "physical viewpoint", and the context of the 1907 and 1911 papers, the physical properties of global time in static fields are self-evident, whereas they are in need of justification from a more modern vantage point.

## 2.7   Conclusion: coordinates and time

In 1921, five years after his review article on the completed general theory of relativity, Einstein gave the Stafford Little Lectures at Princeton University, in which he introduced and reviewed both the special and the general theory. The text of these lectures was published in 1922 as *The Meaning of Relativity* [11]. Coming back to the subject of the behavior of rods and clocks in a gravitational field [11, pp. 90–92], Einstein notes that only in local inertial systems the coordinates can be chosen in such a way that they conform to "naturally measured lengths and times". For the case of the static field generated by a central mass, and natural coordinates usually adopted in this situation, however, a unit measuring rod will not fit exactly in a unit coordinate interval: as Einstein says, its "coordinate length" will be shortened—moreover, as Einstein remarks, this coordinate length, and its dependence on location and orienta-

---

[11]Actually, the sense in which general relativity makes all frames of reference equivalent is controversial at least, see e.g. [2].

tion, will depend on the chosen system of coordinates. So here we seem to be dealing with a coordinate effect, in the modern sense.

However, turning to time, Einstein remarks that the interval between two beats of a unit clock ($ds = 1$) corresponds to a longer "time" ($dx_4 > 1$) "in the unit used in our system of coordinates", but he immediately continues [11, p. 92]:

> The rate of a clock is accordingly slower the greater is the mass of the ponderable mass in its neighbourhood. We therefore conclude that spectral lines which are produced on the sun's surface will be displaced towards the red, compared to the corresponding lines produced on the earth, by about $2.10^{-6}$ of their wave-lengths.

This is the exact same argument as in the 1916 paper, which, as we have seen, becomes difficult to understand if we think of $dx_4$ as a completely arbitrary coordinate, and of the slowing down as a pure coordinate effect.

Evidently, in 1921 Einstein was aware of the in principle arbitrary character of coordinates, as shown by his discussion about the behavior of measuring rods. Nevertheless, he clung to his earlier strategy according to which the concept of time should be made as physical as possible. It had been this strategy that had helped him decisively in creating his special theory of relativity, and in taking the first steps towards general relativity. Moreover, in special cases (static gravitational field, the presence of symmetries, etc.) this same strategy is helpful even in the finished theory of general relativity. Awareness of this methodological motif that runs through Einstein's early work makes much of this work more easily understandable. This applies to the content and validity of his early relativity papers, but also to appreciating the difficulties Einstein encountered in coming to grip with relativity in its most general form, in which global time is generally no longer a sensible physical concept.

# References

[1] Dieks, D., The Adolescence of Relativity: Einstein, Minkowski and the Philosophy of Space and Time. Chapter 9 in: Petkov, V. (ed.), *Minkowski Spacetime: A Hundred Years Later*, Springer, 2010. See also: Dieks, D., Time in Special Relativity. Chapter 6 in: Ashtekar, A. and Petkov, V. (eds.), *The Springer Handbook of Spacetime*, Springer, 2014.

[2] Dieks, D., Another Look at General Covariance and the Equivalence of Reference Systems, *Studies in History and Philosophy of Modern Physics*, **37**, 2006, 174–191.

[3] Earman, J. and Glymour, C., The Gravitational Red Shift as a Test of General Relativity: History and Analysis, *Studies in History and Philosophy of Science*, **11**, 1980, 175–214.

[4] Eddington, A.S., *Report on the Relativity Theory of Gravitation*. London: Fleetwood Press, 1918; 2nd edition, 1920. Reprinted by the Minkowski Institute Press, Montreal, 2014.

[5] Eddington, A.S., *Space, Time and Gravitation*. Cambridge: Cambridge University Press, 1923. Reprinted by the Minkowski Institute Press, Montreal, 2017.

[6] Einstein, A., Zur Elektrodynamik bewegter Körper, *Annalen der Physik* **17**, 1905, 891–921. Reprinted in [17], pp. 275–310. English translation in [12], pp. 35–65.

[7] Einstein, A., Über das Relativitätsprinzip und die aus demselben gezogenen Folgerungen, *Jahrbuch der Radioaktivität und Elektronik* **4**, 1907, 411–462. Reprinted in [17], pp. 433–488. English translation in: Schwartz, H.M., Einstein's comprehensive 1907 essay on relativity, *American Journal of Physics*, **45**, 1977, 512–517, 811–817, 899–902.

[8] Einstein, A., Berichtigungen zu der Arbeit "Über das Relativitätsprinzip und die aus demselben gezogenen Folgerungen", *Jahrbuch der Radioaktivität und Elektronik* **5**, 1908, 98–99. Reprinted in [17], pp. 493–495.

[9] Einstein, A., Über den Einfluß der Schwerkraft auf die Ausbreitung des Lichtes, *Annalen der Physik* **35**, 1911, 898–908. English translation in [12], pp. 97–108. Reprinted in [13], pp. 485–497.

[10] Einstein, A., Die Grundlage der allgemeinen Relativitätstheorie, *Annalen der Physik* **49**, 1916, 769–882. Reprinted in [14], pp. 283–339. English translation in [12], pp. 100–164.

[11] Einstein, A., *The Meaning of Relativity*. Princeton: Princeton University Press, 1922.

[12] Einstein, A., Lorentz, H.A., Weyl, H., and Minkowski,, H., *The Principle of Relativity*. New York: Dover Publications, 1952.

[13] Klein, M.J., et al. (eds), *The Collected Papers of Albert Einstein, Vol. 3*. Princeton: Princeton University Press, 1993.

[14] Kox, A.J., et al. (eds.), *The Collected Papers of Albert Einstein, Vol. 6*. Princeton: Princeton University Press, 1996.

[15] Pais, A., *Subtle is the Lord. The Science and Life of Albert Einstein*. Oxford, Oxford University Press, 1982.

[16] Schilp, P.A. (ed.), *Albert Einstein, Philosopher-Scientist*. La Salle: Open Court, 1949.

[17] Stachel, J. (ed.),*The Collected Papers of Albert Einstein, Vol. 2*. Princeton: Princeton University Press, 1989.

# 3 The Meaning of Background Independence and the Nature of Spacetime: A Critical Review from a Constructivist Perspective

Tian Yu Cao

**Abstract**   ???

## 3.1   Introduction

In the discussion of quantum gravity, "background independence" is one of most frequently mentioned notions. Philosophers as well as most physicists take it as a crucial constraint any intended formulation of quantum gravity has to satisfy to be acceptable. As to the exact meaning of the notion, however, it is subject to interpretation. In this talk, I will first list some popular readings of BI, which are very different in their justification and implications. Then I will move to the origin of this notion, that is, to Einstein's deliberations on the so-called hole argument in the years leading to his 1915 formulation of the general theory of relativity and their philosophical implications. These implications, especially those concerning the nature of space and time, will be critically as well as historically and scientifically examined in the context of the dominant views on space and time held by Newton, Leibniz, Kant, Engels, March, and those suggested by recent developments in the studies of quantum gravity.

My discussion, especially those about the ontological status of manifold, naturally leads to bringing Kant back to the stage. The talk will end with some claims about the Kantian perspective in our conception of quantum gravity.

## 3.2   What is meant by background independence?

In publications on quantum gravity, the notion of background independence, or BI, is understood differently by different people. Roughly, they fall into three

C. Duston, M. Holman (Eds), *Spacetime Physics 1907 - 2017. Selected peer-reviewed papers presented at the First Hermann Minkowski Meeting on the Foundations of Spacetime Physics, 15-18 May 2017, Albena, Bulgaria* (Minkowski Institute Press, Montreal 2019). ISBN 978-1-927763-48-3 (softcover), ISBN 978-1-927763-49-0 (ebook).

categories. First, it is understood as independence of special coordination system. For example, Ed Witten in his 1993 discussion of a quantum version of BI, claimed that a BI formulation of string theory should be an intrinsic one, no dependence on a chosen base point, which is equivalent to a special coordinate from which one started with. More generally, he claimed that BI means all initial conditions should be determined uniquely by a fundamental theory. Later string theorists claimed that the M theory based on duality is a BI theory because it does not depend on any of the five versions of perturbative string theories. Some physicists, Moshe Rozali (2008) for example, take this kind of BI as an aesthetic requirementwithout physical content. It is just like in differential geometry, Rozali argues, one demands that equation should be written in a form independent of the choice of charts and coordinate embeddings. Thus a formulation is manifest or implicit BI becomes an issue of taste, not a big deal in physics.

Secondly, BI is understood as independence of not only specific coordinate systems, but more importantly of a background spacetime with a fixed geometry. It demands that in a BI dynamic theory, different geometries should be its solutions, rather thanbeing presumed as a fixed background. This interpretation of BI has been widely accepted. But it has also been widely expanded as well. Lee Smolin, for example, has advocated on various occasions and in many publications that a BI theory should not rely on any physical circumstances that do not originate from constructs within the theory itself, all has to be derived from inside the theory. Thus, even the M theory and the Ads/CFT conjecture in the string approach don't qualify as BI because of the imposition of the extra dimensions and asymptotic anti-de Sitter boundary conditions, which are by definition outside of theory itself. If Smolin were right, then the only possible BI theory would be a cosmological one about the entire universe with a spatially compact topology without boundaries. Many philosophers, Gordon Belot for example, follow suit and claim that the only proper context for the discussion of BI is the entire universe with a spatially compact topology. Tying BI to cosmology would make BI irrelevant, if not impossible, to any practical effort in constructing a theory of quantum gravity, even if we ignore the issue of how to derive the dimensionality of the universe from any dynamic theory.

Another understanding of BI is based on a separation of background from fixed parameters. While background refers to those dynamically changeable quantities, such as different geometries in a fundamental theory such as GTR, fixed parameters, as part of the data needed to specify dynamic problems, and thus as non-dynamic aspect of the theory, can be used to define superselection sectors in theoretical discourse. For example, the asymptotic Ads geometry and flat one cannot be mutually changed into each other by any physical processes; they just define two different theories, two different superselection sectors. According to this understanding of BI, taking non-dynamical initial conditions, boundary conditions, fundamental parameters, such as dimensionality, topological and differential structures, which are outside of dynamics, would not violate the principle of BI. Such a lenient understanding would allow more formulations to claim BI without compromising the core claim of BI, that is, dynamically changeable geometry should not be taken as a fixed stage on which physical processes are described without itself being affected.

## 3.3    The implications of the hole argument

To my knowledge, the origin of the whole BI industry originated from John Stachel's authoritative interpretation of Einstein's hole argument and its sequels leading to the genesis of the general theory of relativity. The story is influential and well-known, [1] thus a brief summary would be enough for later discussions.

The hole argument, according to Stachel, says that general covariance has deprived manifold points' individuality if causality is not to be compromised. This implies that points would have no spatial temporal meaning until they are enmeshed into a relational structure as placeholders, whose identity is determined by their places and roles in the structure. The relational structure referred here is constituted by metric tensor fields, which are solutions to the field equation. A further implication is that the states of the metric field are not labeled by manifold points (or supported by manifold points, as the substantivalist philosopher of spacetime John Earman [2] frequently claims), but can be characterized, for example, by field's internal structural features, the first and second fundamental forms defining scalar and external curvatures. The argument's implication for the nature of spacetime being dynamic and relational is contained in Einstein's 1952 claim that spacetime has no separate existence, but only as the structural quality of the field. [3] This claim has made physicists and philosophers busy in digesting and explaining. Lee Smoling and Carlo Rovelli have stressed its relationist implications. [4] Stachel's reading is more complicated. He has succinctly summarized the lesson learnt from the hole argument: no kinematics without dynamics. This claim, I think, should be read properly without improper inflation or deflation. Geometry certainly sits at the core. But what about dimension, boundary conditions and other fundamental parameters? Should we aspire to derive all of them from a dynamic theory? Answers to these questions are closely related with one's understanding of the nature of spacetime, which is the subject of the nest section.

## 3.4    Traditional views on the nature of spacetime

Traditionally, the nature of spacetime is understood in two ways. [5] The substantivalist school presumes an ontologically independent existence of spacetime, which serves as the stage for the physical entities, events and processes. The major arguments for this position, according to Earman, are, first, the necessity of a reference framework for accommodating absolute motions, and second, the necessity for any field theory to have an ontological support. These two arguments will be addressed soon. According to the anti-substantivalist school, spacetime is a derivative structure, which has to be explained by or derived from ontologically primary physical entities, relations and processes.

Newton's absolute space is a substantival entity. It functions as a reference frame and is supposed to be the precondition for absolute motion to be conceivable. It was conceived by Mach as a monster because is acted on physical objects without being acted upon by them. [6] It can be argued, as Earman did, as an ontological foundation for field theories. The argument is valid for classical field theories as well as relativistic quantum field theories based on the Minkowskian spacetime. But the validity collapses once we take general theory of relativity into consideration, as the hole argument has indicated.

For Leibniz, space and time were not ontologically independent entities, but just the totality of spatial and temporal relations among objects. Here the relations were not conceived as ontologically self-subsistent non-supervenient relations without being anchored on relata, as Paul Teller [7] and ontic structural realists Steven French and his followers [8] have frequently claimed, but presumed the existence of material objects. Leibniz's relationist version of anti-substantivalism was based on two principles: the principle of sufficient reason and that of the identity of the indiscernible or PII. There is no sufficient reason, however, for having to have a sufficient reason for everything. If you push it hard enough, it will reveal itself as just a disguised version of theology. As to PII, anyone who gets acquainted with quantum physics would discard it immediately.

Mach's anti-substantivalist view of space was different. While Newton insisted that the ultimate reference framework, with which absolute motions can be defined, should be conceived as an absolute space, Mach assumed that the required reference framework was in fact provided by the configuration of physical degrees of freedom in the universe. Thus we have obtained a functional absolute space without having committed to an ontologically independent absolute space. [9]

Mach's conception of the reference framework, opened the door for Einstein to introduce an agent, the gravitational field, which, due to its universal coupling to all physical degrees of freedom in the whole universe, is able to provide the required reference framework in specifying spatial and temporal relations without itself being the spacetime. That is, this agent functions as Mach's totality of the physical degrees of freedom in the universe. The implications of introducing the gravitational field, which is dynamic in nature, into the understanding of spacetime are rich and deep, and have been exploited by Einstein first, and many others afterwards.

A crucial difference between Einstein's view of the spatial and temporal relations and the relationist view held by Leibniz, and all the way down to Adolf Grunbaum and many others these days, [10] is that these relations, in Einstei's case, are not externally specified, but rather, they are intrinsically constituted by the metric tensor field. That is why Einstein claimed that spacetime is only a structural quality of the field. Starting from this, we can develop a structural, constitutive and constructive view of spacetime, to which I now turn.

## 3.5   A Constructivist Approach

If the metric tensor field plays such a crucial constitutive role in Einstein's view of spacetime, a problem naturally arises is how to characterize these fields. Are they existing in spacetime? Where to define the equation for the metric tensor field? Surely we need a manifold. Even in more fundamental theories the same question remains. In many theories of quantum gravity, such as those of spin foams, causal sets, causal dynamical triangulations and in geometrogenesis, no spacetime is presumed, it is only emerged in the classic limit from the quantum theory. In these theories, all the spatial-temporal kinematic structures can be deduced from local dynamic processes. But the dynamic processes still have to be conceptualized in terms of a manifold. In this respect, a leader in loop quantum gravity, Abhay Ashtekar, once argued against Julian Barbour, a fundamentalist in the relationist school, in the late 1980s, that "one cannot even write the constraint equations if one does not have a manifold as such." [11]

So we have to start form a manifold, without which no field equation would be formularizable, and no sensible conception of spacetime that is informed by best available scientific theories would be possible. But is it legitimate to turn this epistemic necessity of using a manifold into an ontological argument in support of the so-called manifold substantivalism, as Earman has argued? What is exactly the ontological status of a manifold?

A reasonable answer to this question can be based on the assumption that the totality and structures of the points in a manifold can be viewed as *isomorphic* to those of the totality of physical events in the whole universe. It may even have the status of a pre-spacetime, similar to Kant's space and time as *a priori* forms of our sensibility which can be used to structure our experience and make sense of our experiences. One possible argument in support of this claim is that the dimensionality of a four-dimensional manifold is deeply anchored in our perceptual structure.

Even though a minimally structured manifold (with only a global topological structure and dimensionality) is the starting point for further construction of the reality of spacetime, this acknowledgment of its epistemic necessity cannot be exploited in support of manifold substantivalism for two reasons. First, as we have already pointed out, the manifold points have no direct spatio-temporal meaning, and that their spacetime meaning is constituted by the chromo-geometrical structures (metric and closely related Riemann tensor), which endow the points with individuality either through imposing non-reflexive metrical relationships upon manifold points as argued by Simon Saunders in 2003, [12] or by characterizing the points with four or more invariants of the Riemann tensor, as indicated by Bergmann in 1957; [13] Komar in 1958; [14] and Stachel in 1993. For this reason, the points of a manifold enjoy no ontological priority over the chromo-geometrical structures. Second, as Stachel indicated long time ago, even the global structures of a manifold are not fixed and globally defined, but as a result of the maximal compatible extension of local solutions to a dynamic theory, thus a loophole to a manifold substantivalism is closed.

The above mentioned two reasons can be rephrased in terms of the relationship between a field and a manifold. First, if in classical field theories or in special relativistic field theories, such as conventional quantum field theories, fields are anchored on the points of a manifold, this is not the case in general theory of relativity. Here, as Einstein stressed in 1952, it is not that fields situated in a spacetime manifold, but rather these fields are spatially extended and temporary endured, and endowed the manifodl points with spatial and temporal meanings. Thus the ontological priori status of manifold, let alone its spacetime connotation, is deprived by the field. A manifold is constituted by certain fields into a spacetime with certain structural features. Second, if the global structure of a manifold as a spacetime is the result of the global extension of local solutions to a field equation, a manifold can only be legitimately viewed as the mathematical expression of a field's spatial and temporal features.

A further construction of the reality of spaetime is facilitated by a clarification of the relationship between the kinematic and dynamic structures. Here we have to notice that the kinematic structures, mainly the chromo-geometrical structures are not substantial but purely relational. Thus substantial entities have to be found to ontologically support them, so that the dynamics of these structures can be materialistically rather than Platonically understood. The gravitational field, as a substantial physical entity represented by the connection field, is the required dynamical entity, whose relational aspects are mathematically represented by the metric tensor, which represents the chromo-geometrical structure. The two are inseparably connected to each other through the geodesic equation as the compatibility condition. Thus although the spatio-temporal relations are specified by the metric, the metric itself is ontologically supported by the inertio-gravitational field (the connection). Once the relationship between metric and connection is thus clarified, the mystery about the dynamicity of the metric (as a purely relational structure), or the dynamic nature of spacetime in general, as is convincingly put in display through the general theory of relativity, is dispelled: it is only an *epiphenomenon* of the dynamical behavior of the connection, as a substantial physical structure or entity.

Both the metric and connection are taken to be holistic structures that enjoy ontological priority over their components. For the metric, manifold points are only placeholders for the spatio-temporal relations it stipulates; for the connection, it is both a substantial entity with universal couplings with all kinds of physical entities, and a holistic structure stipulating the possible behaviors of test bodies if they are put somewhere in the field and interacting with parts of the field (the local values of the field) there. Clearly, this structural, constitutive and constructive view of spacetime, a new version of anti-substantivalist view of spacetime, distinguishes itself from the externalist relationism with its intrinsic constitution of spatial temporal relations by the metric, and with its clarification of the ontological relationship between the substantial gravitational field and the metric field as the mathematical expression of its relational structure.

## 3.6 Kant and quantum gravity

In the above discussion about the necessity of manifold, we already mentioned Kant. What is the relevance of Kant to quantum gravity?

It is well known that Kant rejected both Newton's view and Leibniz's view of space and time as objective entities or relations. Rather, he took them as *a prior* condition of our sensibility, as *a priori* requirement for our sensory-cognitive faculties to which all things must conform. For Kant, space and time are parts of a systematic framework we use to structure our experience, and make sense of our experiences.

It is also well known that Kant's *a priori* view of space, time and geometry is a thing of the past, foundered with the rise of non-Euclidean geometry since the mid-$19^{th}$ century. With the rise of quantum physics, which entails the collapse of any possibility of visualizability in the microscopic world, as vigorously and convincingly argued by Niels Bohr , Kant's view of space and time was deeply sunk into a disgraceful situation, was completely rejected as one of the victories of quantum physics.

Is this true? Yes, but not quite. In one crucial sense, Bohr revived Kantianism by claiming the necessity, a kind of Kantian transcendental necessity, of having macroscopic measuring instruments, which are necessarily anchored in classical spacetime, to make sense of quantum physics. One may escape from Bohr's Kantianism by taking many worlds seriously. Granted. But even within quantum physics itself, the Kantian notion of space and time or space-time as the precondition for quantum physics is unavoidable. The very notion of quantization, either in the canonical formulation, or in the path integral formulation, presumes the existence of spacetime. How can one to have canonical commutation relations without a metric, or have path integral without a notion of path? But both the metric and paths are part of a background spacetime.

Of course, everybody knows that conventional quantum physics, quantum field theory included, is a background dependent theory, underpinned by a Minkowskian spacetime, or a curved but non-dynamic manifold. But what about a BI theory of quantum gravity, which is not definable at the scale at which QFT is definable, but is definable only at a much smaller scale, perhaps at the Planck scale? In such a theory, what is rejected is not just any specific background spacetime, but the very notion of spacetime itself: spacetime has to emerge from physical activities described by a theory of quantum gravity, as a classic limit of something which has no reference to spacetime at all. In this case, Kant is completely rejected.

Then how to describe what happens to this quantum gravitational field? Since no spacetime can be or should be involved, we have to find other ways than the Kantian one of conceptualizing the situation. However, a question remains. Do we still need something to make sense of what happens in the quantum gravity regime, from which a classical spacetime can emerge, or the inverse problem can be solved? In other words, do we still need a generalized Kantian perspective?

Recent developments in loop quantum gravity, in spin foams, causal sets, causal dynamical triangulations and in geometrogenesis, have arguably shown that kinematic structures, spatial and temporal and causal structures, can be derived from local dynamical processes, although all these processes have to be conceptualized with a notion of manifold.

Surely any recovery of classical limit, such as geometry and material degrees of freedom emerged from the dynamical processes of the same underlying field proposed by geometrogenesis approach, would have to go through chains of phase transitions. That is, as the result of heterogenerous emergence, they are qualitatively novel, different from what happens in the quantum gravity regime. Still, since phase transitions are intrinsically determined by the dynamic processes, which is describable only when a manifold is available, Kant is still with us even in the deepest layer of the physical world, in the sense that a parameter spacetime is needed to make sense of physical events there.

## 3.7   A dialectical ascendancy from the abstract to the concrete

The major steps in our construction of the reality of spacetime can be summarized as follows. First, we started from a Kantian form of sensibility, an unstructured bare manifold as a pre-spacetime, whose points could be used to represent physical point-coincident events, although these points themselves have no spatial and temporal meanings. In a sense, these points have quiddity of spacetime in a generalized Kantian sense. For example, the dimensionality of the manifold is closely related with human perceptive structure. But these points do not have haecceity of spacetime before being enmeshed with a metric tensor field, that is, they cannot differentiate themselves from each other, and thus are unable to support spatial-temporal relationship which can only be constituted by by manifold points with haecceity. This lack of haecceity of manifold points is only a mathematical expression of the fact that the individuality of these points as the anchor of the spatial-temporal events is constituted by a gravitational field. With the introduction of the metric tensor field, not only the Leibnizian relations, but also the functions of the Newtonian absolute spacetime can be effectively and convincingly derived or reconstructed.

Second, with the clarification of the ontological relationship between the metric field and the gravitational field, we get a clearer sense that the reality of spacetime is ultimately constituted by the physically substantial gravitational field. The introduction of this dynamical agent into the constitution of spacetime has an additional benefit for the constructivist view as against the substantivalist view because according to the later, the absolute spacetime can only be an empty reference framework without any source of dynamicity.

Once a series of substantial and dynamical fields are introduced for the constitution of various real and concrete incarnations of spacetime, the door for the exploration of the richness of spacetime is opened. One may think about Elie Cartan and torsion. [15] But I mainly have quantum gravity in mind. When quantum gravity is brought into the consideration, the advantage of the constitutive approach becomes supreme. The emergence of classical spacetime from quantum regime through chains of phase transitions is heterogeneous in nature. This means that qualitatively novel, different spacetimes can result due to different dynamic processes which intrinsically determine those phase transitions. So the concrete and rich reality of spacetime can be investigated through the investigations of concrete physical processes connecting the quantum regime and classical regimes. [16]

Third, this constitutive view has brought us from Kant's form of sensibility as the starting point of construction of the reality of spacetime to Friedrich Engels's conception of space and time as forms of existence of matter, [17] Engels's notion of matter here can be properly understood as physical degrees of freedom. Engels's form seems to have affinity to Kant's form. But there is a leap from the subjective to the objective. The dialectics of form and content and the unity of form and content would allow us to explore the roles of physical entities and processes in constituting the reality of spacetime, which exploration I have just surveyed moments ago.

Now the various stages in the constitution are coincided with the steps in our construction of the reality of spacetime, from very abstract Kantian form of pre-spacetime, to more and more concrete reality of spacetime. These steps can be viewed as the steps in the ascendance from the abstract to the concrete in Hegel's dialectical logic, which was used by Karl Marx in his construction of social reality. Since what are involved in the constructive ascendancy described above are not the logical categories, but physical entities, events and processes, this dialectical constructive approach to understanding the nature of spacetime should be regarded as an application of Marx's dialectics rather than Hegel's.

Then what is our concrete reality of spacetime we have constructed so far? In my view, spacetime as a means in structurally describing the behavior of physical degrees of freedom under the influence of gravitation, is essentially an ontological expression of gravity's effects. Turning attention from spacetime to the dynamical issue of gravity was the major contribution of Einstein. It is, to my knowledge, also a consensus among physicists and most philosophers of spacetime.

## 3.8   Concluding remark

The above examination suggests that the reality of spacetime is the result of stepwise human construction: from very abstract Kantian form of pre-spacetime to more and more concrete reality of spacetime. On the other hand, although these constructive steps can be viewed as the steps in the ascendance from the abstract to the concrete in Hegel's dialectical logic, what are involved in the constructive ascendancy are not the Hegelian logical categories, but *physical entities, events and processes*. For this reason, the dialectical constructive approach to understanding the nature of spacetime should be regarded as an application of Marx's materialistic dialectics rather than Hegel's idealist one; that is, the truth of spacetime is a human construction, which, however, is based on objective reality. This conclusion seems in line with Einstein's "reli-

48

gious belief in super-human objectivity."

# References

[1] Stachel, J (1980) "Einstein's search for general covariance, 1912B1915," a paper presented to the 9th International Conference on General Relativity and Gravitation, Jena, 1980; later it is printed in Einstein and the History of General Relativity (edited by D. Howard and J. Stachel; Boston, Birkhauser), 63-100; Satchel, J. (1993) "The meaning of general covariance," in Philosophical Problems of the Internal and External Worlds (eds. John Earman et al., University of Pittsburgh Press), 129-160.

[2] Earman, J. (1989). World-Enough and Space-Time (MIT Press, Cambridge, MA).

[3] Einstein, A. (1952): "Relativity and the problem of space," appendix 5 in the 15th edition of Relativity: The Special and the General Theory (Methuen, London, 1954), 135–157.

[4] See Smolin, L (2001) Three Roads to Quantum Gravity (Weidenfeld and Nicolson and Basic Books, London and New York) and Rovelli, C (2004) Quantum Gravity (Cambridge University Press).

[5] The traditional views are briefly yet precisely summarized in Earman (1989)

[6] Mach, E. (1883), Die Mechanik in ihrer Entwicklung. Historisch-Kritisch dargestellt (Brockhaus, Leipzig); see also Cao, T. Y. (1997) Conceptual Development of 20th Century Field Theories (Cambridge University Press), section 4.3.

[7] Teller, P. (1986): "Relational Holism and Quantum Mechanics," British Journal forthe Philosophy of Science, 37:71-81.

[8] French, S. and Ladyman, J. (2003a) "Remodelling structural realism: Quantum physics and the metaphysics of structure," Synthese 136 (1): 31-56; (2003b): "Between Platonism and phenomenalism: Reply to Cao" Synthese, 136: 73-78.

[9] See Cao (1997) section 4.3 on Mach.

[10] Grunbaum, A. (1977) "Absolute and relational theories of pace and space-time," in Foundations of Space-Time Theories (edited by J. Earman, C. Glymore and J. Stachel; University of Minnesota Press), 303-373; see also Cao (1997), P.121.

[11] More on this and relevant references can be found in Cao (2006).

[12] Saunders, S. "Indiscernibles, general covariance, and other symmetries: the case for non-reductive relationalism," in Revisiting the Foundations of Relativistic Physics (eds. Abhay Ashtekar et al. Kluwer, Dordrecht), 151-173.

[13] Bergmann, P. G. (1957): "Topics in the theory of general relativity," in BrandeisUniversity Summer Institute of Theoretical Physics, 1-44.

[14] Komar, A. (1958) "Construction of a complete set of independent observables in the general theory of relativity," Physical Review 111: 1182-87.

[15] Cartan, E. (1922). 'Sur une géneralisation de la notion de courbure de Riemann et les éspaces à torsion', Compt. Rend. Acad. Sci. Paris, 174: 593–595.

[16] Cao, T. Y. (2007): "Conceptual Issues in Quantum Gravity Invited lecture at the 13th International Congress of Logic, Methodology and philosophy of Science, August 9 – 15, 2007, Beijing China; unpublished.

[17] Engels (1883): Dialectics of Nature (International Publishers Co., April 1968).

# 4 PROPER-TIME MEASUREMENT IN ACCELERATED RELATIVISTIC SYSTEMS

URI BEN-YA'ACOV

**Abstract**    Separate constituents of extended systems measure proper-times on different world-lines. Relating and comparing proper-time measurements along any two such world-lines requires that common simultaneity be possible, which in turn implies that the system is linearly-rigidly moving so that momentary rest frames are identifiable at any stage of the system's journey in space-time.

Once momentary rest-frames have been identified, clocks moving on separate world-lines are synchronizable by light-signal communication. The synchronization relations for two clocks are explicitly computed using light-signals exchanged between them. Implications for the clock hypothesis are included. Also, since simultaneity is frame-dependent, incorrect usage of it leads to pseudo-paradoxes. Counter-examples are discussed.

## 4.1  Introduction

In classical Newtonian mechanics any set of particles may be grouped to form a "system". Time is absolute, independent of the referenc-frame, therefore common to all the chosen constituents, and a centre-of-mass (CM) may be unambiguously defined.

Relativity theory is closer to reality than Newtonian mechanics, telling us that time measurement is reference-frame-dependent. Then, except for the trivial case of a system whose constituents all move inertially with the same velocity, there is no common time for the system as a whole.

For point-like particles, proper-time is measured along their world-line, which may also be used for their age. Our originating research question is then *to what extent it is possible to assign the concept of common or characteristic proper-time to spatially extended systems*; in particular, with the wish to use this common proper-time for the *age* of the system.

For a system whose constituents all move inertially with the same velocity,

C. Duston, M. Holman (Eds), *Spacetime Physics 1907 - 2017. Selected peer-reviewed papers presented at the First Hermann Minkowski Meeting on the Foundations of Spacetime Physics, 15-18 May 2017, Albena, Bulgaria* (Minkowski Institute Press, Montreal 2019). ISBN 978-1-927763-48-3 (softcover), ISBN 978-1-927763-49-0 (ebook).

the common proper-time is identified with the time reading of the common rest-frame. But if the system is not inertial the issue becomes far from trivial.

If the system is closed so that its total energy-momentum is constant and the CM frame of the system is inertial, then the common proper-time may be discussed in relation to dynamical quantities. This was done recently, to some extent, in refs.[1, 2]. Here we wish to consider kinematically the case of non-inertial systems.

For a point-like particle moving on the world-line $x^\mu = (t, \vec{r}(t))$ the proper-time lapse between two events, say A and B, is computed, as is well-known [3, 4], by the integral

$$\Delta \tau_{AB} = \int_{t_A}^{t_B} \sqrt{dt^2 - d\vec{r}^2} = \int_{t_A}^{t_B} \sqrt{1 - \vec{v}^2} dt \qquad (4.1)$$

If the particle is inertial this is the reading of a clock attached to the particle's rest-frame. Otherwise, this is the cumulative reading of clocks relative to which the particle is momentarily at rest.

Extended systems may be regarded as being composed of a number of point-like constituents. If the system accelerates then different constituents, moving on distinct world-lines, measure the proper-time differently. Then, in order to approach the issue of assigning a common proper-time to the whole system, the issue of *relating and comparing proper-times measured at distinct constituents* must be sorted out first. This is the purpose of the present talk.

Proper-time lapses measured along separate world-lines may be easily compared if the world-lines intersect twice, but then only between the intersections. This is the case with various versions of the so called 'twins paradox'. Otherwise, in the absence of intersections, relating and comparing proper-time measurements at any two points of an extended system requires that some kind of simultaneity be possible. Therefore, to mimic in such cases the proper-time measurement for an inertial system, it follows that *momentary rest frames, common to the whole system, must be identifiable at any stage of the system's journey in space-time.* The last statement in italics is recognized as characterizing rectilinear relativistic rigid motion [5, 6], which is possible for arbitrary (also time-dependent) accelerations (taking into account necessary differential accelerations between different points). Therefore, linearly-rigidly accelerated systems are used in the following to study comparative proper-time measurement in extended systems.

Since proper-times are Lorentz invariant quantities they should be treated in a Lorentz covariant manner. Linear relativistic rigid motion with general (not-necessarily constant) accelerations is discussed in the following Lorentz covariantly, allowing to relate accelerations, velocities and proper-times of arbitrarily different points along the moving system.

Simultaneity is then used to link and compare the time evolution of different parts of the system. For accelerating systems, the rapidity $\eta = \tanh^{-1}(v)$ [7] is a very convenient parameter to identify the momentary rest-frames. Once simultaneity has been thus defined, clocks moving on separate world-lines are synchronizable by communicating light-signals, either between them or emitted from a source in between the clocks. The synchronization relations are computed and found to depend on the rapidity difference between emission and reception of the signals and the spatial separation of the clocks, and much less significantly on the details of the acceleration. These results are then used to support the relativistic clock hypothesis.

When momentary rest-frames are required, simultaneity can only be relative to the system as a whole, and should be referred to, accordingly, as *proper simultaneity*. Proper simultaneity is possible only for rigidly moving systems, and, since simultaneity is frame-dependent, incorrect application of it leads to wrong conclusions and appearance of so-called 'paradoxes'. To emphasize this aspect of non-inertial motion, the article is finalized with demonstration of the ambiguity of proper-time comparison in non-rigid motion and discussion of two pseudo-paradoxes, Bell's spaceships 'paradox' [8] and Boughn's 'identically accelerated twins' [9].

*Notation.* The entire article is confined to Special Relativity only, referring to events $x^\mu = \left(x^0, x^1, x^2, x^3\right)$ in flat Minkowski space-time. With the convention $c = 1$ and metric tensor $g_{\mu\nu} = \text{diag}\left(-1, 1, 1, 1\right)$, $\mu, \nu = 0, 1, 2, 3$, for any 4-vectors $a^\mu = (a^0, \vec{a})$ and $b^\mu = (b^0, \vec{b})$ the inner product is $a \cdot b = -a^0 b^0 + \vec{a} \cdot \vec{b}$.

## 4.2  Time measurement on spatially extended systems

Composite systems consist of a number (small or large) of points, each point moving on a separate world-line in space-time. At any such point, a virtual clock may be placed. Synchronization of separated clocks requires that simultaneity be established between them. Simultaneity is defined relative to a particular reference frame. If the clocks are inertial and relatively at rest then synchronization is naturally carried out in their common rest frame. Otherwise, for non-inertial clocks, common momentary rest frames identifying common momentary simultaneity hyper-planes must be found. Common momentary rest-frames are characteristic of linear rigid motion. We now review the notion of simultaneity, then consider accelerated linear motion and introduce into it the rigidity condition.

### 4.2.1  Simultaneity and rigidity

Let $x_A^\mu (\theta_A)$ and $x_B^\mu (\theta_B)$ designate the world-lines of two separate point-like entities, A and B, with $\theta_A$ and $\theta_B$ general time-like evolution parameters. Simultaneity is established between the two world-lines when there is a space-like displacement vector orthogonal to both : For each event $x_A^\mu (\theta_1)$ on A's world-line there is a unique event $x_B^\mu (\theta_2)$ on B's world-line so that the displacement vector $\Delta^\mu (\theta_1, \theta_2) \equiv x_A^\mu (\theta_1) - x_B^\mu (\theta_2)$ is orthogonal to $x_A^\mu (\theta_A)$ at $\theta_A = \theta_1$. In general $\Delta^\mu$ is not orthogonal to $x_B^\mu (\theta_B)$ at $\theta_B = \theta_2$. Only when $\Delta^\mu$ is orthogonal to both $x_A^\mu (\theta_A)$ and $x_B^\mu (\theta_B)$ with the conditions

$$\Delta (\theta_1, \theta_2) \cdot \frac{dx_A}{d\theta_A} (\theta_1) = 0 \quad \& \quad \Delta (\theta_1, \theta_2) \cdot \frac{dx_B}{d\theta_B} (\theta_2) = 0 \qquad (4.2)$$

may A and B be regarded *simultaneous*. If the simultaneity condition (4.2) is continuously maintained along these world-lines then $\Delta^\mu(\theta_1, \theta_2)$ is of constant length[1] $\sqrt{\Delta \cdot \Delta}$ and the motion is necessarily rigid.

The existence of a common rest frame, even momentarily, is a requisite : If two clocks are relatively moving, so they do not have a common rest frame, then there is no clear definition of simultaneity, even if the clocks themselves are inertial. Choosing different reference frames, in particular the rest frames of each clock, to determine simultaneity, determines different ratios between the time scales measured by the clocks (see Section 4.5).

## 4.2.2 Proper-time measurement in rectilinear accelerated motion

The world-line of a point particle moving along the $x$-direction may be written, relative to some inertial reference frame S, as

$$x^\mu = \left(t, \xi^1(t), \xi^2, \xi^3\right) \tag{4.3}$$

with constant $\xi^2, \xi^3$. Its unit 4-velocity is $u^\mu = \gamma(v)\,(1, v, 0, 0)$, with $v(t) = d\xi^1/dt$ and $\gamma(v) = \left(1 - v^2\right)^{-1/2} = dt/d\tau$. The acceleration 4-vector at the point is

$$a^\mu = \frac{du^\mu}{d\tau} = \gamma^4\,(v)\,\frac{dv}{dt}\,(v, 1, 0, 0) = a n_1^\mu\,, \tag{4.4}$$

with $a = \gamma^2(v)\,(dv/d\tau)$ the proper acceleration and $n_1^\mu = \gamma(v)\,(v, 1, 0, 0)$ is the space-like unit 4-vector orthogonal to $u^\mu$ indicating the spatial direction of motion.

For linearly accelerating bodies it is very convenient to use the *rapidity* $\eta(v) \equiv \tanh^{-1}(v)$ – the additive quantity in the superposition of co-linear velocities [7] – as the evolution parameter. It satisfies $d\eta = \gamma^2(v)dv = a d\tau$, so that the basic relation between the proper acceleration, the proper-time and the rapidity

$$a = \frac{d\eta}{d\tau}\,, \tag{4.5}$$

which holds for all rectilinear motion, is obtained [10, 11].

In terms of the rapidity, the particle's unit 4-velocity and the spatial unit vector in the direction of motion are, respectively,

$$u^\mu(\eta) = (\cosh\eta, \sinh\eta, 0, 0) \quad , \quad n_1^\mu(\eta) = (\sinh\eta, \cosh\eta, 0, 0)\,. \tag{4.6}$$

The world-line $x^\mu(\eta)$ then satisfies

$$\frac{dx^\mu}{d\eta}(\eta) = \frac{u^\mu(\eta)}{a(\eta)} \tag{4.7}$$

---

[1]It is emphasized that the constancy of the distance $\Delta$ is due to the special relativistic context; this wouldn't necessarily be the case in a general relalivistic context.

which upon integration yields

$$x^\mu(\eta) = \left( \int^\eta \frac{\cosh\eta}{a(\eta)} d\eta, \int^\eta \frac{\sinh\eta}{a(\eta)} d\eta, 0, 0 \right). \tag{4.8}$$

It should be noted that under Lorentz transformations in the direction of motion the rapidity changes by an additive constant. Therefore, while the proper acceleration and rapidity differences are Lorentz invariant, $\eta$ as a variable is not, so the function $a(\eta)$ is necessarily frame-dependent.

## 4.2.3 Rectilinear rigid motion

We now recall the essentials of rectilinear rigid motion [11]. Born's rigidity condition [5] implies that in rectilinear rigid motion there is always, continuously, a momentary rest frame that is common to all the constituents of the system. The Herglotz-Noether theorem [6] then verifies that such motion is possible with arbitrary accelerations.

The existence always of momentary rest frames common to the whole system implies that all its constituents move with the same (varying) spatial velocity $v$ and rapidity $\eta = \tanh^{-1}(v)$ relative to any inertial reference frame. Hence, the foregoing analysis in Section 4.2.2 is equally valid to all points and $\eta$ may be used as a common evolution parameter for the whole system.

In order to describe and analyze the motion of the system in space-time, some reference point must be initially chosen within it. This point defines a reference world-line $x^\mu = x_o^\mu(\eta)$ with $\tau_o(\eta)$ its proper-time. Following the foregoing discussion an orthonormal tetrad $\{u^\mu, n_i^\mu\}$ is chosen that is carried along the reference world-line, with $u^\mu(\eta)$ and $n_1^\mu(\eta)$ given by (4.6) and the other two (constant) unit vectors $n_i^\mu$ ($i = 2, 3$) corresponding to displacements perpendicular to the spatial direction of motion. The tetrad $\{u^\mu, n_i^\mu\}$, corresponding to the common momentary rest-frames, is common to the whole system. The motion of the whole system is therefore completely determined by that of the reference world-line and the spatial triad $\{n_i^\mu(\eta)\}$.

The triad $\{n_i^\mu(\eta), i = 1, 2, 3\}$ spans the $\eta$-simultaneity hyperplanes relative to $x_o^\mu(\eta)$. These are the simultaneity hyperplanes relative to which synchronization of the clocks of the system is possible. Since all the points of the system share the same simultaneity hyperplanes, the choice of the reference point is arbitrary.

Any other point in the system may be defined relative to the reference world-line by a set of 3 constant distance parameters $\{\zeta^i\}$ relative to the triad $\{n_i^\mu(\eta)\}$, with the world-line

$$x^\mu(\zeta, \eta) = x_o^\mu(\eta) + \zeta^i n_i^\mu(\eta) \tag{4.9}$$

Let $a_o(\eta)$ be the proper acceleration of the reference point. From eqs. (4.7) and (4.9) then follows

$$\frac{d}{d\eta} x^\mu(\zeta, \eta) = \left( \frac{1}{a_o} + \zeta^1 \right) u^\mu(\eta). \tag{4.10}$$

Since all unit 4-velocities are parallel they must be identical, $u^\mu(\zeta, \eta) = u^\mu(\eta)$ (*i.e.*, just saying that all the constituents move with the same velocity). The proper acceleration at $x^\mu(\zeta, \eta)$ is then identified as

$$\frac{1}{a(\zeta, \eta)} = \frac{1}{a_o(\eta)} + \zeta^1 . \tag{4.11}$$

Eq. (4.11) implies limitation on the spatial extension of the system via the condition $1 + \zeta^1 a_o > 0$, since otherwise $x^\mu(\zeta, \eta)$ would enter into the Rindler horizon relative to the reference point.

Since the accelerations are point-dependent, so are also the proper-times. Combining eqs. (4.5) and (4.11) then yields

$$d\tau(\zeta, \eta) = d\tau_o(\eta) + \zeta^1 d\eta , \tag{4.12}$$

so the relation between proper-time lapses and rapidity difference between any two simultaneity hyperplanes $\eta = \eta_1$ and $\eta = \eta_2$ is

$$\Delta\tau(\zeta, \eta_1 \to \eta_2) = \Delta\tau_o(\eta_1 \to \eta_2) + \zeta^1 \Delta\eta . \tag{4.13}$$

($\Delta\eta = \eta_2 - \eta_1$). For any two points A and B then follows the relation between the proper-time lapses measured along their corresponding world-lines

$$\Delta\tau_B(\eta_1 \to \eta_2) = \Delta\tau_A(\eta_1 \to \eta_2) + \left(\zeta_B^1 - \zeta_A^1\right)\Delta\eta , \tag{4.14}$$

independently of the choice of the reference point $x_o^\mu$ and the particular details of the acceleration. Recalling that rapidity differences and proper-times are Lorentz invariants verifies the Lorentz invariance of these results.

## 4.3 Synchronization with light signals

Spatially extended systems consist of a number of points moving on separate world-lines. Attaching a virtual clock to each point, proper-times may be measured at the points. If these clocks are synchronizable, proper-times measured at different points of the system may be linked and compared.

Synchronization may be understood here as "adjusting the timing of two clocks". When the clocks are separated, some mediating mechanism is required to link their time readings. As originally suggested by Einstein, this may be done by sending light signals, either from a common source situated in between to both clocks or by directly communicating light signals between them.

Let two clocks be located at points A, B along the $x$-axis in a rigidly accelerated system, with proper distance $L$, so that in the notation of Section 4.2.3 $\zeta_B^1 - \zeta_A^1 = L$. At a certain moment a light signal is simultaneously sent from A towards B and from B to A (Figure 4.1). If the system were inertial then it takes equal times for both signals to arrive to their destinations, allowing synchronization to be achieved and maintained. But if the system accelerates, time differences ensue. In the present section we discuss these differences and their dependence on the acceleration.

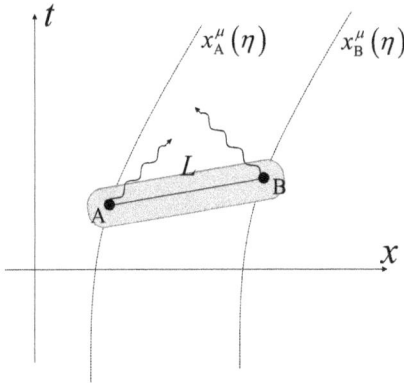

**Figure 4.1:** Space-time diagram showing A & B's world-lines, both A and B on a simultaneity hyperplane, and the signals emitted from both points towards each other.

In terms of the rapidity the clocks' world-lines are given by eq. (4.8)

$$x_i^\mu(\eta) = \left( \int^\eta \frac{\cosh\eta}{a_i(\eta)} d\eta, \int^\eta \frac{\sinh\eta}{a_i(\eta)} d\eta, 0, 0 \right) \qquad i = \mathrm{A}, \mathrm{B}, \qquad (4.15)$$

satisfying $x_\mathrm{B}^\mu(\eta) = x_\mathrm{A}^\mu(\eta) + L n_1^\mu(\eta)$ with the proper accelerations related by eq. (4.11),

$$\frac{1}{a_\mathrm{B}(\eta)} = \frac{1}{a_\mathrm{A}(\eta)} + L \qquad (4.16)$$

Let us consider a signal emited from A at $\eta = \eta_1$ and arriving to B at $\eta = \eta_2$. The important quantity here is the rapidity difference $\Delta\eta = \eta_2 - \eta_1$ which is Lorentz invariant. The 4-vector displacement of the signal is $\Delta x_\mathrm{AB}^\mu = x_\mathrm{B}^\mu(\eta_2) - x_\mathrm{A}^\mu(\eta_1)$, and the light-cone condition on $\Delta x_\mathrm{AB}^\mu$ implies either

$$\int_{\eta_1}^{\eta_2} \frac{\cosh\eta}{a_\mathrm{A}(\eta)} d\eta + L\sinh\eta_2 = \int_{\eta_1}^{\eta_2} \frac{\sinh\eta}{a_\mathrm{A}(\eta)} d\eta + L\cosh\eta_2 \qquad (4.17)$$

or, equivalently,

$$\int_{\eta_1}^{\eta_2} \frac{\cosh\eta}{a_\mathrm{B}(\eta)} d\eta + L\sinh\eta_1 = \int_{\eta_1}^{\eta_2} \frac{\sinh\eta}{a_\mathrm{B}(\eta)} d\eta + L\cosh\eta_1 \qquad (4.18)$$

which yield

$$L = \int_{\eta_1}^{\eta_2} \frac{e^{(\eta_2 - \eta)} d\eta}{a_\mathrm{A}(\eta)} = \int_{\eta_1}^{\eta_2} \frac{e^{-(\eta - \eta_1)} d\eta}{a_\mathrm{B}(\eta)} \qquad (4.19)$$

The light-cone condition (4.19) defines the relation between $L$, $\eta_1$ and $\eta_2$.

Let $\Delta\tau_A(\eta_1, \eta_2)$, $\Delta\tau_B(\eta_1, \eta_2)$ be the proper-time lapses between signal emission and arrival as measured at A and B. From eq. (4.5) it follows that

$$\Delta\tau_i(\eta_1, \eta_2) = \int_{\eta_1}^{\eta_2} \frac{d\eta}{a_i(\eta)} \qquad i = \text{A, B} \tag{4.20}$$

and from (4.16) it follows that

$$\Delta\tau_B = \Delta\tau_A + L\Delta\eta \tag{4.21}$$

Eq. (4.19) may now be used, in conjunction with (4.20), to get useful inequalities. First we get

$$L = \int_{\eta_1}^{\eta_2} \frac{e^{(\eta_2-\eta)}d\eta}{a_A(\eta)} > \int_{\eta_1}^{\eta_2} \frac{d\eta}{a_A(\eta)} = \Delta\tau_A(\eta_1, \eta_2) \tag{4.22}$$

and

$$L = \int_{\eta_1}^{\eta_2} \frac{e^{-(\eta-\eta_1)}d\eta}{a_B(\eta)} < \int_{\eta_1}^{\eta_2} \frac{d\eta}{a_B(\eta)} = \Delta\tau_B(\eta_1, \eta_2) \tag{4.23}$$

Combining (4.22) and (4.23) yields the inequality

$$\Delta\tau_A(\eta_1, \eta_2) < L < \Delta\tau_B(\eta_1, \eta_2) , \tag{4.24}$$

so that the proper-time lapse for the signal transmission as measured at the emitter, $\Delta\tau_A$, is shorter than $L$, while the corresponding lapse measured at the detector, $\Delta\tau_B$, ia larger than $L$, in accordance with (4.21). Then, using (4.21), (4.24) may be turned over to yield

$$(1 - \Delta\eta)L < \Delta\tau_A(\eta_1, \eta_2) < L \tag{4.25}$$

and

$$L < \Delta\tau_B(\eta_1, \eta_2) < (1 + \Delta\eta)L \tag{4.26}$$

If the signal is emited from B at $\eta = \eta_1$, arriving to A at $\eta = \eta_2$, then the light-cone condition (4.19) changes to

$$L = \int_{\eta_1}^{\eta_2} \frac{e^{(\eta-\eta_1)}d\eta}{a_A(\eta)} = \int_{\eta_1}^{\eta_2} \frac{e^{-(\eta_2-\eta)}d\eta}{a_B(\eta)} , \tag{4.27}$$

slightly different from (4.19) but leading to the same inequalities. In either case, these relations, applicable for arbitrary accelerations, are Lorentz invariant since rapidity differences are Lorentz invariant. The various proper-time differences $\Delta\tau$ depend on the details of the acceleration, but they are bounded by quantities that are acceleration-independent; it may therefore be appreciated that the dependence of the synchronization relations on the details of the acceleration is relatively limited.

As a specific example, for constant acceleration it follows from (4.19) and (4.27) that for signals in either direction (A $\to$ B or B $\to$ A)

$$\Delta\eta = \ln(1 + a_A L) = -\ln(1 - a_B L) \tag{4.28}$$

with the proper-time lapses

$$\Delta\tau_A = \frac{L\Delta\eta}{e^{\Delta\eta} - 1} \quad , \quad \Delta\tau_B = \frac{L\Delta\eta}{1 - e^{-\Delta\eta}} \tag{4.29}$$

Therefore, signals in both directions emitted simultaneously will also arrive simultaneously. This is not the case for varying acceleration.

These results are easily extended for signals emitted simultaneously from a common source situated in between A and B.


## 4.4   Accelerating clocks

The foregoing results bear consequences regarding the properties of accelerating clocks:

A clock is a device with an intrinsic periodic mechanism. It is convenient, whenever possible, to regard clocks ideally as point-like, because then their time evolution is confined to a single world-line. The idea of inertial point-like clocks is necessary for the construction of the Minkowski space-time continuum. Time is then assumed to be measured on non-inertial point-like clocks according to (4.1), independently of the details of the acceleration [3, 4]. The last assertion is known as the *clock hypothesis*.

So far there is no physical evidence to question the validity of the clock hypothesis, which was thus un-mentionally assumed at the beginning of the paper. Yet, this validity was questioned several times (see, *e.g.*, [12, 13, 14, 15]). Our results add, theoretically, an argument in favour of the clock hypothesis : Real physical clocks are composite, spatially extended systems. Being regarded as composed of point-like constituents moving on different world-lines, these offer a continuum of proper-times within which the proper-time characterizing the clock must be identified. These proper-times may be related only when the clock is linearly-rigidly accelerating, and then the relation (4.14) between the proper-times is indeed independent of the details of the acceleration. In practice, even the largest difference between proper-time readings, which is $L\Delta\eta$, where $L$ is the spatial dimension of the clock, is very minute [11], so the readings are very close : Let $L = 10$m, $v_1 = 0$ and $v_2 = 0.9c$. Then $\Delta\eta = \tanh^{-1}(0.9) = 1.47$ and $L\Delta\eta/c = 4.9 \times 10^{-8}$s. The discrepancy is real, but macroscopically hardly noticeable.

As is evident from the previous section, the details of the acceleration may enter only in the process of synchronization, or initial linking, of the time measurement along these world-lines. If the clock starts inertial and accelerates only after synchronization has been completed, then the consequent relations between the time-measurements are independent of the details of the acceleration. It is only when synchronization is attempted while the clock is already

60

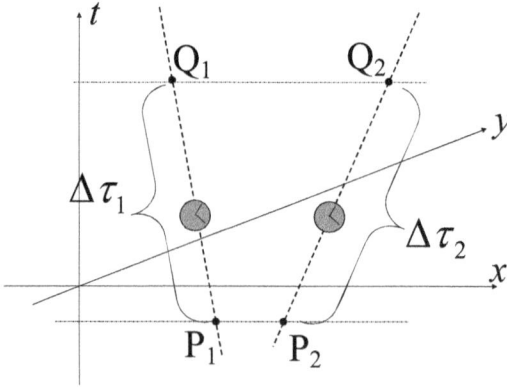

**Figure 4.2:** Space-time diagram showing the world-lines of two relatively moving inertial point-like clocks. Proper-time lapses are measured or computed between simultaneity hyperplanes $P_1P_2$ and $Q_1Q_2$ relative to the external observer.

accelerating that time-measurements depend on the acceleration, but here, again, the discrepancies are minute, of the order of $L\Delta\eta$.

## 4.5 The ambiguity of proper-time comparison in non-rigid motion

The proper-time lapse computed relative to an external inertial observer is given by (4.1). This proper-time lapse is Lorentz-invariant when computed between two fixed events (*i.e.*, independent of the external observer), but if the limiting events are determined by the external observer the situation changes, even for inertial motion.

Consider two point-like clocks moving with different constant velocities (Figure 4.2). The relation between the proper-time lapses, measured along their world-lines between events simultaneous relative to the same external observer, are

$$\frac{\Delta\tau_1}{\Delta\tau_2} = \frac{\sqrt{1-v_1^2}}{\sqrt{1-v_2^2}} \tag{4.30}$$

Using the rapidities for both velocities, with $\eta_{1,2} = \eta(v_{1,2})$ and $\Delta\eta = \eta_2 - \eta_1$,

$$\begin{aligned}\frac{\Delta\tau_1}{\Delta\tau_2} &= \frac{\cosh\eta_2}{\cosh\eta_1} = \frac{\cosh(\eta_1+\Delta\eta)}{\cosh\eta_1} = \cosh(\Delta\eta) + \frac{\sinh\eta_1}{\cosh\eta_1}\sinh(\Delta\eta) = \\ &= \cosh(\Delta\eta) + v_1\sinh(\Delta\eta)\end{aligned} \tag{4.31}$$

The rapidity difference $\Delta\eta$ is Lorentz-invariant, but $v_1$ is observer-dependent.

Since the range of possible $v_1$ values is $-1 < v_1 < 1$, then follows the inequality

$$e^{-\Delta\eta} < \frac{\Delta\tau_1}{\Delta\tau_2} < e^{\Delta\eta} \qquad (4.32)$$

which defines the measure of ambiguity of the proper-times ratio, which is clearly observer-dependent. The inequality (4.32) is Lorentz invariant due to the Lorentz invariance of the rapidity difference $\Delta\eta$.

## 4.6 The case of two spaceships – an example for the correct usage of simultaneity

As has already been pointed out, comparison of proper-time lapses requires introducing proper simultaneity; and since simultaneity is frame dependent and not preserved by Lorentz transformations, much care has to be taken at this point. In the present section we discuss an example that stresses and highlights the need for correct choice of the simultaneity hyperplanes :

Two small (so that they may be regarded point-like) spaceships, A and B, are at rest in the (inertial) home-base, distance $L$ apart. At a certain moment a signal is received causing the engines of the spaceships to ignite and they both simultaneously embark into space following what seems to be the same journey plan relative to the home-base reference frame $S_H\{(x^\mu)\}$

$$x_A^\mu = (t_A, \xi(t_A), 0, 0) \qquad , \qquad x_B^\mu = (t_B, \xi(t_B) + L, 0, 0) \qquad (4.33)$$

where $\xi(t)$ describes some non-uniform motion.

Two questions may be asked :

1. A stretchable string connects the two spaceships. Will the string stretch and eventually be torn apart ?

2. The astronauts in both spaceships are twins. Will they still be of the same age after the journey ?

Both questions have been raised up in the literature and discussed to some extent. Question (i) is the basis for the so-called 'Bell's spaceships paradox' [8]. Question (ii) was raised by Boughn [9], referring to the astronauts as 'identically accelerated twins'. Both questions require comparison, either for the relative distance or for the proper-time difference, between the spaceships. Comparison, as we have already seen, requires a common platform, which must be a proper simultaneity hyperplane. The quality of the answers depends, therefore, on the correct choice of the reference frames which define the correct simultaneity hyperplanes :

The scenario (4.33) requires that the relative distance between the spaceships remain constant relative to the home-base reference frame during the journey; but this simultaneity is irrelevant – the astronauts can measure time, distances, velocities and accelerations only relative to themselves. If the spaceships are connected by a string, any information regarding the state of its stretching can only be obtained by direct measurement relative to the space-ships themselves.

Therefore, the concept of *proper distance* must be introduced – the distance between two points which are simultaneously at rest. Without requiring proper simultaneity the whole concept of length or distance between two world lines is completely ambiguous.

The need for proper simultaneity is clear when we consider that the space-ships need to operate their engines in order to accelerate. The engines are operated by the computers of the spaceships which need to be programmed accordingly. The operation program necessarily uses time for the different stages of the journey. Whose time is it ? it can only be the proper-time of the spaceship, which is the time provided by the spaceship's clock (which, quite reasonably, constitutes a component of the computer). The accelerations of the spaceships are therefore determined relative to their own proper-times, and may be compared unambiguously only on proper simultaneity hyperplanes. Therefore, the twins in (4.33) are not really *identically* accelerated.

Therefore, for questions that concern proper distances and proper times the correct simultaneity hyperplanes must be relative to the spaceships themselves, *i.e.*, in which both spaceships are momentarily at rest.

Moving so that there is always a common momentary rest frame for both spaceships, implies that the motion must be rigid, with world-lines given by (4.15). The equal-$\eta$ simultaneity hyperplanes correspond to the common motion of the two spaceships at the string's ends. On the other hand, the motion proposed in (4.33) is certainly non-rigid, since constant relative distance is assumed relative to the home-base rather than between the spaceships.

It is easy to demonstrate explicitly that there is no common rest frame (therefore no common simultaneity hyperplane) in Bell's and Boughn's scenario (4.33) if the spaceships accelerate : It suffices to assume hyperbolic motion (constant proper acceleration). Writing the home-base scenario (4.33) in terms of the rapidities $\eta_{A,B}$,

$$x_A^\mu = (\rho \sinh \eta_A, \rho \cosh \eta_A, 0, 0) \ , \ x_B^\mu = (\rho \sinh \eta_B, \rho \cosh \eta_B + L, 0, 0) \quad (4.34)$$

$(a = \rho^{-1}$ is the common proper acceleration), and taking into account that the momentary velocities are $v_{A,B} = \tanh \eta_{A,B}$, the spaceships have a common rest frame at the home-base only for $\eta_A = \eta_B = 0$, just before launching into their journey. Lorentz transforming to another inertial reference frame $S_R \{(\bar{x}^\mu)\}$ moving with velocity $V_R = \tanh \eta_R$ relative to the home-base, the world-lines (4.34) become

$$
\begin{aligned}
\bar{x}_A^\mu &= (\rho \sinh (\eta_A - \eta_R), \rho \cosh (\eta_A - \eta_R), 0, 0) \ , \\
\bar{x}_B^\mu &= (\rho \sinh (\eta_B - \eta_R) - L \sinh \eta_R, \rho \cosh (\eta_B - \eta_R) + L \cosh \eta_R, 0, 0)
\end{aligned}
$$
$$(4.35)$$

$\bar{\eta}_{A,B} = \eta_{A,B} - \eta_R$ are the rapidities relative to $S_R$. Both spaceships are momentarily at rest in $S_R$ when $\eta_A = \eta_B = \eta_R$, but these two events are not simultaneous – they correspond to different $S_R$-times – $\bar{x}_A^0 = 0$ and $\bar{x}_B^0 = -L \sinh \eta_R$.

Consequently, if the spaceships accelerate, it is impossible to find in any inertial reference frame two simultaneous events in which both space-ships are momentarily together at rest. Simultaneity could be achieved only with rigid motion, with

$$
\begin{aligned}
x_{\mathrm{A}}^{\mu} &= (\rho\sinh\eta_{\mathrm{A}}, \rho\cosh\eta_{\mathrm{A}}, 0, 0) \quad, \\
x_{\mathrm{B}}^{\mu} &= ((\rho+L)\sinh\eta_{\mathrm{B}}, (\rho+L)\cosh\eta_{\mathrm{B}}, 0, 0)
\end{aligned}
\qquad (4.36)
$$

which coincides with (4.34) only at the home-base rest-frame, with $\eta_{\mathrm{A}} = \eta_{\mathrm{B}} = 0$.

Therefore, when the motion is not rigid the whole problem is ill-posed right from the start : To be able to measure the string's proper length it must be at rest, even momentarily, relative to some inertial reference frame which defines a simultaneity hyperplane; but in a non-rigid motion the two spaceships do not share any common simultaneity hyperplane. Therefore, the mere concept of proper length is meaningless in non-rigid motion. Similarly, for world-lines in non-rigid motion there is no way to even compare the ages.

Both questions above are answerable *only* if the motion is rigid. The answers are then immediate :

1. The string maintains constant proper length, with differential acceleration along it. The components of the string feel stresses, but these are only required to maintain the accelerated rigid motion and they don't change the string's length.

2. The astronauts' ages are determined by the proper-time lapses along their world-lines. Comparison of the ages requires simultaneity hyperplanes, which exist only for rigid motion. Then it follows from (4.21) that the ages differ by $L\Delta\eta$, *i.e.*, depending on the proper distance between the twins and the rapidity difference between the home-base and the end station.

The two questions coincide with a recent discussion of Bell's 'paradox' by Franklin [16], who, among other things, also compared the Minkowskian times of the right and left spaceships (or brothers) which are obviously the same in any instantaneous rest frame. However, the *ages* of the brothers are determined not by the Minkowskian times but by the proper-times measured along their (separate) space-time trajectories. If the end station moves relative to the home-base (so that $\Delta\eta \neq 0$ between the initial and final states) then the brothers do indeed end up with different ages, simply because of siting in separated spaceships.

## 4.7 Concluding remarks

The ages or proper-times measured at different constituents of an extended system may be related and compared only if momentary simultaneity hyperplanes may be identified along the system's journey in space-time. The relation of proper-time lapses at two distinct points of an accelerating system is then uniquely determined, Lorentz covariantly, only for rectilinear relativistic rigid motion. Rectilinear rigid motion may therefore serve to model comparative proper-time measurement in accelerated relativistic systems. This modelling was used here to reflect upon and discuss the clock hypothesis and the correct use of simultaneity.

Besides being the characterizing property that allows proper-time comparison, rigidity has a value of its own also for the following reason : The space-time picture of extended systems is of a congruence of world-lines. What makes this congruence a "system", more than just only a collection of world-lines ? Such a collection becomes a "system" – a whole that is more than just the sum of its constituents – when there is a property which does not pertain to the individual constituents but characterizes the group as a whole. Rigidity is such a property.

In addition, we point out the useful use of the rapidity $\eta$ as the parameter of evolution for linearly accelerated systems.

Finally, we also make note of the fact that while the continuum picture of Minkowski space-time uses point-like clocks, these are only idealizations of real clocks which are necessarily spatially-extended. We may therefore conclude that the continuum picture of Minkowski space-time is only approximately self-consistent, even without taking into account gravitation and quantum mechanics.

# References

[1] Ben-Ya'acov U, Internal time observable of classical relativistic system *J. Physics A : Math. Gen.* **39**, 667-83

[2] Ben-Ya'acov U, Internal-time and dilatations in classical relativity *J. Physics: Conference Series* **66**, 012009

[3] Misner C W, Thorne K P and Wheeler J A 1973 *Gravitation* (San Francisco : Freeman) p. 393

[4] Rindler W 2006 *Relativity : Special, General, and Cosmological* (2nd ed., Oxford : Oxford University Press)

[5] Born M 1909 Die Theorie des starren Elektrons in der Kinematik des Relativitätsprinzips (The Theory of the Rigid Electron in the Kinematics of the Principle of Relativity) *Annalen der Physik* **335** (**11**) 1-56

[6] Herglotz G 1910 Über den vom Standpunkt des Relativitätsprinzips aus als starr zu bezeichnenden Körper (On bodies that are to be designated as "rigid" from the standpoint of the relativity principle) *Annalen der Physik* **336** (**2**) 393-415
Noether F 1910 Zur Kinematik des starren Körpers in der Relativtheorie *Annalen der Physik* **336** (**5**) 919-44

[7] Rhodes J A and Semon M D 2004 Relativistic velocity space, Wigner rotation and Thomas precession *Am. J. Phys.* **72** **(7)** 943-60

[8] Bell B S 1993 *Speakable and Unspeakable in Quantum Mechanics* (Cambridge : Cambridge University Press) pp 67-8
Dewan E and Beran M 1959 Note on Stress Effects due to Relativistic Contraction *Am. J. Phys.* **27** 517-8

[9] Boughn S P 1989 The case of the identically accelerated twins *Am. J. Phys.* **57** **(9)** 791-3

[10] Minguzzi E 2005 Differential aging from acceleration: An explicit formula *Am. J. Phys.* **73** **(9)** 876-80

[11] Ben-Ya'acov U 2016 The 'twin paradox' in relativistic rigid motion *Eur. J. Phys.* **37** 055601

[12] Mashhoon B 1990 The hypothesis of locality in relativistic physics *Phys. Let. A* **145** 147-53

[13] Mainwaring S R and Stedman G E 1993 Accelerated clock principles in special relativity *Phys. Rev. A* **47** 3611-9

[14] Kowalski F V 1996 Accelerating light clocks *Phys. Rev. A* **53** 3761-6

[15] Fletcher S C 2013 Light Clocks and the Clock Hypothesis *Found Phys* **43** 1369-83

[16] Franklin J 2010 Lorntz contraction, Bell's spaceships and rigid body motion in special relativity *Eur. J. Phys.* **31** 291-8

# 5 RIGID BODY MOTION IN SPECIAL RELATIVITY

JERROLD FRANKLIN

**Abstract**    We study the acceleration and motion of a rigid body in special relativity. We first show that the definition of 'rigid body' in relativity differs from the usual classical definition, so there is no difficulty in dealing with rigid bodies in relativistic motion. We then describe

1. the motion of a rigid body undergoing constant acceleration to a given velocity.

2. the motion of a rigid body due to an applied impulse.

3. the motion of a relativistic spaceship.

## 5.1   Introduction

How can we talk about rigid bodies in special relativity when some authorities deny their existence in special relativity? For instance, Pauli[1] wrote "the concept of a *rigid body* has no place in relativistic mechanics," while Panofsky and Phillips[2] state that special relativity "precludes the existence of the 'ideal rigid body'." Most other textbooks do not mention the words 'rigid body' in connection with special relativity.

Yet, in his 1905 paper[3], Einstein writes the phrases "Let there be given a stationary rigid rod ...", and "We envisage a rigid sphere...", and four years later Born[4] gave conditions for rigid body motion in relativity. Thus rigid bodies are at the heart of special relativity, yet some authorities deny their existence.

Can we resolve these statements? Although the previous quote of Pauli is often referred to, he went on to add "it is nevertheless useful and natural to introduce the concept of a *rigid motion* of a body." What does he mean by this? Pauli's (and others'[5, 6, 7]) objection to use of the term"rigid body" in special relativity was that the rotation of a rigid body could not be described in relativity. That seems to be asking too much of it, since special relativity includes only Lorentz transformations and no (nonstatic) rotational transfor-

C. Duston, M. Holman (Eds), *Spacetime Physics 1907 - 2017. Selected peer-reviewed papers presented at the First Hermann Minkowski Meeting on the Foundations of Spacetime Physics, 15-18 May 2017, Albena, Bulgaria* (Minkowski Institute Press, Montreal 2019). ISBN 978-1-927763-48-3 (softcover), ISBN 978-1-927763-49-0 (ebook).

mations for anything. We shall thus discuss the motion of a rigid body only in translational motion with the acceleration parallel to the velocity, and Pauli did countenance that.

The objection of Panofsky and Phillips to a relativistic rigid body is that "its ends would move simultaneously as observed from any frame". (In fact, we will show below that the ends of an accelerating rigid rod do not move simultaneously in any frame.) They also argue that if a truly rigid body were kicked at one end, the other end would move instantly rather than at a retarded time.

This is not only a relativistic objection. The fact that an electromagnetic signal could not propagate faster than $c/\sqrt{\epsilon\mu}$ was shown long before the advent of relativity. Since a rod is held together by electromagnetic forces, the simultaneous motion of the right end if the left end were kicked is ruled out on classical grounds, because of the need to use the retarded time. Actually, of course, $c/\sqrt{\epsilon\mu}$ is an unrealistically fast upper limit to the speed of motion in a material rod. The actual transmission speed of an impulse is really governed by the speed of sound in the rod, which is orders of magnitude less than $c/\sqrt{\epsilon\mu}$.

The argument is sometimes made that, in the nonrelativistic motion of an object, the speed of sound in the object has no limit if Young's modulus is taken to infinity, while special relativity limits the speed of transmission to c. However, the notion of an infinite Young's modulus is not tenable for real physical objects, so any realistic speed of sound is orders of magnitude lower than the speed of light. Consequently, there is no real difference in relativistic or nonrelativistic mechanics in the introduction of the abstraction of an ideal rigid body.

Actually, the cooperative nature of a many body wave function in quantum mechanics includes the possibility of the components of an object moving simultaneously. One example of this is the recoil of a crystal in the Mossbauer absorption of a photon. If no phonons are excited in the absorption, the entire crystal recoils collectively. Not to do so would spoil the Mossbauer effect.

In prerelativistic dynamics, the motion of a rigid body is generally defined as preserving the dimensions of the body during any motion of the body. There are two problems with this definition. First, any actual physical object will have elastic properties, so there must be some distortion during accelerated motion. Second, due to the finite velocity of sound in any real object, one end of a rigid rod will not move until a short time after the other end is struck.

These difficulties are generally dispensed with by assuming that the body is so rigid that the elastic deformation can be ignored, and the speed of sound so fast that the initial delay in the motion of the other end can also be neglected. This leads to the abstraction of an 'ideal rigid body' that is used in all the books and papers treating classical rigid body motion. The finite speed of sound in the object does lead to oscillations that are generally too small to be visible, and only manifest themselves as the 'ping' you hear when you strike a rigid body.

Then again there is the Panofsky and Phillips objection that the 'relativistic length' of a moving object would change as its velocity increases. This violates the classical definition that rigid body motion preserves the dimensions of a body during any motion of the body. But, this is an example of how using a prerelativistic definition for a relativistic phenomenon leads to confusion. In fact, the proper relativistic definition of a rigid body turns the classical definition on its head. If an object retained its length while moving, its length would increase in its rest system. Consequently, we take as our definition of a rigid body,

**A RIGID BODY RETAINS ITS REST FRAME LENGTH WHILE IN MOTION.**[1]

This requires a moving rigid body to change its length in any frame in which it is moving.

## 5.2 Constant acceleration

We consider the motion of a rigid rod of length $L_0$ that starts from rest in a Lorentz system S. We assume a constant acceleration so that we can find explicit trajectory equations for the motion. Thus each point on the rod undergoes a constant acceleration in its instantaneous rest system S′ [2]. We show below that, in order to keep a constant length in its rest system, the front and back ends of the rod must have different constant accelerations, $a'_F$ and $a'_B$, in the rest system.

As the rod's velocity increases in the frame S, the constant acceleration $\mathbf{a}'$ of any point on the rod in its rest system is related to the acceleration $\mathbf{a}$ in frame S where that point is moving with velocity $v$ by[3][8]

$$\mathbf{a}' = (1 - v^2)^{-\frac{3}{2}}\mathbf{a} = \gamma^3\mathbf{a}. \tag{5.1}$$

We use the relation

$$\frac{d}{dt}(\gamma\mathbf{v}) = \frac{d}{dt}\left[\frac{\mathbf{v}}{\sqrt{1-\mathbf{v}^2}}\right] = \gamma\mathbf{a} + \gamma^3\mathbf{v}(\mathbf{v}\cdot\mathbf{a})] = \gamma^3\mathbf{a}, \tag{5.2}$$

which holds for $\mathbf{a}$ parallel to $\mathbf{v}$, to write

$$\mathbf{a}' = \frac{d}{dt}(\gamma\mathbf{v}). \tag{5.3}$$

---

[1] This is called 'Born rigid motion'[4].

[2] By "instantaneous rest system", we mean a Lorentz system moving at constant velocity in which that point on the rod is momentarily at rest.

[3] We use units with $c = 1$.

We solve this differential equation by the following steps:

$$a' = \frac{d(\gamma v)}{dt}$$

$$a't = \gamma v = \frac{v}{\sqrt{1 - v^2}}$$

$$v = \frac{a't}{\sqrt{1 + a'^2 t^2}} = \frac{dx}{dt}$$

$$\int_{x_0}^{x} d\bar{x} = \int_0^t \frac{a'\bar{t}d\bar{t}}{\sqrt{1 + a'^2 \bar{t}^2}}$$

$$x = x_0 + \left(\sqrt{1 + a'^2 t^2} - 1\right)/a'. \tag{5.4}$$

The equation of motion of each end of the rod is given by Eq. (5.4) as

$$x_F = L_0 + \left(\sqrt{1 + a_F'^2 t_F^2} - 1\right)/a_F' \tag{5.5}$$

$$x_B = \left(\sqrt{1 + a_B'^2 t_B^2} - 1\right)/a_B', \tag{5.6}$$

where $t_F$ and $t_B$ are the times at which the front $(x_F)$ and back $(x_B)$ ends of the rod are measured.

Rigid body motion for the rod means keeping the distance between the ends of the rod constant at $L_0$ in their mutual rest system. In order to transform to the rest system of the rod, we have to know $x_F$, $x_B$, $t_F$, and $t_B$ when each end has the same velocity in S. We can do this by using the relations

$$t = \gamma v/a' \quad \text{and} \quad \gamma = \sqrt{1 + a'^2 t^2}, \tag{5.7}$$

which follow from the steps in Eq. (5.4) above. Then, we have

$$x_F = L_0 + (\gamma - 1)/a_F'$$
$$x_B = (\gamma - 1)/a_B' \tag{5.8}$$

for the location of each end of the rod when they have the same velocity $v$. The two times $t_F$ and $t_B$ are now different. These times are given by

$$t_F = \gamma v/a_F'$$
$$t_B = \gamma v/a_B'. \tag{5.9}$$

The condition that the distance between the ends in the rest system be fixed at $L_0$ can be imposed by Lorentz transforming their difference $\Delta x = x_F - x_B$ in system S to the rest system. The space and time differences for the two ends follow from Eqs. (5.8) and (5.9):

$$\Delta x = L_0 + (\gamma - 1)\delta \tag{5.10}$$
$$\Delta t = \gamma v \delta, \tag{5.11}$$

where

$$\delta = \frac{1}{a'_F} - \frac{1}{a'_B}. \tag{5.12}$$

The Lorentz transformation to the rest frame is

$$
\begin{aligned}
L_0 &= \Delta x' = \gamma(\Delta x - v\Delta t) \\
&= \gamma[L_0 + (\gamma - 1)\delta - v^2 \gamma \delta] \\
&= \gamma L_0 + (1 - \gamma)\delta.
\end{aligned}
\tag{5.13}
$$

This equation has the solution

$$L_0 = \delta = \frac{1}{a'_F} - \frac{1}{a'_B}, \tag{5.14}$$

so the acceleration of the back end of the rod is related to that of the front end by

$$a'_B = \frac{a'_F}{1 - a'_F L_0}. \tag{5.15}$$

Thus there is a fixed relation between the constant accelerations of the two ends of the rod in its instantaneous rest system. Maintaining these different rest frame accelerations for each end will keep the rest frame distance, $L_0$, between them constant. The variation in acceleration also holds for any point on the rod, with its rest frame acceleration given by $a'_B$ in Eq. (5.15) with $a'_F$ being the acceleration of the front end and $L_0$ representing the $x$ distance from the front end.

We see that in order to keep a body rigid in its rest frame, the acceleration has to vary throughout the body in a specific way. Although the acceleration varies, there will be no strain because this varying acceleration preserves the rest frame dimensions of the body. Any stress in the body will not be appreciably different than the stress induced by nonrelativistic acceleration of a rigid body. Also, it does not matter where on the rigid body the impetus for acceleration acts. The accelerated motion is a cooperative process with the acceleration of any part of the rigid body being specified by Eq. (5.15)

The results above give the motion of the ends of a rigid rod undergoing continuous constant acceleration. We now relate this to a rod that undergoes constant acceleration from rest that ends when the rod reaches a final velocity $V$. Equation (5.9) shows that, in frame S, the back end will reach the velocity $V$ at a time $T_B = \gamma V/a'_B$, which is earlier than the time $T_F = \gamma V/a'_F$ at which the front end reaches velocity $V$. This means that, starting at $T_B$, the back end will move at constant velocity, while the front end continues to accelerate until $T_F$, at which time each end will continue with the same velocity $V$.

The motion of the rod in frame S is shown in as the solid trajectory in Fig. 1.

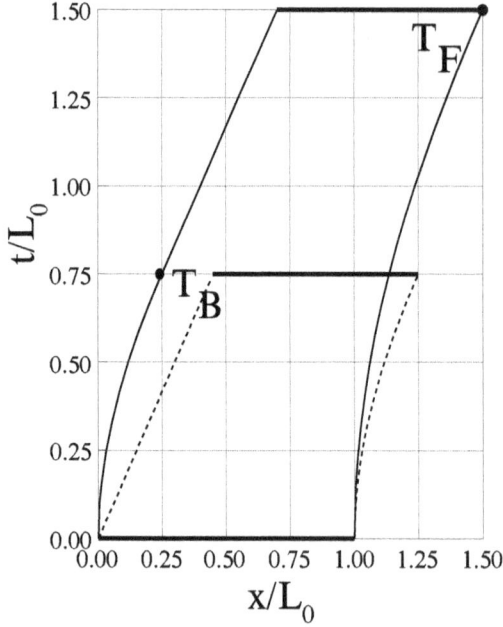

**Figure 5.1:** Constant acceleration of a rigid body. The solid curve is the trajectory for continuous acceleration. The dashed curve is for impulsive acceleration. The time $T_B$ on the solid curve represents the end of acceleration for the back end of the rod, and $T_F$ for the front end.

The figure represents the space-time curve for acceleration in frame S from rest to a final velocity $V = 0.6$, for which $\gamma = 1.25$. We have chosen the rest frame accelerations to be $a'_F = 1/(2L_0)$ and $a'_B = 1/L_0$, which are consistent with Eq. (5.14). The acceleration continues until each end of the rod reaches velocity $V$, which occurs at equal times in the rest system, but at the unequal times $T_F$ and $T_B$ shown on the figure. At the time the acceleration stops in the rest system, the front end of the rod is at a position $X_F$ and the back end is at $X_B$. The difference $X_F - X_B$ is given by Eq. (5.10) to be $\gamma L_0 = (5/3)L_0$. This length is at different times in system S, but it would be the measured length if observers in S made the length measurement when told to by passengers at the front and back ends of the rod.

After the acceleration of the back end stops at time $T_B$, the back end of the rod travels in frame S at constant velocity $V$, while the front end continues to accelerate until time $T_F$. The length[4] of the rod decreases to

$$L = \gamma L_0 - V(T_F - T_B) = \gamma L_0 - \gamma V^2 L_0 = L_0/\gamma, \qquad (5.16)$$

where we have used Eq. (5.9) for the time difference $(T_F - T_B)$. At time $T_F$, both ends of the rod will have the same velocity $V$, and they will continue

---

[4]When we use the word 'length' or the symbol $L$ without a qualifier, we mean the difference $x_F - x_B$ measured at equal times in system S.

to move at that constant velocity. At any time after $T_F$, the rod's length, measured at equal times in frame S, remains a constant length $L = L_0/\gamma$, the usual 'Lorentz contraction'.

During the accelerated motion, the distance between the ends of the rod measured at equal times is given until time $t_B$ by the difference

$$L = x_F - x_B = \sqrt{t^2 + 1/a_F'^2} - \sqrt{t^2 + 1/a_B'^2}, \quad 0 \le t \le t_B, \tag{5.17}$$

where we have used Eqs. (5.5) and (5.6) for $x_F$ and $x_B$. From time $t_B$ until time $t_F$, the distance between the ends is given by the difference of $x_F$ as given by Eq. (5.5) and $x_B$ given by $x_B = V(t - T_B) + X_B$. After some algebra, this results in

$$L = \sqrt{t^2 + 1/a_F'^2} - Vt. \quad t_B \le t \le t_F, \tag{5.18}$$

Although the motion described above keeps the rest frame length of the rod constant, we see that the distance between the ends of the rod, measured in system S at the same time for each end, will decrease. This decrease is shown in Fig. 2, which is a plot of Eqs. (5.17) and (5.18). The equal time length continually decreases from $L_0$ to $L_0/\gamma$ when each end has the final constant velocity $V$. We see that while the classical definition of a rigid body requires it to have a constant length while accelerating, the relativistic definition requires its length to change.

The stopping motion of a rigid rod of moving length $L_0/\gamma$ undergoing constant deceleration from an initial velocity $-V$ to come to rest at $t = 0$ is given by the same equations (5.5) and (5.6) as for acceleration, but with the changes $t \to -t$, $v \to -v$, and the interchange of the subscripts $F$ and $B$. This corresponds to the reverse motion with time going from $-t$ to 0. This can be depicted on Fig. 1, by just moving down the vertical time axis (now thought of as $-t$). The rod moves with velocity $-V$ until time $-t_B$ (located at $T_F$ in the figure), at which time the new back end will start to decelerate. The front end will continue at constant velocity $-V$, until it starts to decelerate at $-t_F$ (located at $T_B$ in the figure). Each end will come to rest at $t = 0$, with the length of the rod now $L_0$.

## 5.3 Impulsive acceleration

Impulsive acceleration occurs when one end of a rod is given an infinite acceleration in an infinitesimal time $\Delta t$ so that, in the limit $\Delta t \to 0$, the product $a\Delta t$ approaches a finite change $\Delta V$ in the velocity of one end of the rod. We consider the case of a rigid rod originally at rest for the which the back end acquires a velocity $V$, and continues to move at that constant rate. It does not matter where on the rod the impulse is exerted. Because of the cooperative nature of rigid body acceleration, it will always be the back end that acquires the instantaneous velocity $V$ with $a_B' \to \infty$.

We see from Eq. (5.14) that, with $a_B' \to \infty$, the front end will have a finite acceleration $a_F' = 1/L_0$. Then, using Eq. (5.5), the front end of the rod will

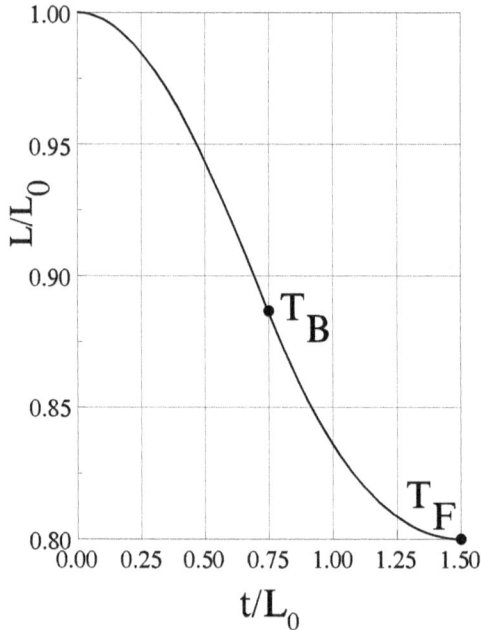

**Figure 5.2:** Equal time length of an accelerating rigid rod. The length decreases from $L_0$ to $L_0/\gamma$. The time $\mathbf{T_B}$ on the solid curve represents the end of acceleration for the back end of the rod, and $\mathbf{T_F}$ for the front end.

follow the trajectory

$$
\begin{aligned}
x_F &= L_0 + \left( \sqrt{1 + t_F^2/L_0^2} - 1 \right) L_0 \\
&= \sqrt{L_0^2 + t_F^2}.
\end{aligned}
\tag{5.19}
$$

This acceleration will continue until the front end reaches the same velocity as the back end. From Eq. (5.9), we see that this occurs at a time

$$
T_F = V\gamma L_0,
\tag{5.20}
$$

after which both ends continue at the constant velocity $V$. This impulsive motion is shown as the dashed trajectory in Fig. 1 for the same final velocity $V = 0.6$ as we used for continuous acceleration.

The inelastic collision of a rigid rod with a brick wall so that the rod comes to rest after impacting the wall corresponds to moving down in time on the dashed trajectory in Fig. 1. The front end of the rod continues at constant velocity $-V$ until it strikes the wall. The back end starts to decelerate at the time shown as $T_B$ in Fig. 1, and follows the equation $x = \sqrt{t^2 + L_0^2}$ with $t^2$ decreasing until it equals zero, and the length of the rod is $L_0$. Viewers in system S may be surprised to see the back end of the rod start to decelerate before the front end hits the wall. However, in the rest system of the rod, the onset of deceleration occurs at the same time for each end. Because the invariant separation of the front and back ends is space-like, the relative time order can be different in other Lorentz frames, but this has no physical significance. The early deceleration of the back end seen by viewers in system S is illusory. We can draw the following conclusions from what the viewers see:

**RELATIVE TIME IS AN ILLUSION**
**FOR SPACE-LIKE SEPARATIONS.**

**THERE IS NO 'BEFORE' OR 'AFTER'**
**FOR SPACE-LIKE SEPARATIONS.**

An elastic collision of a rigid rod with a wall so that the rod rebounds with the same velocity as it approached the wall corresponds to the same approach to instantaneous rest as above, followed by immediate impulsive acceleration. The collision will be elastic if the final velocity has the same magnitude as the approach velocity.

## 5.4   A relativistic spaceship

Let's say that you are the pilot of a spaceship that is too long to fit in its hangar.[5]

The ship can be given a relativistic velocity so that it is shorter than the hangar.

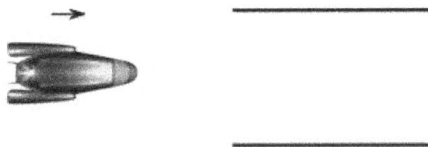

---

[5]This is a space age version of the 'pole and barn paradox', which has been the subject of many conflicting papers and online treatments. I have not seen one of these that treats the situation correctly.

Then the spaceship can enter the hangar,

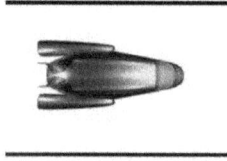

and both doors be closed with the spaceship inside.

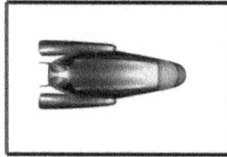

The back door can be opened,

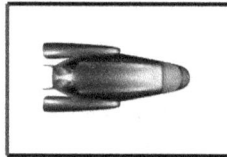

to let the spaceship leave the hangar.

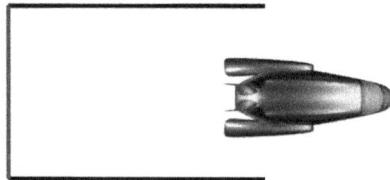

It looks like we have managed to put the spaceship (temporarily) into a hangar that is too short for it, but did that really happen or was it an illusion?

We can get some insight by looking in the rest frame of the spaceship. There, the people in the spaceship see a Lorentz shortened hangar coming toward them with velocity $V$. At first, both doors of the hangar are open. Then, shortly before the hangar approaches the spaceship, the rear door closes briefly and then opens again, so the spaceship can pass through. The front door remains open until the spaceship has completely passed through it, at which point it can close. The rear door remains open until the spaceship has passed completely through it.

The important point is that the two doors that closed simultaneously in the hangar's rest system follow a completely different time ordering in the system in which the hangar is moving. Consequently, nothing unusual happens in the rest system of the spaceship. Passengers in the spaceship would ask spectators on the ground, what made you think we were inside the hangar with both doors closed?

Although people on the ground and those in the spaceship see completely different things, the two situations are connected by the Lorentz transformation between the rest system of the hangar and that of the spaceship. Was the spaceship actually locked temporarily inside the hangar or was that an illusion? We can answer that question by looking at what would happen if the ground crew forgot to reopen the back door of the hangar.

Then, in the spaceship's rest frame, the back door of the hangar would strike the front end of the spaceship, greatly damaging it in what we will assume was a completely inelastic collision. If the hangar and door are strong enough, and are strongly attached to the earth, the spaceship will undergo an impulsive acceleration bringing it up to the speed of the hangar as discussed earlier in section 3. It will follow the dashed path shown in Fig. 1 for impulsive acceleration, becoming Lorentz contracted, but still longer than the Lorentz contracted hangar. Since the rear end of the spaceship will project out beyond the hangar, the front door of the hangar will not be able to close but will be obstructed by the spaceship.

What happens in the rest frame of the hangar follows directly by Lorentz transforming the trajectories in the original rest frame of the spaceship to the rest frame of the hangar. There, when the back door remains closed, the spaceship will come to a halt at the back door just as in the case of an inelastic collision with a fixed wall discussed at the end of section 2. As seen there by going down the curve for impulsive acceleration in figure 1, the rear end of the spaceship decelerates before the collision, so that the spaceship has its full rest frame length when it comes to rest at the back door. Then, just as in the rest frame of the spaceship, the front door is blocked from closing by the spaceship.

The above discussion shows that if the back door does not reopen, the picture

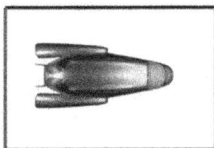

of the spaceship inside the hangar with both doors closed cannot happen. If the backdoor does not reopen, the final picture will look like this:

78

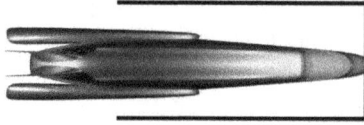

Just as in the rest frame of the spaceship, the front door will be blocked from closing simultaneously with the back door.

There are some more quite unusual pictures to look at. We have already shown the picture

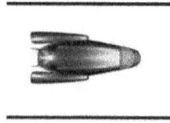

of the spaceship inside the hangar in the case when the back door will close and then reopen. If the back door does not reopen, the picture would be

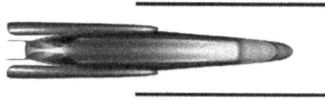

These two pictures are remarkable. In each picture, the hangar has both doors open, but in the top picture the spaceship is Lorentz contracted and in the bottom picture, it is at its full rest frame length. For the top picture, the back door will shut and then quickly reopen. For the bottom picture, the back door will not reopen. Neither the closing nor possible reopening of the back door has yet happened. How does the spaceship know what the door will do before it happens?

The situation seems paradoxical for people watching the events. People on the ground, and even the ground crew, see just by looking at the length of the spaceship, whether or not the ground crew will reopen the back door even before the ground crew decides what to do.

This crazy situation is resolved by the conclusion:

## THERE IS NO 'BEFORE' OR 'AFTER' FOR SPACELIKE SEPARATIONS.

It is not that one or the other of the contradictory pictures is wrong, but rather that each of them is a kind of illusion. In the rest frame of the spaceship, the sequence of events is straightforward. The spaceship is at rest with its full length. If the backdoor is reopened, the hangar passes by the spaceship, which fits through both doors. If the backdoor remains shut, the hangar collides with the nose of the spaceship. In either case, there is no ambiguity, and no confusion about the time sequence of events. This is an example of the fact that:

**WHAT HAPPENS TO AN EXTENDED OBJECT HAPPENS IN ITS REST FRAME.**

As applied to the relativistic spaceship, this means that a simple understanding of the sequence of events acting on the spaceship is only possible in the rest frame of the spaceship. Another result of this conclusion is that application of Newton's first and second laws (in their relativistic version) for forces acting simultaneously on an extended object can only be done in the object's rest frame. A Lorentz transformation would then give the corresponding result in any other reference frame. Trying to apply these laws to moving objects has led to a number of 'paradoxes' in special relativity.

## 5.5   Summary and Conclusions

Using the definition that *a rigid body retains its rest frame length while in motion*, we have discussed the accelerated motion of rigid bodies in special relativity. The trajectories for uniform acceleration and impulsive acceleration of a rigid rod have been shown in figure 1. Deceleration to a stop and collision with a brick wall can be seen in the figure by moving down the time axis, reversing the direction of time. Although these trajectories are for uniform acceleration, the general features would be about the same for any reasonable acceleration. These general features are:

1. Different parts of an accelerating rigid body undergo different accelerations in the rest frame, with passengers at the rear of an accelerating spaceship feeling a greater force than those at the front.

2. In a collision with a brick wall, the rear end of a rigid body decelerates before the collision of the front-end so that the rigid body attains its rest frame length by the time of the collision.

In section 4, we treated the case of a relativistic spaceship flying through the opening and closing doors of a hangar that was shorter than the rest length of the spaceship. Some of the remarkable aspects of this scenario lead us to the following conclusions:

1. **Relative time is an illusion for spacelike separations.**

2. **There is no 'before' or 'after' for spacelike separations.**

3. **What happens to an extended object happens in its rest frame.**

# References

[1] W. Pauli, *Theory of Relativity* p. 132 (Pergamon Press, Oxford, UK) 1958.

[2] W. K. H. Panofsky and M. Phillips, *Classical Electricity and Magnetism, 2nd Ed.* p.287 (Addison-Wesley, Reading, MA) 1962.

[3] A. Einstein, Annalen der Physik, **17**, 891 (1905).

[4] M. Born, Annelen der Physik, **30**, 1 (1909).

[5] P. Ehrenfest, Phys. Z., **10**, 918 (1909).

[6] G. Herglotz, Ann. Phys., Lpg., **31**, 393 (1910).

[7] F. Noether, Ann. Phys., Lpg., **31**, 919 (1910).

[8] See, for instance, Eq. (14.26) in: J. Franklin, *Classical Electromagnetism, 2nd Ed.* (Dover Publications, 2017).

# 6 THIN SHELL WITH FICTITIOUS MOTIONS

HOU Y. YAU

**Abstract**   We show that a spherical thin shell with fictitious radial motions can mimic the gravitational effects of a massive spherical thin shell. These fictitious radial motions (either with a uniform velocity or oscillate about the thin shell) are not physical motions of the thin shell. Instead, we use their hypothetical effects on time and distance measurements to define the spacetime metric on the surface of the timelike thin shell. The system with fictitious motions is spherically symmetric with time translational and time reflection symmetries. The external spacetime outside the thin shell is static and satisfies the Schwarzschild solution for the gravitational field of a spherically symmetric mass.

## 6.1   Introduction

The Schwarzschild metric has a remarkably simple form. Its diagonal time component, $g_{tt}$, is inversely proportional to its diagonal radial component, $g_{rr}$ [1]. Because of this simple form, many attempts have been made to reconcile the metric by invoking the reciprocity of time dilation and length contraction in special relativity [2]. For instance, the Schwarzschild metric can be written as,

$$ds^2 = [1 - \frac{v(r)^2}{c^2}]dt^2 - [1 - \frac{v(r)^2}{c^2}]^{-1}dr^2 - r^2 d\Omega^2, \qquad (6.1)$$

where $v(r) = -(2GM/r)^{1/2}$. It happens that $v(r)$ is the free falling velocity of a particle with zero initial velocity at infinity. The diagonal time and radial metric components can then be expressed in terms of the time dilation and length contraction observed for a moving object with velocity $v(r)$. As advocated in refs. [3, 4, 5, 6, 7, 8], the Schwarzschild metric is obtained from the study of a radially accelerated free falling frame by insinuating the concepts of time dilation, length contraction, equivalence principle and Newtonian gravity

C. Duston, M. Holman (Eds), *Spacetime Physics 1907 - 2017. Selected peer-reviewed papers presented at the First Hermann Minkowski Meeting on the Foundations of Spacetime Physics, 15-18 May 2017, Albena, Bulgaria* (Minkowski Institute Press, Montreal 2019). ISBN 978-1-927763-48-3 (softcover), ISBN 978-1-927763-49-0 (ebook).

in the derivation. These results have led to a false belief that it is possible to derive the Schwarzschild metric without any reference to the field equations of general relativity.

The result that Schwarzschild metric can be obtained by combining special relativity, equivalence principle and Newtonian gravity is a mere coincidence. As demonstrated by Gruber et al. [9], two gravitational effects are necessary to specify the two *a priori* diagonal components $g_{tt}(r)$ and $g_{rr}(r)$. The first piece of gravitational information they considered is the free falling radial inward velocity $v(r)$ of a particle measured by an observer stationary at a coordinate $r$. With the knowledge of this velocity, one can obtain the total mass-energy of a free falling particle measured by the same stationary observer. The diagonal component $g_{tt}(r)$ can then be directly inferred, for example, from a gedanken experiment with photons (Appendix in ref. [9]).

To obtain the diagonal radial component $g_{rr}(r)$, Gruber et al. chose to study the gravitational force law, $f(r) = -d^2r/d\tau^2$, where $\tau$ is the proper time measured by the free falling observer. This is 'a plausible analogy of New-tonian radial acceleration that is at the same time relativistically palatable'. With both $v(r)$ and $f(r)$ known, $g_{rr}(r)$ can be inferred. For example, in the Newtonian gravitational theory, $v(r) = -(2GM/r)^{1/2}$ and $f(r) = GM/r^2$ for a gravitating body of mass $M$. The diagonal metric components obtained are $g_{tt} = 1 - 2GM/rc^2$ and $g_{rr} = [1 - 2GM/rc^2]^{-1}$, which happen to be the exact solutions we are looking for. However, when we extend our considera-tions to relativistic gravity, we expect $v(r)$ and $f(r)$ will have to be corrected to capture the relativistic gravitational effects. This can be done by writing $v(r) = -(2GM/r)^{1/2}U(GM/rc^2)$ and $f(r) = (GM/r^2)W(GM/rc^2)$, where $U(GM/rc^2)$ and $W(GM/rc^2)$ are functions with expansions of the dimension-less quantity $GM/rc^2$. (Note that $U = W = 1$ in Newtonian gravity.) The relativistic corrections are supposed to be captured in the dimensionless func-tions $U$ and $W$. Therefore, in order to determine $g_{tt}(r)$ and $g_{rr}(r)$, information from experiments or theory other than the Newtonian gravity must be speci-fied. This is where the Einstein's field equations are needed.

The assumption that all of the higher power of $GM/rc^2$ in functions $U$ and $W$ are zero is unjustified without the Einstein's field equations, albeit it may be true by accident that the particular metric components are already given exactly in the non-relativistic limit. Using only special relativity, equiv-alence principle and Newtonian gravity, it is not sufficient to determine the components of the Schwarzschild metric when relativistic gravitational effects are significant. These arguments, thus, falsify the claims proposed in refs. [3, 4, 5, 6, 7, 8] that the Schwarzschild metric can be obtained without the explicit use of general relativity.

The aforementioned inverse proportionality of the diagonal coefficients is a source of continuing confusions. Based on various reasonings, Schild [10], Rindler [11], Sacks and Ball [12], Kassner [13] have arrived at the same conclu-sion that there is no simple derivation of the Schwarzschild metric. Although it is unreasonable to believe that the Schwarzschild metric can be derived from time dilation and length contraction without general relativity, the erroneous attributions of these features do come up with an answer that looks like the correct one. If so, can there be other constructive ways to use these concepts in the gravitational theory? In this paper, we show that the time dilation and length contraction for the 'fictitious motions' (either a uniform fictitious radial velocity or fictitious oscillations) of a thin shell can be applied to mimic the gravitational properties of a massive thin shell.

In quantum theory, we learned that a particle is an oscillator. However, its amplitude has only a probabilistic interpretation based on Born's postulate [14]. In refs. [15, 16, 17], we demonstrate a possibility that, apart from the classical description of mass [18, 19], a particle can have an intrinsic oscillation in proper time. By allowing matter to vibrate in the time direction, we can reconcile the quantum properties of a matter field. The scalar field describing the vibrations of matter in time satisfies the Klein-Gordon equation and Schrödinger equation. The energy in this system is quantized under the constraint that the energy of mass is on shell. Apart from the proper time oscillation, we can identify another intrinsic property for a particle. The fictitious radial oscillation introduced in this paper has effects that can curve spacetime. A thin shell with fictitious radial oscillations can be contracted to infinitesimal radius that resembles a point mass. In addition, the spacetime around this thin shell with infinitesimal radius is the Schwarzschild field. The fictitious radial oscillation is another intrinsic property that a particle can hypothetically possess.

This paper is organized as follows. Section 2 outlines the properties of a thin shell with a 'uniform fictitious radial velocity'. Time dilation and length contraction derived from this fictitious velocity can be used to define the spacetime metric on the surface of a thin shell. Section 3 shows that a thin shell with 'fictitious radial oscillations" can also produce the same effects obtained in Section 2 based on a time translational symmetry. Section 4 demonstrates that the external spacetime outside the thin shell with fictitious motions is static and satisfies the Schwarzschild solution for the gravitational field of a spherically symmetric mass. Section 5 is reserved for discussions. In the following analysis, we will adopt the units $c = G = 1$.

## 6.2 Thin Shell with a Uniform Fictitious Radial Velocity

The way Eq. (6.1) is written has created a deceptive impression that the Schwarzschild metric can be derived from time dilation and length contraction. As it appears, an observer $O'$ stationary at a particular radial coordinate $r$ is subject to a velocity $v(r)$. However, this velocity is not physical motion of the observer in spacetime. In fact, observer $O'$ is considered stationary in the spatial coordinates. This "fictitious velocity" is what is responsible for the apparent time dilation and length contraction, i.e. $(dt')^2 = [1 - v(r)^2]dt^2$ and $(dr')^2 = [1 - v(r)^2]^{-1}dr^2$. As discussed earlier, it is unreasonable to assume that we can apply these concepts at every point in spacetime to obtain the Schwarzschild metric. Here, we will limit their applications only on the surface of a timelike hypersurface. Our goal is not to derive the Schwarzschild metric but to show that the gravitational properties derived from a spherical thin shell with fictitious radial motions is compatible with those for a spherical thin shell with mass $M$.

Let us consider a coordinate system $(t, r, \theta, \phi)$. The coordinate time $t$ is measured by the clock of a stationary observer $O$ located at spatial infinity. The radial coordinate $r$ is defined as the circumference, divided by $2\pi$, of a sphere centered around the spherical thin shell to be investigated. The angular coordinates $\theta$ and $\phi$ are the usual polar spherical angular coordinates. This coordinate system is the same adopted for the conventional Schwarzschild field.

We will assume an infinitesimally thin spherical shell $\Sigma$ with radius $\breve{R}$ is centered at the origin of this coordinate system. A stationary observer $\breve{O}$ on this thin shell will be subject to a uniform fictitious radial velocity $v_{fm}$ as if it is constantly moving relative to the observer $O$ at spatial infinity. However, the thin shell itself has no physical movement in regular spacetime. The uniform fictitious radial velocity $v_{fm}$ has effects on the time and distance measured at $r = \breve{R}$ while observer $\breve{O}$ remains stationary on the thin shell. Outside this thin shell, the spacetime is a vacuum with no fictitious radial motion. The hypothetical effects of the fictitious velocity on time and distance measurements are used to define the spacetime metric on the surface of the timelike thin shell.

In a Minkowski spacetime, the clock of a stationary observer at any location shall be synchronized with the clock of observer $O$ at spatial infinity. However, this is not the case for observer $\breve{O}$ stationary on the thin shell. For the purpose of our discussions, it will be easier if we adopt the use of a fictitious frame $Q$. In this fictitious frame, it is the clock of the fictitious observer $Q$ that synchronizes with the clock of $O$. In addition, the two observers measure the same length for the same object. We shall consider the effects of the fictitious velocity at $r = \breve{R}$ as if $\breve{O}$ is moving in the fictitious frame of $Q$. An observer $\breve{O}$ on the thin shell will have a uniform fictitious velocity $v_{fm}$ moving relative to the fictitious observer $Q$. Therefore, $\breve{O}$ is under the constant effects of a uniform fictitious velocity while remaining at rest relative to $O$ at spatial infinity.

Let us consider two events in frame $\breve{O}$. We will assume the fictitious velocity is $|v_{fm}| < 1$. The infinitesimal $d\breve{t}$ and $d\breve{r}$ are the differences in the time and radial coordinates between the two events. They can be related to the coordinate increments $dt$ and $dr$ for the same two events observed by $O$,

$$\begin{bmatrix} dt \\ dr \end{bmatrix} = \begin{bmatrix} \Upsilon^t{}_{\breve{t}} & \Upsilon^t{}_{\breve{r}} \\ \Upsilon^r{}_{\breve{t}} & \Upsilon^r{}_{\breve{r}} \end{bmatrix} \begin{bmatrix} d\breve{t} \\ d\breve{r} \end{bmatrix}. \tag{6.2}$$

In the local frames of $O$ and $\breve{O}$, the respective basis vectors in the time and radial directions are orthogonal, i.e. $\vec{e}_t \cdot \vec{e}_r = 0$ and $\vec{e}_{\breve{t}} \cdot \vec{e}_{\breve{r}} = 0$. On the other hand, $\breve{O}$ is stationary relative to $O$. The time and radial basis vectors in frame $O$ are parallel to their counterparts in frame $\breve{O}$, i.e. $\vec{e}_{\breve{t}} \parallel \vec{e}_t$, and $\vec{e}_{\breve{r}} \parallel \vec{e}_r$. Under these conditions, the transformation matrix $\Upsilon$ is diagonal,

$$\Upsilon^t{}_{\breve{r}} = \Upsilon^r{}_{\breve{t}} = 0. \tag{6.3}$$

When $d\breve{r} = 0$, $d\breve{t}$ is a proper time measured by the clock carried by $\breve{O}$. This timelike interval can be Lorentz transformed to the fictitious frame of $Q$,

$$d\underline{t} = \gamma d\breve{t}, \tag{6.4}$$

$$d\underline{r} = \gamma \underline{v}_{fm} d\breve{t}, \tag{6.5}$$

where $\gamma = [1 - (v_{fm})^2]^{-1/2}$. In the fictitious frame, $\breve{O}$ travels a distance $d\underline{r}$ over a time $d\underline{t}$. On the other hand, the clocks of $O$ and $Q$ are synchronized. $O$ shall measures the same time as $Q$,

$$dt = d\underline{t} = \gamma d\breve{t}. \tag{6.6}$$

However, $O$ is physically stationary relative to $\breve{O}$, i.e.

$$dr = 0. \tag{6.7}$$

The underlined quantity in Eq. (6.5) is a fictitious infinitesimal displacement that appears only in the fictitious frame of $Q$. Under the effect of this fictitious velocity, time measured by the clock of $O$ is dilated but there is no physical movement between $O$ and $\breve{O}$. From Eq. (6.6),

$$\Upsilon^t{}_{\breve{t}} = \gamma = [1 - (v_{fm})^2]^{-1/2}. \tag{6.8}$$

Next, we will consider a measuring rod with length $d\breve{r}$ carried by $\breve{O}$. This spacelike interval can be expressed as two events measured at the endpoints of the rod simultaneously, $d\breve{t} = 0$. Again, we can Lorentz transform these two events to the fictitious frame of $Q$,

$$d\underline{t} = \gamma \underline{v}_{fm} d\breve{r}, \tag{6.9}$$

$$d\underline{r} = \gamma d\breve{r}. \tag{6.10}$$

From the viewpoint of $Q$, the rod carried by $\breve{O}$ is moving at a velocity $\underline{v}_{fm}$. To obtain the moving length $d\underline{l}$ of the rod, we shall subtract $d\underline{r}$ by the distance traveled by the rod during $d\underline{t}$,

$$d\underline{l} = d\underline{r} - \underline{v}_{fm} d\underline{t} = \gamma^{-1} d\breve{r}. \tag{6.11}$$

Since the rod is of the same length as measured by $O$ and $Q$,

$$dr = d\underline{l} = \gamma^{-1} d\breve{r}. \tag{6.12}$$

On the other hand, the rod carried by $\breve{O}$ is stationary relative to $O$. The underlined quantities in Eqs. (6.9) and (6.11) are fictitious temporal and spatial infinitesimal displacements that only appear in the fictitious frame of $Q$. Their effects shorten the rod observed in frame $O$ but there is no physical movement between $O$ and $\breve{O}$. The spacelike interval $dr$ representing the length of the rod in frame $O$ is measured simultaneously at the endpoints,

$$dt = 0. \tag{6.13}$$

From Eq. (6.12),

$$\Upsilon^r{}_{\breve{r}} = \gamma^{-1} = [1 - (v_{fm})^2]^{1/2}. \tag{6.14}$$

Therefore, Eq. (6.2) becomes,

$$\begin{bmatrix} dt \\ dr \end{bmatrix} = \begin{bmatrix} [1 - (v_{fm})^2]^{-1/2} & 0 \\ 0 & [1 - (v_{fm})^2]^{1/2} \end{bmatrix} \begin{bmatrix} d\breve{t} \\ d\breve{r} \end{bmatrix}, \tag{6.15}$$

which is based on the results from Eqs. (6.3), (6.8) and (6.14).

## 6.3 Thin Shell with Fictitious Oscillations

In this section, we will show that the same transformation matrix $\Upsilon$ from Eq. (6.15) can be obtained by considering a thin shell that has fictitious oscillations in the radial direction. Let us again consider the infinitesimally thin spherical shell $\Sigma$ with radius $\breve{R}$. We will introduce fictitious radial oscillations on the surface of this thin shell,

$$\underline{r}_f(\underline{t}) = -\Re \cos(\omega_0 \underline{t}), \tag{6.16}$$

$$\underline{v}_f(\underline{t}) = \frac{\partial \underline{r}_f(\underline{t})}{\partial \underline{t}} = \Re \omega_0 \sin(\omega_0 \underline{t}), \tag{6.17}$$

where $\Re$ and $\omega_0$ are the amplitude and angular frequency of the fictitious oscillations respectively. As we have discussed in the last section, we will adopt the use of a fictitious frame $Q$. In this fictitious frame, it is the clock of a fictitious observer $Q$ oscillating about $r = \breve{R}$ that synchronizes with the clock of observer $O$ at spatial infinity, i.e. $t = \underline{t}$. In addition, the two observers measure the same length for the same object, i.e. $l = \underline{l}$. Although an observer $\breve{O}$ on the thin shell has no physical motion relative to observer $O$ at spatial infinity, $\breve{O}$ is subject to a fictitious oscillation. At a particular time, $\breve{O}$ is displaced a distance $\underline{r}_f$ with an instantaneous velocity $\underline{v}_f$ relative to $Q$ in the fictitious frame. The fictitious oscillation does not carry an observer through regular spacetime. There is no physical motion of matter. Instead, the fictitious oscillation is used to define the geometry of spacetime at $r = \breve{R}$. Furthermore, the system is spherically symmetric with fictitious oscillations in the radial direction only. The region outside the thin shell is a vacuum which is source free. There is no fictitious vibration in the vacuum spacetime.

From Eqs. (6.16) and (6.17), the fictitious displacement and instantaneous velocity at $\underline{t} = \underline{t}_m = \pi/(2\omega_0)$ are,

$$\underline{r}_f(\underline{t}_m) = 0, \tag{6.18}$$

and

$$v_f(\underline{t}_m) = v_{fm} = \Re\omega_0, \qquad (6.19)$$

where we will assume $v_{fm} = \Re\omega_0 < 1$ in this section. $\breve{O}$ is traveling with a velocity $v_{fm}$ in the fictitious frame with no displacement relative to $Q$. (Note that we equate the peak fictitious velocity $v_f(\underline{t}_m)$ with the uniform fictitious velocity $v_{fm}$ developed in the previous section to show later that both the use of fictitious oscillations and the uniform fictitious velocity can produce the same gravitational effects.) As we shall recall, the properties at $r = \breve{R}$ with fictitious velocity $v_{fm}$ are already discussed in Section 2. However, in the fictitious oscillating system, apart from the instantaneous velocity $v_f$, $\breve{O}$ also has a displacement $r_f$ relative to the fictitious observer $Q$. As a simple harmonic oscillating system, we expect both the fictitious displacement and its instantaneous velocity can have effects on $\breve{O}$.

Let us rewrite the transformation matrix $\Upsilon$ from Eq. (6.15) in terms of a constant $\breve{I}$ for the thin shell with fictitious radial oscillations, i.e.

$$\begin{bmatrix} dt \\ dr \end{bmatrix} = \begin{bmatrix} (1-\breve{I})^{-1/2} & 0 \\ 0 & (1-\breve{I})^{1/2} \end{bmatrix} \begin{bmatrix} d\breve{t} \\ d\breve{r} \end{bmatrix}, \qquad (6.20)$$

where

$$\breve{I} = \omega_0^2[r_f(\underline{t})]^2 + [v_f(\underline{t})]^2 = v_{fm}^2 = \Re^2\omega_0^2. \qquad (6.21)$$

At $\underline{t} = \underline{t}_m = \pi/(2\omega_0)$, $\breve{I} = v_{fm}^2$. The oscillation has an instantaneous velocity $v_{fm}$ with no displacement relative to the fictitious observer $Q$ as discussed earlier. On the other hand, the constant fictitious velocity from Section 2 is replaced by a fictitious oscillation in Eqs. (6.20) and (6.21). The constant $\breve{I}$ is the summation of two parts analogous to the "potential" and "kinetic" components of a classical oscillating system. At a particular instant, the fictitious displacement $r_f(\underline{t})$ and the fictitious instantaneous velocity $v_f(\underline{t})$ satisfy Eq. (6.21) which is a typical solution for a simple harmonic oscillating system. In addition, $\Upsilon$ is a constant matrix. The effects of the fictitious oscillation on $\breve{O}$ at any instant is equivalent to those produced by the fictitious velocity $v_{fm}$ at $\underline{t} = \pi/(2\omega_0)$. The same transformation matrix $\Upsilon$ from Eq. (6.15) can be obtained by replacing the constant fictitious radial velocity with fictitious oscillation in the radial direction.

As we shall recall, the total energy of a simple harmonic oscillating system is typically conserved. This is what we expect for the system with fictitious oscillations. Based on the Noether's theorem, the system we are considering shall have a time translational symmetry. This fact will be confirmed below. As a result, the combined effects from the fictitious displacement $r_f$ and fictitious velocity $v_f$ on observer $\breve{O}$ shall be a constant over time. Therefore, there is no surprise that $\Upsilon$ is a constant matrix for the thin shell with fictitious radial oscillations.

The fictitious oscillations in the system are entirely radial. It is spherically symmetric and there is no rotational motion. Thus, the line element at $r = \breve{R}$ can be written as [20],

$$ds^2 = g_{tt}(\breve{R})dt^2 + 2g_{tr}(\breve{R})dtdr + g_{rr}(\breve{R})dr^2 - \breve{R}^2d\Omega^2, \qquad (6.22)$$

where $\Omega$ is the metric induced on each 2-sphere using the radial coordinates of our reference system.

The temporal (radial) coordinate in frame $O$, and its counterpart in frame $\breve{O}$, are of different scale. From Eq. (6.20), we can relate the basis vectors in frames $O$ and $\breve{O}$,

$$\vec{e}_{\breve{t}} = \vec{e}_t(1 - \breve{I})^{1/2}, \tag{6.23}$$

$$\vec{e}_{\breve{r}} = \vec{e}_r(1 - \breve{I})^{-1/2}. \tag{6.24}$$

As a result, the metrics at $O$ and $\breve{O}$ are different. From Eqs. (6.23) and (6.24),

$$g_{tt}(\breve{R}) = \vec{e}_{\breve{t}} \cdot \vec{e}_{\breve{t}} = (1 - \breve{I})\vec{e}_t \cdot \vec{e}_t = 1 - \breve{I}, \tag{6.25}$$

$$g_{rr}(\breve{R}) = \vec{e}_{\breve{r}} \cdot \vec{e}_{\breve{r}} = (1 - \breve{I})^{-1}\vec{e}_r \cdot \vec{e}_r = -(1 - \breve{I})^{-1}, \tag{6.26}$$

$$g_{tr}(\breve{R}) = g_{rt}(\breve{R}) = \vec{e}_{\breve{t}} \cdot \vec{e}_{\breve{r}} = \vec{e}_t \cdot \vec{e}_r = 0, \tag{6.27}$$

where $\vec{e}_t \cdot \vec{e}_t = 1$, $\vec{e}_r \cdot \vec{e}_r = -1$, and $\vec{e}_t \cdot \vec{e}_r = 0$. Therefore, the line element at $r = \breve{R}$ is,

$$ds^2 = [1 - \breve{I}]dt^2 - [1 - \breve{I}]^{-1}dr^2 - \breve{R}^2 d\Omega^2. \tag{6.28}$$

Apart from the time translational symmetry, the spacetime at $r = \breve{R}$ is also invariant under time reflection symmetry ($t \to -t$).

## 6.4 Schwarzschild Spacetime

From Eq. (6.21), both the uniform fictitious radial velocity $v_{fm}$ and the fictitious oscillation with amplitude $\Re$ can give rise to the same line element in Eq. (6.28). Using these fictitious radial motions (either a uniform fictitious radial velocity or fictitious oscillations) to define the geometry of spacetime at $r = \breve{R}$, the metric derived on the surface of a thin shell is different from the assumed flat spacetime at spatial infinity. The geometry of spacetime at these two distant locations are different. If the spacetime manifold outside the thin shell is smooth and continuous, its structures cannot be flat. The fictitious motions have effects that can curve spacetime.

Eq. (6.28) is the line element of the Schwarzschild metric on the surface of a thin shell with total mass $M$ if we set

$$\breve{I} = \frac{2M}{\breve{R}}, \tag{6.29}$$

or

$$M = \frac{\breve{R}\Re^2\omega_0^2}{2} = \frac{\breve{R}v_{fm}^2}{2}. \tag{6.30}$$

From general relativity, the vacuum space–time $v^+$ outside this spherical thin shell $\Sigma$ (a time-like hypersurface) is the Schwarzschild spacetime, i.e.

$$ds^2 = [1 - \frac{\breve{R}\Re^2\omega_0^2}{r}]dt^2 - [1 - \frac{\breve{R}\Re^2\omega_0^2}{r}]^{-1}dr^2 - r^2 d\Omega^2, \tag{6.31}$$

or

$$ds^2 = [1 - \frac{\breve{R}\underline{v}_{fm}^2}{r}]dt^2 - [1 - \frac{\breve{R}\underline{v}_{fm}^2}{r}]^{-1}dr^2 - r^2 d\Omega^2. \tag{6.32}$$

The spacetime around the thin shell with fictitious motions is static with time translational and time reflection symmetries.

The Birkhoff's theorem [21, 22] states that the gravitational field of any spherically symmetric vacuum region is necessarily static, and its metric is that of the Schwarzschild spacetime. This applies to the external field of any non-rotating, spherical, uniform thin shell whether the shell is static, fluctuating or collapsing. Applying the same principle, the time-like hypersurface $\Sigma$ can be expanded (or contracted) by 'carrying' the fictitious oscillations along geodesics orthogonal to the original surface to a new sphere $\Sigma'$. As long as mass $M$ given in Eq. (6.30) is remaining constant during this transformation, the metric and curvature of the external field will not be affected. Under this condition, the amplitude of the radial oscillation is,

$$\Re = \sqrt{\frac{2M}{\breve{R}\omega_0^2}}, \tag{6.33}$$

or the uniform fictitious radial velocity is,

$$\underline{v}_{fm} = \sqrt{\frac{2M}{\breve{R}}}. \tag{6.34}$$

The metric from Eqs. (6.31) and (6.32) will encounter a coordinate singularity when the shell is contracted to a radius $\breve{R} = 2M$ (the event horizon). Although the fictitious instantaneous velocity on the shell at event horizon can reach the speed of light (i.e. $\underline{v}_{fm} = 1$ from Eqs. (6.19), (6.33) and (6.34) when $\breve{R} = 2M$), it is not physical motion of matter. As we shall recall, the fictitious motions have effects on the time and distance measurements of a stationary observer. The spacetime metric at $r = \breve{R}$ can be inferred from these fictitious motions. However, there is no physical movement by the stationary observer or the thin shell. The fictitious motions can curve spacetime but will not produce superluminal transfer of energy even for a shell that is inside the event horizon with $\underline{v}_{fm} > 1$.

The amplitude $\Re$ from Eq. (6.33), the uniform fictitious velocity $\underline{v}_{fm}$ from Eq. (6.34), and the related spacetime curvature tensors derived from the metric (e.g. the coordinate independent Kretschmann invariant [23], $R^{\alpha\beta\gamma\delta}R_{\alpha\beta\gamma\delta} = 48M^2/r^6 = 12\breve{R}^2\Re^4\omega_0^4/r^6$), are well defined as the shell is contracted until it reaches a radius $\breve{R} \to 0$ where it meets a true singularity [24, 25]. At this point, the shell is infinitely small but has infinitely large amplitude of oscillations, $\Re \to \infty$, or infinitely large fictitious radial velocity, $\underline{v}_{fm} \to \infty$. As predicted by Birkhoff's theorem, the metric around this infinitely small shell is the Schwarzschild spacetime. An infinitesimally small thin shell with fictitious motions can mimic the gravitational properties of a point mass in relativity.

Before we conclude this section, let us address a question about the fictitious oscillations. From Eq. (6.34), the uniform fictitious radial velocity $\underline{v}_{fm}$ can be established directly from the mass $M$. Therefore, there is no ambiguity about what value $\underline{v}_{fm}$ shall be used when applying to a thin shell with mass $M$. However, this is not so trivial for the amplitude of fictitious oscillations $\mathfrak{R}$. In order to determine $\mathfrak{R}$, we also need to specify the angular frequency $\omega_0$ in Eq. (6.33). This is ambiguous when we apply the fictitious oscillations for a macroscopic mass since there is no known 'distinguished' angular frequency that we can associate with a macroscopic object at rest. The angular frequency can therefore be assigned arbitrarily. On the contrary, there is one reasonable choice when the application is for a quantum particle. As de Broglie has conjectured in his 1924 thesis [26], a particle possesses an internal clock. This hypothesis has been recently tested, and the experimental data obtained are found to be compatible with the theory [27]. Considering the particle as a point mass, the angular frequency $\omega_0$ specified in our formulation can be taken as the frequency of the de Broglie's internal clock. This frequency is unique for the particle. However, it is unreasonable to extend this idea of an internal clock for a macroscopic object. Quantum effects are supposed to be negligible in the classical energy level and there is no analogy of an internal clock for a macroscopic object. Therefore, to avoid ambiguity, we will limit our application of the fictitious oscillations for quantum systems only.

## 6.5  Conclusions and Discussions

The misconception that time dilation and length contraction can be applied to derive the Schwarzschild metric without general relativity is a source of continuing confusions. Repeated attempts have been made despite this idea has been refuted many times by others. Yet the simple form of this concept makes its application very appealing. In this paper, based on a better understanding of what limits its applications, we borrow the idea and apply it to derive the metric on the surface of a thin shell. One major difference from the previous attempts is that the outside spacetime is maintained as vacuum without fictitious motions in our study. Our goal is not to derive the Schwarzschild metric without general relativity. Instead, this application helps to paint a simple picture on how matter can be connected to spacetime. A massive thin shell exerts fictitious motions on its surface. This alters the spacetime metric on the surface of the thin shell and curves the surrounding external spacetime. In turn, the curved spacetime tells other matters how to react in the presence of the thin shell. The previously unsuccessful idea can have another application in the gravitational theory.

The fictitious radial oscillation is different from the proper time oscillation of a particle studied in refs. [15, 16, 17]. In these referenced papers, we show that a particle can have an intrinsic oscillation in proper time. As a part of spacetime, this oscillation in time shall have effects on its surrounding spacetime. Here, we demonstrate another intrinsic property possible for a particle. The fictitious radial oscillations with de Broglie's frequency on a thin shell with infinitesimal radius can give rise to the Schwarzschild spacetime. Unlike the proper time oscillation of a particle, the fictitious radial oscillations are not motions of matter in the time direction (nor in the spatial direction). Instead, their effects can curve spacetime, These fictitious radial oscillations can be what we need to connect the proper time oscillation of a particle with the external spacetime. The proper time oscillator is the source of the particle's mass-energy which can be the driving force for the fictitious radial oscillations. It is not implausible that a particle can have additional intrinsic properties that are linked more directly with spacetime.

# References

[1] T. Jacobson, "When is $g_{tt}g_{rr} = -1$?", Class. Quant. Grav. **24**, 5717 (2007).

[2] K. Kassner, "A physics-first approach to the Schwarzschild metric", Advanced Studies in Theoretical Physics **11**, 179 (2017) - Abrief summary of the attempts to reconcile the Schwarzschild metric from special relativity, equivalence principle and Newtonian gravity can be found in the Introduction of this paper.

[3] W. Lenz, 1944 unpublished work cited in A. Sommerfeld, *Electrodynamics (Lectures on Theoretical Physics Vol. 3)*, (Academic Press, New York, 1967).

[4] L. Schiff, "On experimental tests of the general theory of relativity", Am. J. Phys. **28**, 340 (1960).

[5] M. Harwit, *Astrophysical Concepts*, (Wiley, New York, 1973).

[6] P. Rowlands, "A simple approach to the experimental consequences of general relativity", Physics Education **32**, 49 (1997).

[7] J. Czerniawski, "The possibility of a simple derivation of the Schwarzschild metric", *Preprint* gr-qc/0611104.

[8] R. Cuzinatto, B. Pimental, and P. Pompeia, "Schwarzschild and de Sitter solution from the argument by Lenz and Sommerfiled', Am. J. Phys. **79**, 662 (2011).

[9] R. Gruber, R. Price, S. Matthews, W. Cordwell, and L. Wagner, "The impossibility of a simple derivation of the Schwarzschild metric", Am. J. Phys. **56**, 265 (1988).

[10] A. Schild, "Equivalence Principle and Red-Shift Measurements", Am. J. Phys. **28**, 778 (1960).

[11] W. Rindler, "Counterexample to the Lenz-Schiff Argument", Am. J. Phys. **36**, 540 (1968).

[12] W. Sacks and J. Ball, "Simple derivations of the Schwarzschild metric", Am. J. Phys. **36**, 240 (1968).

[13] K. Kassner, "Classroom reconstruction of the Schwarzschild metric", Eur. J. Phys. **36**, 065031 (2015).

[14] M. Born, "On the Quantum Theory of the Electromagnetic Field", Proc. Roy. Soc. Lond. A **143**, 410 (1934).

[15] H. Y. Yau, " Emerged quantum field of a deterministic system with vibrations in space and time", AIP Conf. Proc. **1508**, 514 (2012).

[16] H. Y. Yau, "Probabilistic nature of a field with time as a dynamical variable", Lect. Notes Comp. Sci. **10106**, 33 (2016).

[17] H. Y. Yau, "Temporal vibrations in a quantized field", in *Quantum Foundations, Probability and Information* pg. 269, ed. A. Khrennikov and B. Toni, (Springer, Verlag, 2018).

[18] M. Jammer, *Concepts of Mass in Contemporary Physics and Philosophy*, (Princeton University Press, Princeton, 2009).

[19] L. Okun, *Energy and Mass in Relativity Theory*, (World Scientific, Singapore, 2009).

[20] R. Wald, *General Relativity*, (University of Chicago Press, Chicago, 1984).

[21] G. Birkhoff, *Relativity and Modern Physics*, (Harvard University Press, Cambridge-Massachusetts, 1923).

[22] H. Schmidt, "The tetralogy of Birkhoff theorems", Gen. Rel. Grav. **45**, 395 (2013).

[23] C. Cherubini, D. Bini, S. Capozziello, and R. Ruffini, "Second order scalar invariants of the Riemann tensor: applications to black hole spacetimes", Int. J. Mod. Phys. D **11**, 827 (2002).

[24] S. Hawking and R. Penrose, "The singularities of gravitational collapse and cosmology", Proc. Roy. Soc. Lond. A **314**, 529 (1970).

[25] S. Hawking and G. Ellis, *The Large-Scale Structure of Space–Time*, (Cambridge University Press, Cambridge, 1973).

[26] L. de Broglie, Ph. D. thesis, Universite De Paris (1924).

[27] P. Catillon et al., "A search for the de Broglie particle internal clock by means of electron channeling", Found. Phys. **38**, 659 (2008).

# 7  MOTION, TIME AND GRAVITY

NADJA S. MAGALHAES

**Abstract**   In this work we investigate the concept of time within the context of how physicists use it, based on an operational definition. We analyze the fundamentals of a physical clock relating the time measure that it yields with the measurement of distances in space. In this approach time primarily plays the role of a parameter whose values are obtained by measurements of a dynamical variable of a clock, which is a measure of distance. As a consequence, dynamics establishes correlations between displacements instead of between displacements and time, and velocity is dimensionless. The fact that the ordering parameter (time) relates physically to a system other than the focus of the observation implies that the spacetime metric contains elements of two distinct physical systems. The relation between gravity and spacetime appears naturally through the analysis of the 3D motion of a light clock. We show that gravity distorts spacetime due to its role in changing textcolorblueparticles' trajectories.

## 7.1   Introduction

The meaning of time in Physics is still matter of debate. Associated to this unsettled definition is the ignorance on whether the world is a 3-dimensional one that evolves with time or a 4-dimensional spatiotemporally extended object. This question relates to the standard mathematical framework of special relativity as well as to the quantization of gravity, evidencing the importance of it.

In the context of Physics Aristotle pondered on the nature of time, analyzed it in relation to motion and recognized the difficulties about its attributes [2]. Centuries later Newton adopted the concept of "absolute, true and mathematical time", which had a nature such that it would flow equably without regard to anything external (Newton's substantivalism). He recognized also

C. Duston, M. Holman (Eds), *Spacetime Physics 1907 - 2017. Selected peer-reviewed papers presented at the First Hermann Minkowski Meeting on the Foundations of Spacetime Physics, 15-18 May 2017, Albena, Bulgaria* (Minkowski Institute Press, Montreal 2019). ISBN 978-1-927763-48-3 (softcover), ISBN 978-1-927763-49-0 (ebook).

the relative, apparent and common time: some sensible and external measure of duration by the means of motion, which was commonly used instead of true time [12]. To describe physical systems he assumed that the three-dimensional spatial geometry of the universe was independent of one-dimensional absolute time.

Contemporary to Newton, Leibniz described space and time as systems of relations that exist between objects (Leibniz's relationism) which have no existence apart from the existence of those objects [7]. After much debate Newton's substantivalist views of space and time prevailed and although his idea of absolute time is useful to investigate many systems it is does not suffice to explain some physical phenomena. With his theory of relativity Einstein [5] changed the Newtonian paradigm of absolute time focusing on the idea of relative time, which followed from his theory's two postulates: (1) The laws of physics are invariant in all inertial systems; and (2) The speed of light in a vacuum is the same for all observers, regardless of the motion of the light source.

As a consequence of assuming these postulates, the four dimensions were joined together. In 1908 Minkowski [11] presented a geometric interpretation of special relativity that fused time and the three spatial dimensions of space into a single four-dimensional continuum now known as Minkowski space, or spacetime.

Spacetime has successfully being applied to explain classical and relativistic systems. Nevertheless, since 1908 the reality of the four-dimensional spacetime has been under discussion [11]. This issue is a major motivation of this work and is related to a profound question: what is time?

More recently, in the effort to quantize gravity other ways of dealing with time have emerged, including shape dynamics [3, 4, 8], a gravity theory that completely dismisses time. Notwithstanding, time is a concept - or, according to some, a real object [14, 18] - that perhaps will never cease to be used [9].

In this work we investigate the concept of time within the context of how physicists use it, based on an operational definition. We analyze the fundamentals of a physical clock, C, relating the time that it yields with the measurement of distances in space. Necessarily, the knowledge of time depends on the knowledge of the movement of things [1], which happens in 3-space. In the process of defining motion relative to an observer we argue that the acquisition and storage of a spacetime event is a discrete process. As well, by using a light clock to yield time we conclude that one of the possible Minkowski metrics $(+ - - -)$ leads to an intuitive picture regarding time.

In this approach time primarily plays the role of a parameter whose values are obtained through measurements of a dynamical variable of a clock, which in turn can be considered a monotonically perfect clock: for some choice of initial state, its observed values increase monotonically, a needed characteristic for a parameter that allows the description of evolution. The use of dynamical variables to describe time apparently has limitations in quantum theory [19], so we restrict ourselves to the classical (non-quantum and yet relativistic) one.

Since we show that the dynamical variable that yields time relates to a measure of distance, the average speed is not obtained by dividing a displacement by a time interval but by another displacement. Therefore, kinematic relations are established between displacements instead of between displacements and time. The displacements that result from the clock's readings are uniquely related to values of the parameter time, which is a tool for ordering events related to an observed particle, $P$. As a consequence, we show that the speed of light in natural units ($c = 1$) indicates that time is being measured with the aid of a light clock.

The fact that the ordering parameter (time) relates physically to a system ($C$) other than the focus of the observation ($P$) implies that the spacetime metric contains elements of two distinct physical systems that may or may not be physically correlated. As a consequence we discuss the implications of this fact to gravity in the context of general relativity and we argue that the 4D spacetime seems to be a mathematical construct rather than a physical entity. The effect of gravity can be translated into spacetime dynamics through the 3D motion of a light clock. This is how we explain the common view that "gravity is spacetime" in general relativity.

This paper is organized as follows: in the next Section we relate kinematics to the need for time; in Section 7.3 we introduce our definition of a clock. Section 7.4 presents the definition of velocity and how the use of a light clock yields $c = 1$. In Section 7.5 we discuss spacetime and a physical meaning for the interval $ds$ when the metric $(+ - - -)$ is adopted. The metric of a curved spacetime is deduced in Section 7.6 based on the motion of a light clock. Our final comments are presented in Section 7.7.

## 7.2 Motion

We begin this approach to the study of motion assuming the *relativity principle* (all inertial frames are equivalent for the performance of all physical experiments) as well as the axiom of *invariance of causality* (if an event P causes an event Q in an inertial frame, then P must cause Q in any inertial frame) [15] and Einstein's postulate of the universality of the speed of light (the speed of light relative to any unaccelerated observer is c~ $3\times 10^8$ m s$^{-1}$ regardless of the motion of the light's source relative to the observer) [17].

Consider an uncharged point-like particle, $P$, as the physical system being observed by an observer, $O$, in the absence of gravity. For Einstein [10] an observer is equivalent to an inertial reference frame that has a rigid ruler plus a clock at each point of space. We will adopt this equivalence following the formalism by Schutz [17]. We add the condition that $O$ *must be able to store and analyze* the data collected. All these characteristic imply that the observer is not a quantum system.

An *observation* comprises the simultaneous determination, by $O$, of both the position of $P$ and the instant of time in which the position is being measured. The position $\vec{x} = (x^1, x^2, x^3)$ (in a 3D Euclidean space) is measured relative to $O$ with the ruler, while the instant of time time, $x^0$, is obtained from a clock positioned at $\vec{x}$. Following Minkowski, these values - which are *physically* connected by the same spatial position and by being simultaneously observed[1] - are connected *mathematically* through an *event*, $x^\mu$, which is a four-vector that belongs to a 4D spacetime, with $\mu = 0, 1, 2, 3$: $x^\mu \equiv (x^0, \vec{x})$. The observation may take a while, and the time interval needed for the observation will be denoted by $\epsilon_{obs}$.

After the observation takes place, $O$ must store the data collected for later analysis. The *event collection process* involves not only the simultaneous determination of $x^0$ and $\vec{x}$, but also the *storage* of these data. After the event is determined, a time interval $\epsilon_{sto}$ is needed for $O$ to store it. We will assume, for simplicity, that only after the storage of an event the next event can be observed, avoiding multichannel observation.

The total time needed to collect and store an event characterizes the *sampling time*, $\epsilon_{samp} \equiv \epsilon_{obs} + \epsilon_{sto}$. Therefore, ultimately the event collection process is a *discrete* process whose level of detail, or accuracy, depends on how small $\epsilon_{obs}$ and $\epsilon_{sto}$ can be. In this work we consider this process practically continuous, as our focus is on classical relativity. This means that both $\epsilon_{obs}$ and $\epsilon_{sto}$ are infinitesimally small.

The observer should be able to determine multiple consecutive events. The *data set* is the set of events $\{x^\mu\}$ collected by $O$ during a time interval $\Delta x^0_{if} = x^0_f - x^0_i$, with $\Delta x^0_{if} > \epsilon_{samp}$.

This data set allows the observer to determine the state of *motion* of $P$. If $O$ limited the observations to only one event it would be impossible to determine whether $P$ was moving or not, as such sole event would appear a still picture of $P$. In order to investigate whether $P$ has moved or not, $O$ has to determine at least two events, $x^\mu_i$ and $x^\mu_f$. The observer can conclude that $P$ has moved when the initial spatial position is different from the final one, or: $\Delta \vec{x_{if}} \equiv \vec{x_f} - \vec{x_i} \neq 0$.

The time variable is needed by $O$ in the study of motion first to *order* the observations allowing, for instance, the conclusion that causes precede their effects and, second, to quantify *changes* in the state of motion. Such quantification requires a standard of motion, which is related to the concept of clock and will be discussed shortly. On the other hand, in order to move the observed particle has no need whatsoever of time as measured by the observer.

## 7.3   The clock and the description of motion

The description of $P$'s motion by $O$ depends on how the events are ordered in time, and such ordering must imply that $\Delta \vec{x_{if}}$ is causally consistent.

Physically time is determined from the motion of any system that $O$ chooses to order its observations [6], generically known as the clock (C). We note that time, while expected to be used as an ordering parameter, results from the observation of the motion of a system that is *not* $P$. It has been argued elsewhere that time may have a distinct role, being a component of an actual object [14, 18], a vision that differs from ours, as we argue below.

---

[1]It is impossible for an actual clock, the ruler and $P$ to be simultaneously at the same point in space when $O$ performs the observation. However, we will assume that both the clock and the ruler are close enough to $P$ to yield an event with good accuracy.

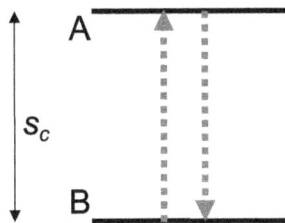

**Figure 7.1:** Schematics of a light clock that belongs to the reference frame of an observer $O$ (not depicted), being at absolute rest in that frame. The parallel mirrors A and B are separated by a constant distance $s_c$ and a light ray (dashed arrow) bounces back and forth, forever trapped.

As an ordering parameter time has arbitrary starting point (origin) and, for simplicity it is convenient that it is given in real, non-negative numbers. Physically, time is related to a dynamical variable of some preferred system, indicating that our framework is non-quantum as the use of dynamical variables to describe time apparently has limitations in quantum theory [19].

Periodic systems normally provide reasonably reliable, predictable values for dynamical variables that can be used to define the time variable. This is so for the motion of the Sun or the Moon in the sky, perhaps the most primitive of the clocks. One particularly useful periodic system is the light clock [6, 13], which is independent of the structure of matter. Such clock can be designed using two parallel mirrors in free motion with a constant distance, $s_c$, between them and letting light bounce back and forth (Figure 7.1). The unit of time would be the interval between two "ticks" of this clock, one tick when light leaves one mirror and the second tick when it returns to that mirror, which corresponds to a total distance of $2s_c$.

Therefore, in a fundamental level an instant in time corresponds to a distance, $x^0 = as_c$, where $a$ is a dimensionless real, non-negative number [6]. This is consistent with $O$'s need for a *standard of motion* (which, in this case, is the steady bouncing of the light ray) in order to determine the motion of $P$ through comparison.

Consequently, a spacetime event $x^\mu$ consists of a *distance* $x^0$ that is traveled by a *standard* object (the trapped light ray) plus the three space *coordinates* of $P$ obtained simultaneously to the determination of $x^0$. This implies that at every two observations the change in $P$'s position can be compared to the respective distance traveled by light in the standard system, C.

Since the value of $x^0$ by definition always increases, then events can be ordered as they are observed. Such ordering allows the observer to determine which event happened before or after a given one. Time then fulfills its role as a convenient parameter used by O to label the measurements performed and to keep them in the sequence in which they were measured. From this data the observer may, for instance, investigate causal relationships between events.

It is then evident that time does *exist* in the sense that it is physically measurable, and it exists as a *distance*[2]. As such, its initial value is arbitrary and it always sums up to positive, real numbers. These properties are those needed for any parameter that labels observations in such a way that the latter can be sequentially ordered. Moreover, by using periodic systems as clocks time also yields a standard of motion, for the clock's uniform motion is in principle well known and predictable.

The 4D spacetime, therefore, appears as fundamentally a 4D *space*: 1D that harbors the displacement of the standard of motion plus 3D that harbor the coordinates of the position of $P$. It is a convenient mathematical construct that allows $O$ to organize simultaneous observations of two *different* systems: the standard of motion (C) and the object of the observation (the particle $P$).

## 7.4   Velocity

We turn back to the end of Section 7.2, where the determination of the state of motion of $P$ was shown to depend on the value of

$$\Delta \vec{x_{12}} \equiv \vec{x_2} - \vec{x_1}. \tag{7.1}$$

These vectors belong to different events: $x_1^\mu = (x_1^0, \vec{x_1})$ and $x_2^\mu = (x_2^0, \vec{x_2})$, with event 1 preceding event 2, or $x_1^0 < x_2^0$.

If $\Delta \vec{x_{12}} = 0$ then $O$ can claim that $P$ did not move while $O$'s standard of motion (the clock's light ray) moved a distance

$$\Delta x_{12}^0 \equiv x_2^0 - x_1^0. \tag{7.2}$$

However, if $\Delta \vec{x_{12}} \neq 0$ then $P$ did move. The question now is: *how* was this motion? A simple way to describe this motion is to compare it to a known, simple motion, as the one that happens inside C. To this end it is convenient to define the ratio, $\vec{v}$, between the two simultaneous observables: the displacement of $P$ ($\Delta \vec{x_{12}}$) and the distance traveled by the light ray that belongs to C ($\Delta x_{12}^0$):

$$\vec{v} \equiv \frac{\Delta \vec{x_{12}}}{\Delta x_{12}^0}. \tag{7.3}$$

This ratio is the definition of the *average velocity* of $P$. In this approach to the study of motion, velocity is naturally dimensionless since it compares two lengths.

One consequence of this definition is that if $P$ is a light ray then equation (7.3) yields a speed $|\vec{v}| = 1$. Consequently, in this approach the speed of light is given directly in natural units. This result is consistent, since both light rays ($P$ and the one in the clock) must travel the same distances due to the postulate of the universality of the speed of light.

Therefore, this approach not only yields the speed of light in natural units, but it also provides a clear physical meaning for $c \equiv |\vec{v}| = 1$: when this happens time measurements are being made with the aid of a light clock.

---

[2]Evidently, time can be given in seconds or any of the usual units of time if $x^0$ is redefined as $t$ according to $t \equiv x^0/V$, where $V$ is a constant that changes second into, say, meter.

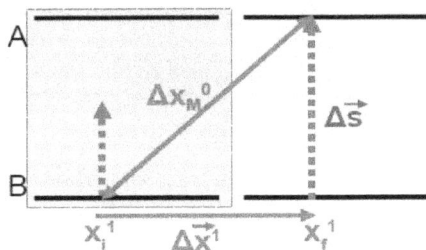

**Figure 7.2:** A moving light clock (highlighted in the grey box) travels to the right (along the $x^1$ axis) relative to an observer $O$ (not depicted) from point $x_i^1$ to point $x_f^1$. Its light ray (whose motion relative to the mirrors is shown with dotted arrows) as well as the mirrors A and B are subjected to simultaneous observations. $\Delta \vec{x^1}$ is the mirrors' displacement measured by $O$, who simultaneously measured $\Delta x_M^0$, which is the distance that $O$ observed that the light ray traveled. The distance $\Delta s$ is traveled by the light ray in the mirrors' frame during $O$'s observations.

## 7.5   Spacetime

The geometry of spacetime is determined by the *squared interval, $ds^2$*, defined by:

$$ds^2 = g_{\mu\nu}dx^\mu dx^\nu, \tag{7.4}$$

where $g_{\mu\nu}$ is the metric tensor and $x^\mu = (x^0, x^1, x^2, x^3)$, with Einstein's summation convention adopted.

A region devoid of any physical influence is described by Minkowski spacetime, which admits two metric tensors that have all components equal to zero except the diagonal ones:

Metric 1:   $g_{\mu\nu} = (+---) \Rightarrow ds^2 = (dx^0)^2 - (dx^1)^2 - (dx^2)^2 - (dx^3)^2$ (7.5)

or

Metric 2:   $g_{\mu\nu} = (-+++) \Rightarrow ds^2 = -(dx^0)^2 + (dx^1)^2 + (dx^2)^2 + (dx^3)^2.$
$$\tag{7.6}$$

While these metrics are mathematically equivalent, one of them is interesting related to time as follows. Let us assume that $O$ is observing a special system, $M$, that consists of the moving clock of Figure 7.2 where the mirrors travel at the speed $v_M$ while the light ray travels at the speed $c$. The clock C at rest in Figure 7.1 belongs to the reference frame of $O$. For simplicity $M$ moves only in the $x^1$ direction.

When observations are close enough to each other, the averaged elements in the figure, $\Delta x_M^0$ and $\Delta \vec{x^1}$, can be respectively described by the infinitesimals $dx_M^0$ and $d\vec{x^1}$. As well, light ray displacements (perpendicular to the mirrors) can be given by $ds$ *both* for C and for $M$, simultaneously. The postulate of the constancy of the speed of light ensures that when $O$ measures a time $ds$ with

C, then the distance $ds$ in Figure 7.2 has the same length, implying that $ds$ is an invariant for all inertial observers: it is the proper time.

By applying the theorem of Pythagoras to the vectors displayed in Figure 7.2 we find

$$(dx_M^0)^2 = ds^2 + (\vec{dx^1})^2. \tag{7.7}$$

This result is sometimes used to illustrate time dilation because since $v_M = \frac{dx^1}{dx_M^0}$ then equation (7.7) can be rewritten as

$$dx_M^0 = ds \left(1 - v_M^2\right)^{-1/2}, \tag{7.8}$$

implying $dx_M^0 > ds$. Thus the "tick" of M as measured by $O$ ($dx_M^0$) is longer than as measured in M's frame ($ds$).

In particular, if the clock in Figure 7.2 ceases to move in relation to $O$ then equation (7.8) yields $ds^2 = (dx^0)^2$, indicating that in this case M marks $O$'s proper time.

Note that we can rewrite equation (7.7) in another manner:

$$ds^2 = (dx_M^0)^2 - (\vec{dx^1})^2. \tag{7.9}$$

The invariant distance ($ds$) thus results from a combination of the two distances that $O$ simultaneously measures. As this equation must be obeyed by the data collected by any inertial observer, then an infinite set of ($dx_M^0$, $\vec{dx^1}$) events could be created with data from observers measuring different $v_M$ speeds. The geometrical representation of these events in the form of a 2-space seems a natural way to organize them. For such space equation (7.9) indicates that the squared interval is $ds^2$ and the metric is $(+ -)$.

Note that equation (7.9) is exactly Metric 1 limited to 2 dimensions. Had we considered that M was moving in 3 spatial dimensions then Metric 1 would be complete in the above equation. Therefore, we showed that Metric 1 can be **deduced** from the observation of a moving clock by an inertial observer that follows the postulates of special relativity.

Evidently, both Metric 1 and 2 can effectively describe spacetime mathematically. Nevertheless, Metric 1 may be particularly enlightening as in its context time can be consistently treated as a distance, which is a main claim of this work. Henceforth we will adopt Metric 1.

We turn back to Figure 7.2 to investigate the case $ds = 0$ which, from equation (7.9), implies $dx_M^0 = dx^1$. Considering that in the figure $\vec{ds}$ is the displacement vector of the light ray in M, then $ds = 0$ indicates that the light ray has zero motion in the *direction* of $\vec{ds}$. If one considers that the light ray that $O$ observes is instead *moving in the direction of* $\vec{dx^1}$ then $ds$ can be null. In this case the result $dx_M^0 = dx^1$ implies that the distance that $O$ sees the light ray cover ($dx^1$) equals the distance that $O$ regards as the time that M measures ($dx_M^0$). Evidently, in this scenario there are no mirrors in M as massive objects are unable to reach the speed of light in the theory of relativity. This conclusion in consistent with equation (7.8) because when $ds = 0$ the only way to obtain a finite value for $dx_M^0$ is when $v_M = 1$, i.e., M has the speed of light. The case $ds = 0$ therefore refers to the motion of a free light ray.

The case $ds^2 < 0$ is a mathematical possibility but it is unphysical in the context of Figure 7.2 because $ds$ is a distance described by real numbers. If $dx^0_M$ and $d\bar{x}^1$ were physically interchangeable so that M could move in the direction of a hypothetical $d\bar{x}^0_M$ then $ds^2 < 0$ would be a physical possibility. The fact that $dx^0$ is a scalar (a counting parameter) instead of a vector is a fundamental asymmetry between space and time.

In the context of the twin paradox [16], the slower aging of the twin that travels $(P)$ can be explained by the longer distances that, as seen by the twin on Earth $(O)$, interacting objects must travel before interacting, as does the light ray of $P$'s clock. Such longer distances postpone interactions that otherwise would be happening at the same pace as those that occur on Earth. When $P$ decelerates to return home, such distances are shortened and the interactions resume their paces closer to normality. The deceleration/acceleration processes are not experienced by $O$ and such asymmetry between the twin's motions is indeed the key for the twins to age at different rates, as is normally believed. As $P$ accelerates again to speeds near $c$ in his/her way back, distances between interactions increase again relative to those on Earth, and even aging happens in a slower rate. When $P$ decelerates to meet $O$, the physical changes that happened at slower paces do not vanish and the twins age at the same rate again, with $P$ younger than $O$.

## 7.6   Gravity

The study above can be generalized to the case of $P$ moving in the presence of a gravitational field. Its proper time would be marked by the clock $M$, whose mirrors' path may now change in space in a complex way due to gravity; ideally, the relative position between the bouncing ray and the mirrors should be kept constant, as well as the separation $s_c$, to ensure the clock's functionality. This condition is achieved by keeping the ray's path perpendicular to the mirrors at all times, as shown in Fig. 7.3, where $M$ moves in 2D. Mathematically this implies $\hat{n}_s \cdot \hat{n}_m = 0$, where $\hat{n}_s$ and $\hat{n}_m$ are unit vectors in the directions of the light ray motion and the mirror's motion, respectively.

We hypothesize that the study of the motion of the mirrors and the light ray across a gravitational field should yield a map of this field in spacetime, unraveling the structure of the metric tensor of Equation (7.4). As a simple illustration, for the situation in Figure 7.3 geometry (law of cosines) yields

$$\Delta s^2 = (\Delta x^0_M)^2 + (\Delta \bar{x})^2 - \Delta x^0_M \, |\Delta \bar{x}| \, \cos \theta$$

where, in general, the terms on the RHS of this equation can be a function of position. For observations infinitesimally close to each other this equation becomes

$$ds^2 = (dx^0_M)^2 + (d\bar{x})^2 - dx^0_M \, |d\bar{x}| \, \cos \theta. \tag{7.10}$$

(For 3D motion of $M$ the angle $\theta$ can be expressed in terms of the three Euler angles and $(d\bar{x})^2 = dx_i dx^i$, $i = 1, 2, 3$.)

As complex gravitational fields are expected to make the terms in equation (7.10) equally complex, under a general gravitational field equation (7.10) is

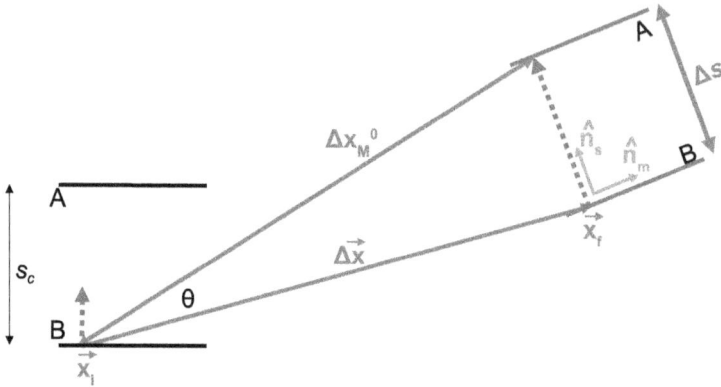

**Figure 7.3:** A moving light clock travels in two dimensions relative to an observer $O$ (not depicted) from point $\vec{x}_i$ to point $\vec{x}_f$. Its light ray, whose motion relative to the mirrors is shown with dotted arrows, and the mirrors A and B always move perpendicularly to each other. $\Delta\vec{x}$ is the mirrors' displacement measured by $O$, who simultaneously measured $\Delta x_M^0$, which is the distance that $O$ observed that the light ray traveled. The distance $\Delta s$ is traveled by the light ray in the mirrors' frame during $O$'s observations. The versors $\hat{n}_s$ and $\hat{n}_m$ are unitary vectors.

expected to take the form of equation (7.4). Therefore, *the effect of gravity on the motion of P is naturally translated into spacetime geometry through the analysis of the 3D motion of P's light clock.*

In this approach gravity is not spacetime but it is related to it through the study of motion, with the aid of a time variable obtained with a clock. What would gravity thus be? To answer this question we first recall what time is. We have shown that time is an auxiliary variable used to study motion, helping in the ordering of events and in the quantification of changes in motion. It is obtained through the use of a standard of motion (the clock) and it is actually a distance.

If motion could be studied without the need for a standard of motion, then time would probably cease to be needed. Investigations in this direction were published in the literature [8, 20]. Our approach shows that time does not bare a physical existence as a particular entity. This leads us to the conclusion that the world is three-dimensional, while the 4D spacetime is a mathematical construct convenient for the analysis of observations.

We then understand gravity not as spacetime but as a physical phenomenon that generates interactions between existing objects due to their mass and/or energy. Gravity changes motion under the postulates of the theory of relativity and for this reason its effects can be described through the squared interval, as argued above, making spacetime a useful tool to study gravity.

## 7.7 Concluding remarks

In this work we investigated motion based on measurement procedures. Through the physical understanding of the meaning of mathematical objects we expect to render greater clarity to general relativistic concepts.

We showed that time is fundamentally a distance. As a result, the speed of light is naturally given as $c = 1$ not due to the conventional use of "natural" units, but because time is ultimately a measure of length.

We argued that the perception of motion precedes the definition of time, indicating that a more fundamental question is "why does motion happen?", which is still to be answered.

We showed how the 4D spacetime is mathematically built in such a way that it allows a connection between two different physical systems, namely the position of the observed system and the clock of the observer at such position. The real world, on the other hand, appears as three dimensional.

The relation between gravity and spacetime resulted as a natural consequence of the analysis of the 3D motion of a light clock. Gravity is not spacetime but the former seems to distort the latter as gravity changes trajectories of objects.

In this work we clarified the meanings of time, spacetime and gravity. Future work could, for instance, determine the paths of light clocks in specific gravitational fields using our approach and then comparing them to the corresponding spacetime squared intervals. Alternatively, spacetime metrics could be deduced from the motion of light clocks in gravitational fields.

## Acknowledgment

The author acknowledges Sao Paulo Research Foundation (FAPESP) for support under the thematic project # 2013/26258-4.

## References

[1] Augustine St: "Confessions," trans. Henry Chadwick, Oxford Univ. Press, Book XI, P 30 (1991)

[2] Aristotle: "Physics", trans. R. P. Hardie and R. K. Gaye, The Internet Classics Archive by Daniel C. Stevenson, Book IV, Part 10. Available at "http://classics.mit.edu//Aristotle/physics.html". Access in 08/04/2018 (2018)

[3] Barbour J B: "The timelessness of quantum gravity: I. The evidence from the classical theory". *Class. Quantum Grav.* **11** 2853-2873 (1994)

[4] Barbour J B: "The end of time". Oxford University Press, New York (2000)

[5] Einstein, A: "Zur Elektrodynamik bewegter Korper". *Annalen der Physik* **322** 891-921 (1905)

[6] Friedmann A A: "The world as spacetime". Minkowski Institute Press, Montreal (2014). English translation from the Russian edition (1923).

[7] Futch, M: "Leibniz's Metaphysics of Time and Space". Springer, New York (2008)

[8] Gomes H, Gryb S, Koslowski T: "Einstein gravity as a 3D conformally invariant theory" *Class. Quantum Grav.* **28** 045005 (24pp) (2011)

[9] Gryb S, Thébault P Y: "Time remains" *Brit. J. Phil. Sci.* **67** 663-705 (2016)

[10] Lorentz, H A, Einstein A, Minkowski, H: "Das Relativitatsprinzip". $6^{th}$ ed. B. G. Teubner, Stuttgart (1958)

[11] Minkowski, H: "Space and Time - Minkowski's Papers on Relativity", trans. Fritz Lewertoff and Vesselin Petkov. Minkowski Institute Press, Montreal (2012)

[12] Newton, I: "Philosophiae Naturalis Principia Mathematica", trans. Andrew Motte, Daniel Adee, New York. Available at "https://ia902706.us.archive.org/0/items/newtonspmathema00newtrich/newtonspmathema00newtrich.pdf". Access in 08/04/2018 (1846)

[13] Ohanian, H C: "Gravitation and spacetime". Norton, New York (1976)

[14] Petkov, V: "On the reality of Minkowski space" *Found. Phys.* **37** 1499-1502 (2007)

[15] Rindler, W: "Essential relativity". 2nd edn. Springer-Verlag, New York (1977)

[16] Rindler, W: "Relativity - Special, general and cosmological". 2nd edn. Oxford Univ. Press, Oxford (2006)

[17] Schutz, B F: "A first course in general relativity". $2^{nd}$ ed. Cambridge University Press, Cambridge (2009)

[18] Smolin L: "Temporal naturalism" arXiv:1310.8539v1 (2013)

[19] Unruh W G, Wald R M: "Time and the interpretation of canonical quantum gravity" *Pys. Rev.* D **40** 2598-2614 (1989)

[20] Verlinde E P: "Emergent gravity and the dark universe" *SciPost Phys.* **2** 016 (2017)

# Part II

# Alternatives to General Relativity

C. Duston, M. Holman (Eds), *Spacetime Physics 1907 - 2017. Selected peer-reviewed papers presented at the First Hermann Minkowski Meeting on the Foundations of Spacetime Physics, 15-18 May 2017, Albena, Bulgaria* (Minkowski Institute Press, Montreal 2019). ISBN 978-1-927763-48-3 (softcover), ISBN 978-1-927763-49-0 (ebook).

# 8 Noether Symmetry Approach for

## FOR

## $F(R,T,X,\varphi)$ Cosmology

Myrzakulov Ratbay, Yerzhanov Koblandy,
Bauyrzhan Gulnur, Meirbekov Bekdaulet

**Abstract**     We consider Noether symmetry approach to find out exact cosmological solutions for $F(R,T,X,\varphi)$ – model of gravity. Also we received Euler-Lagrange equations for this model. Founded the general solution for $F(R,T,X,\varphi)$ – function and particular solution for scale factor.

## 8.1   Introduction

In recent years, the interest to the generalized theory of gravitation has increased. It results in the necessity to clarify the origin of the phenomenon of cosmic acceleration (dark energy), discovered by observing the radiation of SNe Ia [1,2]. Observational data that is currently available is insufficient to determine a single model to explain the observed accelerated expansion of the universe. The simplest of all is cosmological constant $\Lambda$ which was proposed by Einstein in due time. Currently, this model is reduced to the standard cosmological $\Lambda CDM$ model. Another example of the cosmological model describes the accelerated expansion of the steel model with a scalar field. The most popular of scalar field models is quintessence. Lagrangian density for this model $L = X - V(\varphi)$, where $V(\varphi)$ is function of scalar field $\varphi$, $X = \frac{1}{2}\nabla_\mu\varphi\nabla^\mu\varphi$ is kinetic term of this field [3]. As generalization of scalar field models k-essence model with Lagrangian density is used in most general form $L = F(\varphi, X)$ [4] based on model of k-inflation [5]. Other examples of some scalar fields is tachyon field, phantom (ghost) field, dilatonic dark energy, Chaplygin gas. For inclusion in the model of more factors to build more realistic models so-called g-essence is used which also includes fermionic field $\psi$ [6,7].

Another example of gravitation theory is $F(R)$-gravity, which can be regarded as a generalization of general relativity. Here, instead of a Ricci scalar $R$ the function of Ricci scalar $R$ is used, which describes the presence of accel-

C. Duston, M. Holman (Eds), *Spacetime Physics 1907 - 2017. Selected peer-reviewed papers presented at the First Hermann Minkowski Meeting on the Foundations of Spacetime Physics, 15-18 May 2017, Albena, Bulgaria* (Minkowski Institute Press, Montreal 2019). ISBN 978-1-927763-48-3 (softcover), ISBN 978-1-927763-49-0 (ebook).

erated expansion. Also as another modified example of modify gravity can be represented by teleparallel gravity $F(T)$ [8], where $T$ is a torsion scalar [9,10]. Another popular model is the modified theory $F(R,T)$ [11,12], with the Lagrangian depending both on the Ricci scalar and torsion scalar [11-13]. Thus, as we can see, popular study summarizing a variety of models, with a view to finding the most general properties of existing models and to determine the most relevant to observational data.

For this reason in the article, we will use a scalar field with a kinetic term $X$ with united Myrzakulov gravity, written in the most general form. For more generalization, we have the most general model in the form $F(R,T,X,\varphi)$ , including scalar term with kinetic term. To study the model we use the Noether symmetry. This approach is widely used in various branches of physics, including cosmology. Here we consider the Friedmann-Lemaitre-Robertson-Walker (FRW) metric with zero curvature. This paper contains the following sections: Chapter II summarized bases of $F(R,T,X,\varphi)$ gravity. In Chapter III obtained Euler-Lagrange equation for this model. In Chapter IV, received a system of equations of the Noether symmetry. In Chapter V presented some solution of the Noether equations and scale factor solution. The conclusion of cosmological equations of $F(R,T,X,\varphi)$ gravity presented in Chapter VI.

## 8.2  $F(R,T,X,\varphi)$ gravity

We consider here FRW metric, with line element $ds^2 = dt^2 - a(t)^2(dx^2 + dy^2 + dz^2)$. The action in FRW metric is

$$
S = 2\pi^2 \int dt\, a^3 (F - \lambda_1 \left[ R - u + 6 \left( \frac{\ddot{a}}{a} + \frac{\dot{a}^2}{a^2} \right) \right] - \lambda_2 \left[ T - v + 6 \left( \frac{\dot{a}^2}{a^2} \right) \right] -
$$
$$
-\lambda_3 \left[ X - \frac{1}{2}\dot{\varphi}^2 \right]). \tag{8.1}
$$

Here $u, v$ -some function of $a, \dot{a}$.

The Euler-Lagrange equations for this model we can write as

$$
\frac{d}{dt}\frac{\partial L}{\partial \dot{a}} = \frac{\partial L}{\partial a}, \qquad \frac{d}{dt}\frac{\partial L}{\partial \dot{R}} = \frac{\partial L}{\partial R}, \qquad \frac{d}{dt}\frac{\partial L}{\partial \dot{T}} = \frac{\partial L}{\partial T},
$$
$$
\frac{d}{dt}\frac{\partial L}{\partial \dot{X}} = \frac{\partial L}{\partial X}, \qquad \frac{d}{dt}\frac{\partial L}{\partial \dot{\varphi}} = \frac{\partial L}{\partial \varphi}, \tag{8.2}
$$

with the energy condition

$$
E_L = \frac{\partial L}{\partial \dot{a}}\dot{a} + \frac{\partial L}{\partial \dot{R}}\dot{R} + \frac{\partial L}{\partial \dot{T}}\dot{T} + \frac{\partial L}{\partial \dot{X}}\dot{X} + \frac{\partial L}{\partial \dot{\varphi}}\dot{\varphi} - L = 0. \tag{8.3}
$$

For this signature is

$$R = u - 6\left(\dot{H} + 2H^2\right), \qquad (8.4)$$

$$T = v - 6H^2, \qquad (8.5)$$

$$X = \frac{1}{2}\dot{\varphi}^2. \qquad (8.6)$$

Hereafter denoted $F(R, T, X, \varphi)$ as $F$. By varying the action with respect to $R$, $T$ and $X$, one obtains

$$\lambda_1 = F_R, \qquad \lambda_2 = F_T, \qquad \lambda_3 = F_X. \qquad (8.7)$$

Here $F_R$ is derivation of $F$ function by $R$, $F_T$ is derivation of $F$ function by $T$ and $F_X$ is derivation of $F$ function by $X$. After an integration by parts, the point-like Lagrangian have the following form

$$L = a^3\left[F - (R-u)F_R - (T-v)F_T\right] + 6\,a\dot{a}^2\left[F_R - F_T\right] + 6a^2\dot{a}[\dot{R}F_{RR} +$$
$$+\dot{T}F_{RT} + \dot{X}F_{RX} + \dot{\varphi}F_{R\varphi}] - a^3 F_X \left[X - \frac{1}{2}\dot{\varphi}^2\right]. \qquad (8.8)$$

## 8.3 Euler-Lagrange equations

Let us now derive the Euler - Lagrange equations by $a$

$$F - AF_R - BF_T - \left(u_{\dot{a}}\frac{a}{3} - 4H\right)\dot{F}_R - \left(v_{\dot{a}}\frac{a}{3} - 4H\right)\dot{F}_T - 2\ddot{F}_R - F_X\left(X - \frac{1}{2}\dot{\varphi}^2\right) = 0, \qquad (8.9)$$

here

$$A = R - u - \frac{a}{3}u_a + u_{\dot{a}}\dot{a} + u_{\dot{a}a}\dot{a}\frac{a}{3} + u_{\dot{a}\dot{a}}\ddot{a}\frac{a}{3} - 4\dot{H} + 6H^2, \qquad (8.10)$$

$$B = T - v - \frac{a}{3}v_a + v_{\dot{a}}\dot{a} + v_{\dot{a}a}\dot{a}\frac{a}{3} + v_{\dot{a}\dot{a}}\ddot{a}\frac{a}{3} - 4\dot{H} - 6H^2, \qquad (8.11)$$

$$\dot{F}_R = \dot{R}F_{RR} + \dot{T}F_{RT} + \dot{X}F_{RX} + \dot{\varphi}F_{R\varphi}, \qquad (8.12)$$

$$\dot{F}_T = \dot{R}F_{TR} + \dot{T}F_{TT} + \dot{X}F_{TX} + \dot{\varphi}F_{T\varphi}, \qquad (8.13)$$

and

$$\ddot{F}_R = \ddot{R}F_{RR} + \ddot{T}F_{RT} + \ddot{X}F_{RX} + \ddot{\varphi}F_{R\varphi} + \dot{R}\dot{F}_{RR} + \dot{T}\dot{F}_{RT} + \dot{X}\dot{F}_{RX} +$$
$$+\dot{\varphi}\dot{F}_{R\varphi}, \qquad (8.14)$$

where

$$\dot{F}_{RR} = \dot{R}F_{RRR} + \dot{T}F_{RRT} + \dot{X}F_{RRX} + \dot{\varphi}F_{RR\varphi}, \qquad (8.15)$$

$$\dot{F}_{RT} = \dot{R}F_{RTR} + \dot{T}F_{RTT} + \dot{X}F_{RTX} + \dot{\varphi}F_{RT\varphi}, \qquad (8.16)$$

$$\dot{F}_{RX} = \dot{R}F_{RXR} + \dot{T}F_{RXT} + \dot{X}F_{RXX} + \dot{\varphi}F_{RX\varphi}, \qquad (8.17)$$

$$\dot{F}_{R\varphi} = \dot{R}F_{R\varphi R} + \dot{T}F_{R\varphi T} + \dot{X}F_{R\varphi X} + \dot{\varphi}F_{R\varphi\varphi}. \qquad (8.18)$$

From equations gived by $R$, $T$ and $X$ we have zero results.

By $\varphi$

$$0 = F_\varphi - 3HF_X\dot\varphi - \left[\dot{R}F_{XR} + \dot{T}F_{XT} + \dot{X}F_{XX} + \dot\varphi F_{X\varphi}\right]\dot\varphi - \\ - F_X\ddot\varphi. \tag{8.19}$$

The energy condition [14,15] gives

$$E_L = a^3[F_X(X + \frac{1}{2}\dot\varphi^2) + F_R(\dot{a}u_{\dot{a}} + 6H^2 + R - u) + \\ + F_T(\dot{a}v_{\dot{a}} - 6H^2 + T - v) + \\ + 6H(\dot{R}F_{RR} + \dot{T}F_{RT} + \dot{X}F_{RX} + \dot\varphi F_{R\varphi}) - F] = 0. \tag{8.20}$$

These equations are difficult to solve.

## 8.4  The Noether Symmetries Approach

We can write Noether symmetry condition in the following form for the Lagrangian

$$XL = 0, \tag{8.21}$$

here

$$X = \alpha\frac{\partial}{\partial a} + \beta\frac{\partial}{\partial R} + \gamma\frac{\partial}{\partial T} + \delta\frac{\partial}{\partial X} + \epsilon\frac{\partial}{\partial\varphi} + \\ + \dot\alpha\frac{\partial}{\partial\dot{a}} + \dot\beta\frac{\partial}{\partial\dot{R}} + \dot\gamma\frac{\partial}{\partial\dot{T}} + \dot\delta\frac{\partial}{\partial\dot{X}} + \dot\epsilon\frac{\partial}{\partial\dot\varphi}. \tag{8.22}$$

The functions $\alpha, \beta, \gamma, \delta, \epsilon$ depend on the variables $a, R, T, X, \varphi$ and then

$$\dot\alpha = \alpha_a\dot{a} + \alpha_R\dot{R} + \alpha_T\dot{T} + \alpha_X\dot{X} + \alpha_\varphi\dot\varphi, \tag{8.23}$$

$$\dot\beta = \beta_a\dot{a} + \beta_R\dot{R} + \beta_T\dot{T} + \beta_X\dot{X} + \beta_\varphi\dot\varphi, \tag{8.24}$$

$$\dot\gamma = \gamma_a\dot{a} + \gamma_R\dot{R} + \gamma_T\dot{T} + \gamma_X\dot{X} + \gamma_\varphi\dot\varphi, \tag{8.25}$$

$$\dot\delta = \delta_a\dot{a} + \delta_R\dot{R} + \delta_T\dot{T} + \delta_X\dot{X} + \delta_\varphi\dot\varphi, \tag{8.26}$$

$$\dot\epsilon = \epsilon_a\dot{a} + \epsilon_R\dot{R} + \epsilon_T\dot{T} + \epsilon_X\dot{X} + \epsilon_\varphi\dot\varphi. \tag{8.27}$$

Using this we can receive:

$$
\begin{aligned}
0 = {}& 6(\alpha\left[F_R - F_T\right] + \beta a\left[F_{RR} - F_{TR}\right] + \gamma a\left[F_{RT} - F_{TT}\right] + \\
& + \delta a\left[F_{RX} - F_{TX}\right] + \epsilon a\left[F_{R\varphi} - F_{T\varphi}\right] + \alpha_a 2a\left[F_R - F_T\right] + \\
& + \beta_a a^2 F_{RR} + \gamma_a a^2 F_{TR} + \delta_a a^2 F_{RX} + \epsilon_a a^2 F_{R\varphi})\dot{a}^2 + \\
& + \dot{R}^2 6\alpha_R a^2 F_{RR} + \dot{T}^2 6\alpha_T a^2 F_{RT} + \dot{X}^2 6\alpha_X a^2 F_{RX} + \\
& + \dot{\varphi}^2(\epsilon_\varphi a^3 F_X + 6\alpha_\varphi a^2 F_{R\varphi} + \alpha\frac{3}{2}a^2 F_X + \beta\frac{1}{2}a^3 F_{XR} + \\
& + \gamma\frac{1}{2}a^3 F_{XT} + \delta\frac{1}{2}a^3 F_{XX} + \epsilon\frac{1}{2}a^3 F_{X\varphi}) + \dot{a}\dot{R}6a(\beta a F_{RRR} + \\
& + \gamma a F_{RRT} + \delta a F_{RRX} + \epsilon a F_{RR\varphi} + 2\alpha_R\left[F_R - F_T\right] + \\
& + (\alpha_a a + \beta_R a + 2\alpha)F_{RR} + \gamma_R a F_{TR} + \delta_R a F_{RX} + \epsilon_R a F_{R\varphi}) + \\
& + \dot{a}\dot{T}6a(\beta a F_{TRR} + \gamma a F_{TRT} + \delta a F_{TRX} + \epsilon a F_{TR\varphi} + \\
& + 2\alpha_T\left[F_R - F_T\right] + (\alpha_a a + 2\alpha + \gamma_T a)F_{TR} + \beta_T a F_{RR} + \\
& + \delta_T a F_{RX} + \epsilon_T a F_{R\varphi}) + \dot{a}\dot{X}6a(2\alpha_X\left[F_R - F_T\right] + a\beta_X F_{RR} + \\
& + a\gamma_X F_{TR} + \beta a F_{RXR} + \gamma a F_{RXT} + \delta a F_{RXX} + \epsilon a F_{RX\varphi} + \\
& + (2\alpha + \alpha_a a + \delta_X a)F_{RX} + \epsilon_X a F_{R\varphi}) + \dot{a}\dot{\varphi}6a(2\alpha_\varphi\left[F_R - F_T\right] + \\
& + a\beta_\varphi F_{RR} + a\gamma_\varphi F_{TR} + \epsilon_a\frac{a^2}{6}F_X + 2\alpha F_{R\varphi} + \beta a F_{R\varphi R} + \\
& + \gamma a F_{R\varphi T} + \delta a F_{R\varphi X} + \epsilon a F_{R\varphi\varphi} + \alpha_a a F_{R\varphi} + \delta_\varphi a F_{RX} + \epsilon_\varphi a F_{R\varphi}) + \\
& + \dot{R}\dot{T}6a^2(\alpha_R F_{TR} + \alpha_T F_{RR}) + \dot{R}\dot{X}6a^2(\alpha_X F_{RR} + \alpha_R F_{RX}) + \\
& + 6\dot{T}\dot{X}a^2(\alpha_X F_{TR} + \alpha_T F_{RX}) + \dot{R}\dot{\varphi}a^2(6\alpha_\varphi F_{RR} + \epsilon_R a F_X + 6\alpha_R F_{R\varphi}) + \\
& + \dot{T}\dot{\varphi}(6\alpha_\varphi a^2 F_{TR} + \epsilon_T a^3 F_X + 6\alpha_T a^2 F_{R\varphi}) + \dot{X}\dot{\varphi}(\epsilon_X a^3 F_X + \\
& + 6\alpha_X a^2 F_{R\varphi} + 6\alpha_\varphi a^2 F_{RX}) + 3a^2\alpha[F - (R - u)F_R - (T - v)F_T + \\
& + \frac{1}{3}a\left(u_a F_R + v_a F_T\right)] + \beta a^3\left[-(R - u)F_{RR} - (T - v)F_{TR}\right] + \\
& + \gamma a^3\left[-(R - u)F_{RT} - (T - v)F_{TT}\right] + \delta a^3[-(R - u)F_{RX} - \\
& - (T - v)F_{TX}] + \epsilon a^3\left[F_\varphi - (R - u)F_{R\varphi} - (T - v)F_{T\varphi}\right] + \\
& + a^3\left[u_{\dot{a}}F_R + v_{\dot{a}}F_T\right](\dot{a}\alpha_a + \dot{R}\alpha_R + \dot{T}\alpha_T + \dot{X}\alpha_X + \dot{\varphi}\alpha_\varphi) - \\
& - (\alpha 3 F_X + \beta a F_{XR} + \gamma a F_{XT} + \delta a F_{XX} + \epsilon a F_{X\varphi})X a^2.
\end{aligned}
$$

$$\tag{8.28}$$

From a Noether symmetry we have:

$\dot{a}^2$ :
$$\alpha\left[F_R - F_T\right] + \beta a\left[F_{RR} - F_{TR}\right] + \gamma a\left[F_{RT} - F_{TT}\right] +$$
$$+\delta a\left[F_{RX} - F_{TX}\right] + \epsilon a\left[F_{R\varphi} - F_{T\varphi}\right] + \alpha_a 2a\left[F_R - F_T\right] +$$
$$+\beta_a a^2 F_{RR} + \gamma_a a^2 F_{TR} + \delta_a a^2 F_{RX} + \epsilon_a a^2 F_{R\varphi} = 0, \qquad (8.29)$$

$\dot{R}^2$ :
$$6\alpha_R a^2 F_{RR} = 0, \qquad (8.30)$$

$\dot{T}^2$ :
$$6\alpha_T a^2 F_{RT} = 0, \qquad (8.31)$$

$\dot{X}^2$ :
$$6\alpha_X a^2 F_{RX} = 0, \qquad (8.32)$$

$\dot{\varphi}^2$ :
$$\epsilon_\varphi a^3 F_X + 6\alpha_\varphi a^2 F_{R\varphi} + \alpha\frac{3}{2}a^2 F_X + \beta\frac{1}{2}a^3 F_{XR} + \gamma\frac{1}{2}a^3 F_{XT} +$$
$$+\delta\frac{1}{2}a^3 F_{XX} + \epsilon\frac{1}{2}a^3 F_{X\varphi} = 0, \qquad (8.33)$$

$\dot{a}\dot{R}$ :
$$2\alpha F_{RR} + \beta a F_{RRR} + \gamma a F_{RRT} + \delta a F_{RRX} + \epsilon a F_{RR\varphi} +$$
$$+2\alpha_R\left[F_R - F_T\right] + \alpha_a a F_{RR} + \beta_R a F_{RR} +$$
$$+\gamma_R a F_{TR} + \delta_R a F_{RX} + \epsilon_R a F_{R\varphi} = 0, \qquad (8.34)$$

$\dot{a}\dot{T}$ :
$$2\alpha F_{TR} + \beta a F_{TRR} + \gamma a F_{TRT} + \delta a F_{TRX} + \epsilon a F_{TR\varphi} +$$
$$+2\alpha_T\left[F_R - F_T\right] + \alpha_a a F_{TR} + \beta_T a F_{RR} +$$
$$+\gamma_T a F_{TR} + \delta_T a F_{RX} + \epsilon_T a F_{R\varphi} = 0, \qquad (8.35)$$

$\dot{a}\dot{X}$ :
$$2\alpha_X\left[F_R - F_T\right] + a\beta_X F_{RR} + a\gamma_X F_{TR} + 2\alpha F_{RX} +$$
$$+\beta a F_{RXR} + \gamma a F_{RXT} + \delta a F_{RXX} + \epsilon a F_{RX\varphi} + \alpha_a a F_{RX} +$$
$$+\delta_X a F_{RX} + \epsilon_X a F_{R\varphi} = 0, \qquad (8.36)$$

$\dot{a}\dot{\varphi}$ :
$$2\alpha_\varphi\left[F_R - F_T\right] + a\beta_\varphi F_{RR} + a\gamma_\varphi F_{TR} + \epsilon_a\frac{a^2}{6}F_X + 2\alpha F_{R\varphi} +$$
$$+\beta a F_{R\varphi R} + \gamma a F_{R\varphi T} + \delta a F_{R\varphi X} + \epsilon a F_{R\varphi\varphi} + \alpha_a a F_{R\varphi} +$$
$$+\delta_\varphi a F_{RX} + \epsilon_\varphi a F_{R\varphi} = 0, \qquad (8.37)$$

$\dot{R}\dot{T}$ :
$$\alpha_R a^2 F_{TR} + \alpha_T a^2 F_{RR} = 0, \qquad (8.38)$$

$\dot{R}\dot{X}$ :
$$\alpha_X a^2 F_{RR} + \alpha_R a^2 F_{RX} = 0, \qquad (8.39)$$

$\dot{T}\dot{X}$ :
$$\alpha_X a^2 F_{TR} + \alpha_T a^2 F_{RX} = 0, \qquad (8.40)$$

$\dot{R}\dot{\varphi}$ :
$$6\alpha_\varphi a^2 F_{RR} + \epsilon_R a^3 F_X + 6\alpha_R a^2 F_{R\varphi} = 0, \qquad (8.41)$$

$\dot{T}\dot{\varphi}$ :
$$6\alpha_\varphi a^2 F_{TR} + \epsilon_T a^3 F_X + 6\alpha_T a^2 F_{R\varphi} = 0, \qquad (8.42)$$

$\dot{X}\dot{\varphi}$ :
$$\epsilon_X a^3 F_X + 6\alpha_X a^2 F_{R\varphi} + 6\alpha_\varphi a^2 F_{RX} = 0, \qquad (8.43)$$

$$3\alpha\left[F - (R - u)F_R - (T - v)F_T + \frac{1}{3}a\left(u_a F_R + v_a F_T\right)\right] +$$
$$+\beta a\left[-(R - u)F_{RR} - (T - v)F_{TR}\right] + \gamma a[-(R - u)F_{RT} -$$
$$-(T - v)F_{TT}] + \delta a\left[-(R - u)F_{RX} - (T - v)F_{TX}\right] +$$
$$+\epsilon a\left[F_\varphi - (R - u)F_{R\varphi} - (T - v)F_{T\varphi}\right] +$$
$$+\dot{a}\alpha_a a\left[u_{\dot{a}}F_R + v_{\dot{a}}F_T\right] + \dot{\varphi}\alpha_\varphi a\left[u_{\dot{a}}F_R + v_{\dot{a}}F_T\right] +$$
$$+\dot{R}\alpha_R a\left[u_{\dot{a}}F_R + v_{\dot{a}}F_T\right] + \dot{T}\alpha_T a\left[u_{\dot{a}}F_R + v_{\dot{a}}F_T\right] +$$
$$+\dot{X}\alpha_X a\left[u_{\dot{a}}F_R + v_{\dot{a}}F_T\right] - (\alpha 3F_X +$$
$$+\beta a F_{XR} + \gamma a F_{XT} + \delta a F_{XX} + \epsilon a F_{X\varphi})X = 0. \qquad (8.44)$$

## 8.5 The Noether Symmetries Solution

From equations for $\dot{R}^2$, $\dot{T}^2$ and $\dot{X}^2$ we are have two possibilities. First for $F_{RR} = F_{RT} = F_{RX} = 0$, and solution of this is a linear equation $F = s_1(\varphi)R + s_2(\varphi)T + s_3(\varphi)X + s_4(\varphi)$.

Second variant for nonlinear solution we have find for $\alpha_R = \alpha_T = \alpha_X = 0$. We can combine equations for $\dot{R}\dot{\varphi}$, $\dot{T}\dot{\varphi}$ and $\dot{X}\dot{\varphi}$ . This will give us next system of equations:

$$\frac{\epsilon_R}{F_{RR}} = \frac{\epsilon_T}{F_{RT}} = \frac{\epsilon_X}{F_{RX}} = -\frac{6\alpha_\varphi}{aF_X}. \tag{8.45}$$

One result we can receive if we take $\alpha_\varphi \neq 0$ as usual taken in some articles. Now we can try to find solution from previous system of equations, that gives us next equations:

$$F_{RR}F_{XT} = F_{RT}F_{RX}, \tag{8.46}$$

$$F_{RT}F_{XX} = F_{RX}F_{XT}, \tag{8.47}$$

$$F_{RR}F_{XX} = F_{RX}^2. \tag{8.48}$$

Last equation here is homogeneous Monge - Ampere equations. After some calculations their solution involving arbitrary constants we can write as:

$$F = (C_1(\varphi)R + C_2(\varphi)T + C_3(\varphi)X)^2 + C_4(\varphi)R + C_5(\varphi)T + C_6(\varphi)X + C_7(\varphi) \tag{8.49}$$

This solution gives us same results as recent observations about the early time inflation, close to $R^2$. Solutions of Monge - Ampere equations involving one arbitrary function will give more general result:

$$F = f(C_1(\varphi)R + C_2(\varphi)T + C_3(\varphi)X, \varphi) + C_4(\varphi)R + C_5(\varphi)T + C_6(\varphi)X + C_7(\varphi) \tag{8.50}$$

Using the equations found above for the square solution $F$ we obtain that for this that $C_1$, $C_2$, $C_3$, $C_4$, $C_5$, $C_6$ constant by $\varphi$. It means that we can combine equations for $\dot{a}R$, $\dot{a}T$ and $\dot{a}X$ as

$$Z(a, \varphi)_R = Z(a, \varphi)_T = Z(a, \varphi)_X = 0. \tag{8.51}$$

Here $Z(a, \varphi) = (2\alpha + \alpha_a a)F_R + \beta aF_{RR} + \gamma aF_{RT} + \delta aF_{RX} + \epsilon aF_{R\varphi}$. This substitution lets us exclude functions $\beta$, $\gamma$, and $\delta$ from calculations. In addition, we can use here Noether charge $\theta$. For $F(R, T, X, \varphi)$ gravity $\theta$ has form:

$$\theta = \alpha\left[a^3(u_{\dot{a}}F_R + v_{\dot{a}}F_T) + 12a\dot{a}(F_R - F_T) + 6a^2\dot{F}_R\right] +$$
$$+ 6a^2\dot{a}[\beta F_{RR} + \gamma F_{RT} + \delta F_{RX} + \epsilon F_{R\varphi}] + \epsilon a^3 F_X\dot{\varphi}. \tag{8.52}$$

Now we can receive from Noether symmetry next particular solutions for Noether symmetry's components:

$$\alpha = \alpha_0\varphi a, \tag{8.53}$$

114

and

$$\epsilon = 6\alpha_0 \frac{C_1}{C_3} \lg F_X + \epsilon_0 a. \tag{8.54}$$

And $Z$ here

$$Z = Z_0 \alpha. \tag{8.55}$$

This give us the general form of $F$:

$$C_7(\varphi) = C_{70}\varphi^2, \tag{8.56}$$

where $C_{70}$ is constant. And now we can write equation for sum of $u$ and $v$:

$$uC_1 + vC_2 = C_8(\varphi)a + C_9, \tag{8.57}$$

were $C_8(\varphi)$ - function and $C_9$ - constant.
If we take $\varphi = \varphi_0 t$, we can determine scale factor as

$$a = a_0 e^{pt^2}, \tag{8.58}$$

where $a_0$, $p$ are constants.

## 8.6 Conclusion

We have considered the model with scalar field and $F(R)$, $F(T)$ - gravity in the most general form $F(R,T,X,\varphi)$ using the method of Noether symmetry. Also for this model received Euler-Lagrange equations. The result shows a number of interesting features. Thus, it is shown that in models that were taken into account the influence of the Ricci scalar component, the torsion scalar and the kinetic term of the scalar field, we must take them equivocally, as a general argument in the form of their sum $C_1(\varphi)R + C_2(\varphi)T + C_3(\varphi)X$. In general our results is Monge - Ampere equation solutions. Involving solutions for arbitrary constants we will get Starobinsky solutions. It is important to note that this model is the simplest solution for the model we are considering, with the exception of the trivial linear solution for F. For this case were obtained the particular solutions of scale factor $a$.

The work was carried out with the financial support of the Ministry of Education and Science of the Republic of Kazakhstan, Grant No. 0118RK00935.

## References

[1] S. Perlmutter, G. Aldering, G. Goldhaber et al. *Measurements of $\Omega$ and $\Lambda$ from 42 High-Redshift Supernovae.* Astroph. J. 1999. Vol.517, No2, P.565-586.

[2] A.G. Riess, A.V. Filippenko, P. Challis et al. *Observational Evidence from Supernovae for an Accelerating Universe and a Cosmological Constant.* Astron. J. 1988. Vol.116, No3, P.1009.

[3] E. Copeland, M. Sami, and S. Tsujikawa. *Dynamics of dark energy.* Int.J.Mod.Phys.D 15 , 1753 (2006).

[4] Roland de Putter, Eric V. Linder. *Kinetic k-essence and Quintessence.* Astropart.Phys.28:263-272,2007.

[5] C. Armendariz-Picon, T. Damour, V. Mukhanov. *k-Inflation.* Phys.Lett.B458:209-218,1999

[6] R. Myrzakulov, K. Yerzhanov, K. Yesmakhanova et al. *g-Essence as the cosmic speed-up.* Astrophysics and Space Science. V.341, I.2 , 681-688

[7] K. Bamba, R. Myrzakulov, O. Razina, K. Yerzhanov. *Cosmological evolution of equation of state for dark energy in G-essence models.* International journal of modern physics D, V. 22 (6), 1350023, 2013

[8] R. Myrzakulov. *Accelerating universe from F(T) gravity* Eur.Phys.J. C71 (2011) 1752

[9] R. Myrzakulov. *Dark Energy in F(R,T) Gravity.* [arXiv:1205.5266]

[10] R. Myrzakulov. *FRW Cosmology in F(R,T) gravity.* The European Physical Journal C, **72**, N11, 2203 (2012). [arXiv:1207.1039]

[11] M. Sharif, S. Rani, R. Myrzakulov. *Analysis of F(R,T) Gravity Models Through Energy Conditions.* Eur. Phys. J. Plus, **128**, N11, 123 (2013). [arXiv:1210.2714]

[12] A. Pasqua, S. Chattopadhyay, R. Myrzakulov. *A dark energy with higher order derivatives of H in the modified gravity f(R,T).* ISRN High Energy Phys. 2014 (2014) 535010. [arXiv:1306.0991]

[13] S. Capozziello, M. De Laurentis, R. Myrzakulov. *Noether Symmetry Approach for teleparallel - curvature cosmology.* [arXiv:1412.1471]

[14] R. Myrzakulov. *F(T) gravity and k-essence.* General Relativity and Gravitation, v.44, N12, 3059-3080 (2012)

[15] E. Gudekli, N. Myrzakulov, K. Yerzhanov, R. Myrzakulov *Trace-anomaly driven inflation in f(T) gravity with a cosmological constant.* Astrophysics and Space Science, (2015) 357: 45

# 9 ON VARIATIONS IN TELEPARALLELISM THEORIES

YAKOV ITIN

**Abstract**    The variation procedure on a teleparallel manifold is studied. The main problem is the non-commutativity of the variation operator with the Hodge dual map. We establish certain useful formulas for variations and restate the master formula due to Hehl and his collaborates. Our approach is different and sometimes easier for applications. By introducing the technique of the variational matrix, we find the necessary and sufficient conditions for commutativity (anti-commutativity) of the variation derivative with the Hodge dual operator. A general formula for the variation of quadratic-type expression is obtained. The described variational technique is then applied to the two viable field-theoretical setups: the electromagnetic Lagrangian on a teleparallel background and the Lagrangian of the translation invariant gravity.

## 9.1 Introduction

One of the biggest challenges of the theoretical physics is to unify the standard model of particles interactions with Einstein's theory of gravity. Both of these theories are in good agreement with observed phenomena. But the fundamental concepts on which they are based, are quite different. Standard model is a quantum field theory on the flat Minkowskian space-time. In contrast to that, the Einstein's description of gravity is a classical field theory essential connected with the geometrical properties of the pseudo-Riemannian manifold. One can not expect the union of these two different theories without substantial modification of each of them. Great efforts were made in recent years to modernize the standard model in accordance with the idea of superstrings, see [1]. The second partner of the pair must also undergo some necessary changes.

C. Duston, M. Holman (Eds), *Spacetime Physics 1907 - 2017. Selected peer-reviewed papers presented at the First Hermann Minkowski Meeting on the Foundations of Spacetime Physics, 15-18 May 2017, Albena, Bulgaria* (Minkowski Institute Press, Montreal 2019). ISBN 978-1-927763-48-3 (softcover), ISBN 978-1-927763-49-0 (ebook).

Einstein's general relativity theory (GR) is almost the only possible theory of gravity within the framework of the pseudo-Riemannian manifold. Thus an essential modification of it can be made only on a basis of some generalization of the underlying geometry. The study of various relevant geometries began immediately after Einstein proposed his theory of general relativity. The most general geometrical theory of gravity is the metric-affine theory [9]. In the present paper we study a rather simpler generalization of the pseudo-Riemann geometry - the theories of teleparallelism. The teleparallel space was introduced for the first time by Cartan [2] and used by Einstein [4], [5] in a certain variant of his unified theory of gravity and electromagnetism. The work of Weitzenböck [6] was the first devoted to the investigation of the geometric structure of teleparallel spaces. Theories based on this geometric structure arise in physics from time to time either to provide an alternative model of the gravity or to describe the spin properties of the matter, see Refs. [7], [8], [11], [12], [13], [14], [19], [21], [22], [20] [24], [25]. The relation between the premetric construction of GR and teleparallelism studied recently in [27].

It is convenient and useful to have a Lagrangian formulation of a field theory. A functional, called the action, is the basis for the Lagrangian formalism. The field equations of the classical field theory are then represented by the critical point condition of the above action functional.

Let us take a brief look at the main properties of the action functionals in the classical relativistic field theory.

- The action functional is the integral over the entire 4-dimensional manifold.

- The integrand is a certain differential 4-form - *Lagrangian density*.

- The Lagrangian density is a scalar invariant under the group of transformations of the system under consideration.

- The Lagrangian density incorporates only the squares of the first-order derivatives of the field variables at the same point - local densities.

We consider the latter condition as a necessary one, because almost all physically meaningful field equations are second-order partial differential equations of the hyperbolic type. The straightforward application of the variational procedure on the teleparallel framework provides various problems, as it was emphasized in [15]. The main source of these problems is connected with the fact that the operators of a certain type, namely the Hodge dual operator, the interior product and the coderivative operator, non-commute with the variational derivative. In the present paper, we study the the variational procedure (free and constrained) on the teleparallel manifold.

The overview of the work is as follows. We start with a brief outline of the variational procedure in the general relativity. In the second section, we describe the preferences that bring the consideration of a teleparallel space instead of a pseudo-Riemannian manifold. The most advantage of such generalization is a possibility to describe the gravity by the quadratic Lagrangians, similarly to the other field theories. Variational procedure on the teleparallel spaces is described in the third section. We prove the commutative relations which coincide in the case of pseudo-orthonormal coframe with the master formula of [15]. The analogous relation described also in [16]. Using the commutative relation we derive a formula for variation of general quadratic-type Lagrangians. The fourth section is devoted to various types of the covariant constrains on a teleparallel structure. We study the relation of such constrains with the commutativity and anti-commutativity of the variational derivative with the Hodge dual map. The last two sections are devoted to the application of our variational formula for viable field theories. We consider the Maxwell Lagrangian on a teleparallel space and the the translation invariant gravity Lagrangian.

## 9.2   Operators on a teleparallel manifold.

The teleparallel spacetime is a 4D-manifold endowed with a smooth field of frames (ordered sets of 4 independent vectors). In this space, two vectors attached to distinct points are parallel when they have the proportional components with respect to the local frames, see [26]. In order to apply the differential forms technique, it is useful to deal with the coframe field instead of the frame one.

**Definition 9.2.1:** *A teleparallel manifold is a pair $(M, G)$, where $M$ is a differential 4D-manifold and $G = \{\vartheta^a(x^\mu)\}$, $a = 0, 1, 2, 3$ is a fixed smooth cross-section of the coframe bundle $FM$.* ∎

At an arbitrary point $x^\mu \in M$, the 4-tuple of 1-forms $\vartheta^a$ constitutes a basis of the covector space $T_x^* M$. This vector space is supplemented by Minkowski's metric $\eta^{ab} = \eta_{ab} = \mathrm{diag}(-1, 1, 1, 1)$. The 1-forms $\vartheta^a$ are assumed to be pseudo-orthonormal relative to this metric. Consequently, the coordinate components of the metric tensor are represented by the components of the coframe field

$$g_{\mu\nu} = \eta_{ab}\vartheta^a_\mu \vartheta^b_\nu\,. \tag{9.1}$$

Let $C^\infty$ be a set of smooth real valued functions on $M$. Denote by $\Omega^p$ the $C^\infty$-modulo of differential $p$-forms on $M$. We recall algebraic operations defined on a $n$-dimensional teleparallel manifold:

- The *exterior (wedge) product* $\wedge : \Omega^p \times \Omega^q \to \Omega^{p+q}$ of two differential forms[1]

$$\alpha = \alpha_{a_1\cdots a_p}\vartheta^{a_1\cdots a_p} \in \Omega^p \quad \text{and} \quad \beta = \beta_{b_1\cdots b_q}\vartheta^{b_1\cdots b_q} \in \Omega^q \tag{9.2}$$

  is defined as

$$\alpha \wedge \beta = \alpha_{a_1\cdots a_p}\beta_{b_1\cdots b_q}\vartheta^{a_1\cdots a_p b_1\cdots b_q}\,. \tag{9.3}$$

  This operation is associative, $C^\infty$-bilinear and not commutative, in general,

$$\alpha \wedge \beta = (-1)^{pq}\beta \wedge \alpha\,. \tag{9.4}$$

---

[1]We abbreviate the wedge product monomials as $\vartheta^{ab\cdots} := \vartheta^a \wedge \vartheta^b \wedge \cdots$

- The *Hodge dual map* $*$ is a $C^\infty$-linear map $* : \Omega^p \to \Omega^{n-p}$, which acts on the wedge product monomials of the basis 1-forms as

$$*(\vartheta^{a_1\cdots a_p}) = \epsilon^{a_1\cdots a_n}\vartheta_{a_{p+1}\cdots a_n}, \cdot \tag{9.5}$$

Here the lower indexed 1-form $\vartheta_a$ is defined as $\vartheta_a = \eta_{ab}\vartheta^b$, while $\epsilon^{a_1\cdots a_n}$ is the totally antisymmetric pseudo-tensor. The square of the Hodge star acting on a $p$-form is given by

$$*^2 = (-1)^{p(n-p)+s}. \tag{9.6}$$

In particular, in four-dimensional Minkowski space, $*^2 = 1$ for forms with odd degree and $*^2 = -1$ for forms with even degree.

- The *inner product* $v \rfloor \alpha$ of a differential form $\alpha \in \Omega^p$ with a vector $v \in T_x M$ is a $C^\infty$-bilinear map $\rfloor : \Omega^p \to \Omega^{p-1}$, which satisfies the following properties:

$$e_a \rfloor \vartheta^b = \delta_a^b, \tag{9.7}$$

$$v \rfloor (\alpha \wedge \beta) = (v \rfloor \alpha) \wedge \beta + (-1)^{\deg\alpha}\alpha \wedge (v \rfloor \beta). \tag{9.8}$$

Here $e_a$ is a basis vector of $T_x M$ and $\vartheta^a$ is a basis 1-form of $T_x^* M$.

We define an *inner product* for a 1-form $w$ and a $p$-form (with $p > 0$) as

$$w \vee \alpha := *(w \wedge *\alpha). \tag{9.9}$$

It is easy to see that this operation is a $C^\infty$-bilinear map $\Omega^p \to \Omega^{p-1}$, which satisfies the same properties as (9.8) and (9.7). In particular, we have

$$\vartheta_a \vee \vartheta^b = \delta_a^b. \tag{9.10}$$

$$w \vee (\alpha \wedge \beta) = (w \vee \alpha) \wedge \beta + (-1)^{\deg\alpha}\alpha \wedge (w \vee \beta). \tag{9.11}$$

Note that the product defined by (9.9) is really a multiplication of two exterior forms thus this type of definition justifies the term "inner product".

We recall also the first-order differential operators defined on a teleparallel manifold:

- The *exterior derivative* operator $d : \Omega^p \to \Omega^{p+1}$ defined as

$$d\alpha = d\left(\alpha_{a_1\cdots a_p}dx^{a_1} \wedge \cdots \wedge dx^{a_p}\right) = d\alpha_{a_1\cdots a_p} \wedge dx^{a_1} \wedge \cdots \wedge dx^{a_p}. \tag{9.12}$$

This is an anti-derivative relative to the wedge product of forms

$$d(\alpha \wedge \beta) = d\alpha \wedge \beta + (-1)^{\deg\alpha}\alpha \wedge d\beta. \tag{9.13}$$

- The *coderivative* operator $d^+ : \Omega^p \to \Omega^{p-1}$ defined by

$$d^+\alpha = *d*\alpha. \tag{9.14}$$

## 9.3 Variational procedure

### 9.3.1 Coframe Lagrangian

Using the operators described above the action functional on the teleparallel manifold can be accepted in the quadratic (Dirichlet) form. The simplest choice of the Lagrangian density can be made in the form of the Yang-Mills Lagrangian:

$$S[\vartheta^a] = \int_M d\vartheta^a \wedge *d\vartheta_a. \tag{9.1}$$

Observe that the action (9.1) satisfies the following conditions:

- It is independent on a particular choice of a local coordinate system.

- It includes only the first order derivatives of the field variables (coframe 1-forms) so the corresponding field equation is of the second order.

- It is a functional of a Dirichlet-type so it provides an equation of a harmonic type.

- It is invariant under the global group of $SO(1,3)$ transformations of the coframe field $\vartheta^a$.

Therefore the generalization of the pseudo-Riemannian structure to a teleparallel structure allows to consider general Lagrangians for gravity similar to the quadratic Lagrangians of the other (already quantized) field theories. In this way one can hope to define the energy of gravity field in a local, covariant form.

### 9.3.2 Variation procedure

The variational procedure on a teleparallel manifold for the action functional $S = S[\vartheta^a]$ can be described as follows: Let $\vartheta^a(\lambda)$ be a smooth 1-parametric family of cross-sections of $FM$, with the initial conditions

$$\vartheta^a(\lambda = 0) = \vartheta^a \qquad \text{and} \qquad \left.\frac{\partial \vartheta^a}{\partial \lambda}\right|_{\lambda=0} = \delta\vartheta^a. \tag{9.2}$$

Due to the least action principle, the field equation of the physical system coincides with the critical point condition for an appropriative action functional. The critical points of the functional $S[\vartheta^a]$ are defined by the condition:

$$\delta S = \left.\frac{dS}{d\lambda}\right|_{\lambda=0} d\lambda = 0. \tag{9.3}$$

The variation operator $\delta$ ($\lambda$-differential) is independent on the space-time coordinates so it satisfies the following rules:

- The ordinary Leibniz rule for the wedge product

$$\delta(\alpha \wedge \beta) = \delta\alpha \wedge \beta + \alpha \wedge \delta\beta. \tag{9.4}$$

- The commutativity with the exterior derivative operator

$$\delta \circ d = d \circ \delta \tag{9.5}$$

In general, the variation operator does not commute with the Hodge star-operator $\delta \circ * \neq * \circ \delta$ and with the coderivative operator $\delta \circ d^+ \neq d^+ \circ \delta$.

The following lemma is useful for the actual calculations of the variations:

**Lemma 9.3.1:** *Variation of the wedge product and of the Hodge dual monomials, respectively, satisfy*

$$\delta \vartheta^{a_1 \cdots a_p} = \delta \vartheta^m \wedge (\vartheta_m \vee \vartheta^{a_1 \cdots a_p}), \tag{9.6}$$

$$\delta * \vartheta^{a_1 \cdots a_p} = \delta \vartheta^m \wedge (\vartheta_m \vee *\vartheta^{a_1 \cdots a_p}). \tag{9.7}$$

**Proof:** Using Leibniz's rule (9.4) we obtain

$$\delta \vartheta^{a_1 \cdots a_p} = \delta \vartheta^{a_1} \wedge \vartheta^{a_2 \cdots a_p} + \vartheta^{a_1} \wedge \delta \vartheta^{a_2} \wedge \vartheta^{a_3 \cdots a_p} + \cdots + \vartheta^{a_1 \cdots a_{p-1}} \wedge \delta \vartheta^{a_p}. \tag{9.8}$$

The property (9.11) of the inner product yields

$$\vartheta_m \vee \vartheta^{a_1 \cdots a_p} = \delta_m^{a_1} \vartheta^{a_2 \cdots a_p} - \delta_m^{a_2} \vartheta^{a_1 a_3 \cdots a_p} + \cdots . \tag{9.9}$$

Comparing (9.8) and (9.9) we obtain (9.6). Since the Hodge dual of a wedge product monomial is also a monomial, (9.7) is a consequence of (9.6). ∎

In particular, we have the following formula for variation of the dual form

$$\delta * \vartheta^a = \delta \vartheta_m \wedge *(\vartheta^m \wedge \vartheta^a) \tag{9.10}$$

Notice that the variation of the dual form is essentially different from the variation $\delta \vartheta^a$ of the basis form $\vartheta^a$.

For the variation of the volume element, formula (9.7) yields

$$\delta * 1 = \delta \vartheta^m \wedge (\vartheta_m \vee *1) = \delta \vartheta^m \wedge *(\vartheta_m \wedge *^2 1) = -\delta \vartheta^m \wedge *\vartheta_m. \tag{9.11}$$

**Proposition 9.3.2:** *For a differential form $\alpha \in \Omega^p$ on a $n$-dimensional manifold, the following commutative relation holds*

$$\boxed{\delta * \alpha - *\delta \alpha = *\left( \delta \vartheta_m \vee (\vartheta^m \wedge \alpha) - \delta \vartheta_m \wedge (\vartheta^m \vee \alpha) \right).} \tag{9.12}$$

**Proof:** Evaluating the $p$-form $\alpha$ in the coframe basis

$$\alpha = \alpha_{a_1 \cdots a_p} \vartheta^{a_1 \cdots a_p}$$

and using the relation (9.7) we derive

$$
\begin{aligned}
\delta * \alpha &= \delta(\alpha_{a_1 \cdots a_p}) * \vartheta^{a_1 \cdots a_p} + \alpha_{a_1 \cdots a_p} \delta * \vartheta^{a_1 \cdots a_p} \\
&= \delta(\alpha_{a_1 \cdots a_p}) * \vartheta^{a_1 \cdots a_p} + \alpha_{a_1 \cdots a_p} \delta \vartheta^m \wedge (\vartheta_m \vee *\vartheta^{a_1 \cdots a_p}) \\
&= \delta(\alpha_{a_1 \cdots a_p}) * \vartheta^{a_1 \cdots a_p} + \delta \vartheta^m \wedge (\vartheta_m \vee *\alpha).
\end{aligned}
$$

On the other hand,

$$\delta\alpha = \delta(\alpha_{a_1...a_p})\vartheta^{a_1...a_p} + \alpha_{a_1...a_p}\delta(\vartheta^{a_1...a_p})$$
$$= \delta(\alpha_{a_1...a_p})\vartheta^{a_1...a_p} + \delta\vartheta^m \wedge (\vartheta_m \vee *\alpha).$$

Therefore,

$$\delta * \alpha - *\delta\alpha = \delta\vartheta_m \wedge (\vartheta^m \vee *\alpha) - *\Big(\delta\vartheta_m \wedge (\vartheta^m \vee *\alpha)\Big)$$
$$= \Big((-1)^{p(n-p)+1}\delta\vartheta_m \wedge *(\vartheta^m \wedge \alpha) - *(\delta\vartheta_m \wedge (\vartheta^m \vee \alpha))\Big)$$
$$= *\Big[\delta\vartheta_m \vee (\vartheta^m \wedge \alpha) - \delta\vartheta_m \wedge (\vartheta^m \vee \alpha)\Big].$$

∎

It is easy to see from the proof of the proposition above that in a particular case of basis forms the following commutation/(anti-)commutation relations hold

$$\delta * \vartheta^{a_1...a_p} \pm *\delta\vartheta^{a_1...a_p} = *\Big(\delta\vartheta_m \vee \vartheta^{ma_1...a_p} \pm \delta\vartheta_m \wedge (\vartheta^m \vee \vartheta^{a_1...a_p})\Big). \quad (9.13)$$

A commutative relation similar to (9.12) was derived for the first time in [15]. A slightly different form is exhibited in [16].

## 9.4   Variations and constrains.

In the ordinary variational problem, the variations of the field variables are considered to be independent – free variations. In order to study a constraint physical system, for instant motion of a particle on a fixed surface, one looks for a critical point of a functional restricted by appropriative constrains. The constrain is usually incorporated by the use of Lagrangian multipliers. Another approach to the constrained variational problem can be proposed by consideration of restricted variations. These variations satisfy some additional conditions. For global $SO(1,3)$ invariant and diffeomorphic covariant models, it is natural to consider covariant constraint conditions.

Let us expand the variation of the basis 1-form $\vartheta^a$ in the coframe basis

$$\delta\vartheta^a = \epsilon^a{}_b\vartheta^b. \quad (9.1)$$

We will refer to the matrix $\epsilon_{ab} = \eta_{ac}\epsilon^c{}_b$ as the *variational matrix*. It admits a natural covariant decomposition

$$\epsilon_{ab} = \epsilon^{(1)}_{ab} + \epsilon^{(2)}_{ab} + \epsilon^{(3)}_{ab}, \quad (9.2)$$

where

$$\epsilon^{(1)}_{ab} = \frac{1}{2}(\epsilon_{ab} - \epsilon_{ba}) \qquad - \text{ antisymmetric variation,} \quad (9.3)$$

$$\epsilon^{(2)}_{ab} = \frac{1}{2}(\epsilon_{ab} + \epsilon_{ba}) - \epsilon\eta_{ab} \qquad - \text{ traceless symmetric variation,} \quad (9.4)$$

$$\epsilon^{(3)}_{ab} = \epsilon\eta_{ab} \qquad - \text{ trace variation.} \quad (9.5)$$

Here

$$\epsilon = \frac{1}{4}\epsilon_{ab}\eta^{ab} = \frac{1}{4}\epsilon^a{}_a \tag{9.6}$$

denotes the scalar part of the variational matrix.

Let us examine under which conditions on the variational matrix the variational operator commutes with Hodge star operator. For a generality, we consider an extended commutativity relation

$$\delta * \vartheta^a = \kappa * \delta\vartheta^a, \tag{9.7}$$

where $\kappa$ is a real parameter. With $\kappa = 1$, we have here the case of commutativity, while $\kappa = -1$ means the anticommutativity of the operators product.

Using (9.10) in the right hand sides of (9.7) we obtain

$$*\delta\vartheta^a = \kappa\delta\vartheta_m \wedge *(\vartheta^m \wedge \vartheta^a). \tag{9.8}$$

In terms of the variational matrix it reads

$$\epsilon^a{}_b\vartheta^b = \kappa(\epsilon d^a_b - \epsilon_b{}^a)\vartheta^b.$$

This conditions holds for arbitrary coframe if and only if the variational matrix satisfies

$$\epsilon^a{}_b = \kappa(\epsilon\delta^a_b - \epsilon_b{}^a). \tag{9.9}$$

Taking the trace in two sides of this matrix equation we obtain a first necessary condition

$$\epsilon = 0. \tag{9.10}$$

Substituting it into (9.9) we have

$$\kappa\epsilon_{ab} + \epsilon_{ba} = 0. \tag{9.11}$$

This equation yields $\kappa^2 = 1$, thus only two cases are available: proper commutativity

$$\kappa = 1, \qquad \epsilon_{(ab)} = 0, \tag{9.12}$$

and anticommutativity

$$\kappa = -1, \qquad \epsilon_{[ab]} = 0. \tag{9.13}$$

The consideration above can be summarized in the following

**Proposition 9.4.1:**
*The variation commutes with the Hodge dual of a basis 1-form*

$$\delta * \vartheta^a = *\delta\vartheta^a \tag{9.14}$$

*if and only if the variational matrix $\epsilon^{ab}$ is antisymmetric.*
*The variation anticommutes with the Hodge dual of a basis 1-form*

$$\delta * \vartheta^a = - * \delta\vartheta^a \tag{9.15}$$

*if and only if the variational matrix $\epsilon^{ab}$ is traceless and symmetric.*

Consider now the variation of the metric tensor

$$\delta g_{\mu\nu} = \delta(\eta_{ab}\vartheta^a_\mu\vartheta^b_\nu) = \eta_{ab}(\delta\vartheta^a_\mu\vartheta^b_\nu + \vartheta^a_\mu\delta\vartheta^b_\nu).$$

The variation of the coframe can be written as

$$\delta\vartheta^a = \epsilon^a{}_b\vartheta^b = \delta(\vartheta^a_\mu dx^\mu) = \delta\vartheta^a_\mu dx^\mu = \vartheta^\mu_b\delta\vartheta^a_\mu\vartheta^b.$$

Therefore,

$$\vartheta^\mu_b\delta\vartheta^a_\mu = \epsilon^a{}_b, \qquad \text{or} \qquad \delta\vartheta^a_\mu = \epsilon^a{}_b\vartheta^b_\mu.$$

Hence the variation of the metric tensor is

$$\delta g_{\mu\nu} = \eta_{ab}(\epsilon^a{}_c\vartheta^c_\mu\vartheta^b_\nu + \epsilon^b{}_c\vartheta^a_\mu\vartheta^c_\nu) = \epsilon_{bc}(\vartheta^c_\mu\vartheta^b_\nu + \vartheta^b_\mu\vartheta^c_\nu). \tag{9.16}$$

Note that the expression in the parentheses is symmetric under the permutation of the indices $c$ and $b$. Thus the variation of the metric tensor $\delta g_{\mu\nu}$ is identically equal to zero if and only if the matrix $\epsilon_{bc}$ is antisymmetric.

**Proposition 9.4.2:**
*The variation of the metric tensor vanishes if and only if the Hodge dual commutes with the variational derivative operator.*

The traceless variations preserve the determinant of the metric tensor $g^{\mu\nu}$. Indeed the relation (9.16) yields:

$$\delta g = g^{\mu\nu}\delta g_{\mu\nu} = 2\epsilon_{bc}\eta^{bc} = 2\epsilon. \tag{9.17}$$

Therefore, the determinant of the metric tensor is preserved if and if the variational matrix is traceless. Notice a connected result. Since $\delta * 1 = \epsilon * 1$ it follows that the volume element is preserved if and only if $\epsilon = 0$. In Table 1, we summarize our conclusions.

**Table 9.1:** Constrained variations and commutativity rules

| Type of variation | Variational matrix | Metric tensor | Variation of volume | Commutativity rule |
|---|---|---|---|---|
| free | $\epsilon_{ab}$ | $\delta g^{\mu\nu} \neq 0$ | $\delta * 1 \neq 0$ | |
| antisymmetric | $\epsilon^{(1)}_{ab}$ | $\delta g^{\mu\nu} = 0$ | $\delta * 1 = 0$ | $\delta * \vartheta^a = *\delta\vartheta^a$ |
| symmetric | $\epsilon^{(2)}_{ab} + \epsilon^{(3)}_{ab}$ | $\delta g^{\mu\nu} \neq 0$ | $\delta * 1 \neq 0$ | |
| traceless | $\epsilon^{(1)}_{ab} + \epsilon^{(2)}_{ab}$ | $\delta g^{\mu\nu} \neq 0$ | $\delta * 1 = 0$ | |
| traceless symmetric | $\epsilon^{(2)}_{ab}$ | $\delta g^{\mu\nu} \neq 0$ | $\delta * 1 = 0$ | $\delta * \vartheta^a = - * \delta\vartheta^a$ |

We have presented the commutativity conditions only for the variations of the basis 1-forms. They can however be extended straightforwardly for forms of arbitrary degree. For instance, for a form $\alpha$ of an arbitrary degree (9.12) yields

$$\begin{aligned}
\delta * \alpha - *\delta\alpha &= *\epsilon_m{}^k\left(\vartheta_k \vee (\vartheta^m \wedge \alpha) - \vartheta_k \wedge (\vartheta^m \vee \alpha)\right) \\
&= *\epsilon_m{}^k\left(\delta^m_k\alpha - \vartheta^m \wedge (\vartheta_k \vee \alpha) - \vartheta_k \wedge (\vartheta^m \vee \alpha)\right) \\
&= *\epsilon\alpha - *\epsilon_{mk}\left(\vartheta^k \vee (\vartheta^m \wedge \alpha) + \vartheta^k \wedge (\vartheta^m \vee \alpha)\right)
\end{aligned}$$

The first term in the last expression vanishes in the case of a traceless variation. The expression in the brackets is symmetric under the interchange of the indices $k$ and $m$. Thus, in the case of an antisymmetric variational matrix $\epsilon_{mk}$, the variation of an arbitrary form commutes with the Hodge dual. We see that the commutativity (anti-commutativity) of the variation with the Hodge dual operator is connected with a specific type of constrained variation of the coframe field. The special case of commutativity was proven in [15].

## 9.5 General quadratic Lagrangian

A typical Lagrangian used in field theory is quadratic. In this section, we study the variational derivative of quadratic-type Lagrangians.

**Proposition 9.5.1:** *On a n-dimensional manifold, the variation of the quadratic expression* $\mathcal{L} = \alpha \wedge *\beta$ *with* $\alpha, \beta \in \Omega^p$ *takes the form*

$$\delta(\alpha \wedge *\beta) = \delta\alpha \wedge *\beta + \alpha \wedge *\delta\beta - \delta\vartheta_m \wedge J^m, \tag{9.1}$$

*where the* $(n-1)$*-form* $J^m$ *is*

$$J^m = (\vartheta^m \vee \beta) \wedge *\alpha - (-1)^p \alpha \wedge (\vartheta^m \vee *\beta). \tag{9.2}$$

**Proof:** Due to Leibniz's rule for the variation (9.4), we write

$$\delta(\alpha \wedge *\beta) = \delta\alpha \wedge *\beta + \alpha \wedge \delta * \beta. \tag{9.3}$$

The formula for commutator (9.12) yields

$$\delta(\alpha \wedge *\beta) = \delta\alpha \wedge *\beta + \alpha \wedge *\delta\beta + \alpha \wedge *\Big(\delta\vartheta_m \vee (\vartheta^m \wedge \beta) - \delta\vartheta_m \wedge (\vartheta^m \vee \beta)\Big). \tag{9.4}$$

Compute the third term on the right hand side of (9.4)

$$
\begin{aligned}
\alpha \wedge *\Big(\delta\vartheta_m \vee (\vartheta^m \wedge \beta)\Big) &= \alpha \wedge *^2\Big(\delta\vartheta_m \wedge *(\vartheta^m \wedge \beta)\Big) \\
&= (-1)^{p(n-p)+1}\alpha \wedge \delta\vartheta_m \wedge *(\vartheta^m \wedge \beta) \\
&= (-1)^p \delta\vartheta_m \wedge \alpha \wedge (\vartheta^m \vee *\beta).
\end{aligned}
$$

The fourth term on the right hand side of (9.4) is

$$\alpha \wedge *\Big(\delta\vartheta_m \wedge (\vartheta^m \vee \beta)\Big) = \delta\vartheta_m \wedge (\vartheta^m \vee \beta) \wedge *\alpha.$$

These expressions prove the statement. ∎

The $(n-1)$ form $J^m$ presents a type of a field current. Due to the symmetry of the expression (9.1) under the permutation of $\alpha$ and $\beta$ indices, we can rewrite $J^m$ in an explicitly symmetric form:

$$
\begin{aligned}
J^m = \frac{1}{2}\Big(&[(\vartheta^m \vee \beta) \wedge *\alpha + (\vartheta^m \vee \alpha) \wedge *\beta] - \\
&(-1)^p[\alpha \wedge (\vartheta^m \vee *\beta) + \beta \wedge (\vartheta^m \vee *\alpha)]\Big). \tag{9.5}
\end{aligned}
$$

In the special case $\alpha = \beta$, we obtain

$$J^m = (\vartheta^m \vee \alpha) \wedge *\alpha - (-1)^p \alpha \wedge (\vartheta^m \vee *\alpha). \tag{9.6}$$

It is instructive to expand the $(n-1)$-form $J^m$ in the coframe basis. We define

$$J^m = J^{mn} * \vartheta_n. \tag{9.7}$$

Using the identity

$$*(\vartheta^k \wedge J^m) = J^{mn} * (\vartheta^k \wedge *\vartheta_n), \tag{9.8}$$

we obtain the explicit expression for the matrix $J^{mn}$

$$J^{mn} = *(\vartheta^n \wedge J^m). \tag{9.9}$$

The two-indexed object $J^{mn}$ can be considered as an analog of the energy-momentum tensor.

**Proposition 9.5.2:** *The tensor $J^{mn}$ is symmetric*

$$J^{mn} = J^{nm}. \tag{9.10}$$

**Proof:** We use the fact that expression (9.2) is actually symmetric for a permutation of the forms $\alpha$ and $\beta$. Using (9.2) and (9.9) we derive

$$
\begin{aligned}
J^{mn} &= *(\vartheta^n \wedge J^m) = *\left[\vartheta^n \wedge \left((\vartheta^m \vee \beta) \wedge *\alpha - (-1)^p \alpha \wedge (\vartheta^m \vee *\beta)\right)\right] \\
&= (-1)^{p-1} * \left((\vartheta^m \vee \beta) \wedge (\vartheta^n \wedge *\alpha) + (\vartheta^n \wedge \alpha) \wedge (\vartheta^m \vee *\beta)\right).
\end{aligned}
$$

Write

$$\vartheta^n \wedge *\alpha = (-1)^{\sigma_1} *^2 \vartheta^n \wedge *\alpha = (-1)^{\sigma_1} * (\vartheta^n \vee \alpha), \tag{9.11}$$

$$\vartheta^m \vee *\beta = *(\vartheta^m \wedge *^2 \beta) = (-1)^{\sigma_2} * (\vartheta^m \wedge \beta), \tag{9.12}$$

where $\sigma_1, \sigma_2$ are two integers. Thus we obtain

$$
\begin{aligned}
J^{mn} = (-1)^{p-1} * \Big( &(-1)^{\sigma_1} (\vartheta^m \vee \beta) \wedge *(\vartheta^n \vee \alpha) + \\
&(-1)^{\sigma_2} (\vartheta^n \wedge \alpha) \wedge *(\vartheta^m \wedge \beta)\Big).
\end{aligned} \tag{9.13}
$$

Now the symmetry of $J^{mn}$ under the permutation of the forms $\alpha$ and $\beta$ yields the symmetry under the permutation of the indices $m$ and $n$. ∎
Calculate the trace of the matrix $J^{mn}$:

$$
\begin{aligned}
J^m{}_m &= *(\vartheta_m \wedge J^m) = *\left[\vartheta_m \wedge \left((\vartheta^m \vee \beta) \wedge *\alpha - (-1)^p \alpha \wedge (\vartheta^m \vee *\beta)\right)\right] \\
&= *\left(\vartheta_m \wedge (\vartheta^m \vee \beta) \wedge *\alpha - \alpha \wedge \vartheta_m \wedge (\vartheta^m \vee *\beta)\right).
\end{aligned} \tag{9.14}
$$

Use the formula for an arbitrary $p$-form $w$

$$\vartheta_m \wedge (\vartheta^m \vee w) = pw. \tag{9.15}$$

The trace of (9.14) takes the form

$$J^m{}_m = p\beta \wedge *\alpha - (n-p)\alpha \wedge *\beta = -(n-2p)*(\alpha \wedge *\beta). \tag{9.16}$$

Thus we obtain

**Proposition 9.5.3:** *For a quadratic Lagrangian*

$$\mathcal{L} = \alpha \wedge *\beta \qquad deg\alpha = deg\beta = p, \tag{9.17}$$

*the matrix $J^{mn}$ is traceless $J^m{}_m = 0$ if and only if the dimensional of the manifold $M$ is even and $2p = n$.*

The variational procedure described above can be straightforwardly extended to a general case of an external field on the manifold $\{M, \vartheta^a\}$. Consider, for instance, the Lagrangian

$$\mathcal{L} = \alpha(\phi, d\phi, \vartheta^a) \wedge *\beta(\phi, d\phi, \vartheta^a), \tag{9.18}$$

where $\phi$ is a certain external (multicomponent) field – differential form of degree $q$. Write the variation of the forms $\alpha, \beta$ as

$$\delta\alpha = \delta\phi \wedge \frac{\delta\alpha}{\delta\phi} + \delta d\phi \wedge \frac{\delta\alpha}{\delta d\phi} + \delta\vartheta^m \wedge \frac{\delta\alpha}{\delta\vartheta^m}, \tag{9.19}$$

$$\delta\beta = \delta\phi \wedge \frac{\delta\beta}{\delta\phi} + \delta d\phi \wedge \frac{\delta\beta}{\delta d\phi} + \delta\vartheta^m \wedge \frac{\delta\beta}{\delta\vartheta^m}. \tag{9.20}$$

The relations (9.1) yields

$$\delta\mathcal{L} = (\delta\phi \wedge \frac{\delta\alpha}{\delta\phi} + \delta d\phi \wedge \frac{\delta\alpha}{\delta d\phi} + \delta\vartheta^m \wedge \frac{\delta\alpha}{\delta\vartheta^m}) \wedge *\beta +$$
$$(\delta\phi \wedge \frac{\delta\beta}{\delta\phi} + \delta d\phi \wedge \frac{\delta\beta}{\delta d\phi} + \delta\vartheta^m \wedge \frac{\delta\beta}{\delta\vartheta^m}) \wedge *\alpha - \delta\vartheta_m \wedge J^m. \tag{9.21}$$

Thus we obtain two Euler-Lagrange field equations

$$\frac{\delta\alpha}{\delta\phi} \wedge *\beta + \frac{\delta\beta}{\delta\phi} \wedge *\alpha - (-1)^q d\Big(\frac{\delta\alpha}{\delta d\phi} \wedge *\beta + \frac{\delta\beta}{\delta d\phi} \wedge *\alpha\Big) = 0, \tag{9.22}$$

$$\frac{\delta\alpha}{\delta\vartheta^m} \wedge *\beta + \frac{\delta\beta}{\delta\vartheta^m} \wedge *\alpha = J^m. \tag{9.23}$$

## 9.6   Maxwell Lagrangian on a teleparallel space

The Lagrangian of Maxwell's theory of electromagnetism can be written as

$$\mathcal{L} = \frac{1}{2}dA \wedge *dA, \tag{9.1}$$

where $A$ is an 1-form of the electro-magnetic potential. For the field $A$ defined on a curved pseudo-Riemannian manifold, the Hodge star operator depends on the metric $g$ on the manifold.

$$\mathcal{L} = \frac{1}{2}dA \wedge *_g dA. \tag{9.2}$$

In this form, the variation of the Lagrangian must be applied by using free variations of the potential $A$ and of the metric $g$.

On a teleparallel manifold, the Lagrangian can be taken in the same form, but now the Hodge star operator depends on the coframe field $\vartheta^a$

$$\mathcal{L} = \frac{1}{2}dA \wedge *_{\vartheta^a} dA. \tag{9.3}$$

The variation of the Lagrangian (9.3) can be calculated by using equations (9.1, 9.2). We have

$$\delta\mathcal{L} = \frac{1}{2}\delta(dA) \wedge *dA + \frac{1}{2}dA \wedge *\delta(dA) - \delta\vartheta_m \wedge J^m, \tag{9.4}$$

where

$$J^m = \frac{1}{2}\left((\vartheta^m \vee dA) \wedge *dA - dA \wedge (\vartheta^m \vee *dA)\right). \tag{9.5}$$

Extracting the total derivative in (9.4) we are left with

$$\begin{aligned}\delta\mathcal{L} &= d(\delta A) \wedge *dA - \delta\vartheta_m \wedge J^m \\ &= d(\delta A \wedge *dA) + \delta A \wedge d*dA - \delta\vartheta_m \wedge J^m,\end{aligned}$$

we obtain two "field equations"

$$d*dA = 0 \tag{9.6}$$

and

$$J^m = 0, \tag{9.7}$$

Equation (9.6) has exactly the same form as the ordinary electro-magnetic field equation, but it is an equation on a curved manifold (operator $*$ depends on the coframe $\vartheta^a$). We can use the strength notation for electro-magnetic field

$$dA = F. \tag{9.8}$$

The first field equation (9.6) takes now the form of the ordinary Maxwell equations

$$\begin{aligned}dF &= 0, \tag{9.9}\\ d*F &= 0. \tag{9.10}\end{aligned}$$

Let us discuss which meaning can be given to Eq.(9.7). Equation (9.5) reads

$$J^m = \frac{1}{2}(\vartheta^m \vee F) \wedge *F - \frac{1}{2}F \wedge (\vartheta^m \vee *F). \tag{9.11}$$

In the components, the 2-form $F$ can be expressed as

$$F = \frac{1}{2} F_{mn} \vartheta^{mn}. \tag{9.12}$$

Notice that the coefficients of $F$ are antisymmetric by definition $F_{mn} = -F_{nm}$. Consider the first term of $J^m$ (9.11)

$$
\begin{aligned}
\frac{1}{2}(\vartheta^m \vee F) \wedge *F &= \frac{1}{8}(\vartheta^m \vee F^{kn}\vartheta_{kn}) \wedge *F_{pq}\vartheta^{pq} \\
&= \frac{1}{4} F^{kn} F_{pq} \delta^m_k \vartheta_n \wedge *\vartheta^{pq} = \frac{1}{2} F^{mn} F_{nq} * \vartheta^q .
\end{aligned}
$$

The second term of (9.11) takes the form

$$
\begin{aligned}
\frac{1}{2} F \wedge (\vartheta^m \vee *F) &= \frac{1}{8} F^{kn}\vartheta_{kn} \wedge (\vartheta^m \vee *F_{pq}\vartheta^{pq}) = -\frac{1}{8} F^{kn} F_{pq}\vartheta_{kn} \wedge *\vartheta^{mpq} \\
&= \frac{1}{8} F^{kn} F_{pq}\vartheta_k \wedge *(\delta^n_m \vartheta^{pq} - 2\delta^n_p \vartheta^{mq}) \\
&= \frac{1}{8} F^{kn} F_{pq} * (2\delta^n_m \delta^k_p \vartheta^q - 2\delta^n_p \delta^k_m \vartheta^q + 2\delta^n_p \delta^k_q \vartheta^m) \\
&= \frac{1}{2} F^{pm} F_{pq} * \vartheta^q + \frac{1}{4} F^{pq} F_{pq} * \vartheta^m
\end{aligned}
$$

Thus the object $J^m$ takes the form

$$J^m = (F^{mn} F_{nk} - \frac{1}{4} F^{pq} F_{pq} \delta^m_k) * \vartheta^k. \tag{9.13}$$

or in the tensor notations

$$J^m{}_k = F^{mn} F_{nk} - \frac{1}{4} F^{pq} F_{pq} \delta^m_k. \tag{9.14}$$

We readily recognize here the standard expression of the energy-momentum tensor for electromagnetic field. The variation procedure yields that it must vanish. This fact is in a clear contradiction with experiments. It is based on the consideration of the non-dynamical coframe field. When the Lagrangian is considered in the form

$$\mathcal{L} = \frac{1}{2} dA \wedge *_{\vartheta^a} dA + \mathcal{L}(\vartheta^a, d\vartheta^a), \tag{9.15}$$

we have the 3-current $J^a$ in the form

$$J^m = (F^{mn} F_{nk} - \frac{1}{4} F^{pq} F_{pq} \delta^m_k) * \vartheta^k - \frac{\delta}{\delta \vartheta^a} \mathcal{L}(\vartheta^a, d\vartheta^a). \tag{9.16}$$

The equation $J^m = 0$ means now the known relation between Noether's and Hilbert's definition of the energy-momentum tensor. a more detailed discussion of this subject is given in [23]. Due to the propositions (5.3-5.4) the traceless condition as well as the symmetric condition are the consequences of the definition of $J_{ab}$. Thus in the case of the Maxwell Lagrangian the object $J^m{}_n$ coincides with the classical expression of the energy-momentum tensor. Note that by the variational procedure described above we obtain it in a symmetric and a traceless form.

## 9.7   Translational Lagrangian in gravity

In the teleparallel approach to gravity, the coframe field $\vartheta^a$ is the basic gravitational field variable. A general Lagrangian density for the coframe field $\vartheta^a$ (quadratic in the first order derivatives) described by the gauge invariant translation Lagrangian [18]. Up to the $\Lambda$-term, it can be written as

$$\mathcal{L} = \frac{1}{2\ell^2} \sum_{I=1}^{3} \rho_I {}^{(I)}V, \tag{9.1}$$

where

$$
\begin{aligned}
{}^{(1)}\mathcal{L} &= d\vartheta^a \wedge *d\vartheta_a, & (9.2)\\
{}^{(2)}\mathcal{L} &= (d\vartheta_a \wedge \vartheta^a) \wedge *(d\vartheta_b \wedge \vartheta^b), & (9.3)\\
{}^{(3)}\mathcal{L} &= (d\vartheta_a \wedge \vartheta^b) \wedge *\left(d\vartheta_b \wedge \vartheta^a\right). & (9.4)
\end{aligned}
$$

and $\ell$ is the Plank length constant.

Note that the Lagrangian (9.1) is a linear combination of three independent terms of the form $\alpha \wedge *\beta$. So we can apply the procedure described above. Using the proposition (5.1) we obtain for the first Lagrangian (9.2)

$$
\begin{aligned}
\delta\left({}^{(1)}\mathcal{L}\right) &= 2\delta(d\vartheta_a) \wedge *d\vartheta^a - \delta(\vartheta_m) \wedge {}^{(1)}J^m \\
&= 2d\left(\delta(\vartheta_a) \wedge *d\vartheta_a\right) + 2\delta(\vartheta_a) \wedge d*d\vartheta^a - \delta(\vartheta_m) \wedge {}^{(1)}J^m, (9.5)
\end{aligned}
$$

where

$$
\begin{aligned}
{}^{(1)}J^m &= (\vartheta^m \vee d\vartheta_a) \wedge *d\vartheta^a - d\vartheta^a \wedge (\vartheta^m \vee *d\vartheta_a) \\
&= 2(\vartheta^m \vee d\vartheta_a) \wedge *d\vartheta^a - \vartheta^m \vee (d\vartheta_a \wedge *d\vartheta^a) & (9.6)
\end{aligned}
$$

Thus the contribution of the Lagrangian ${}^{(1)}\mathcal{L}$ in the field equation is

$$\boxed{2\rho_1\delta(\vartheta_a) \wedge d*d\vartheta^a + 2\rho_1(\vartheta^a \vee d\vartheta_m) \wedge *d\vartheta^m - \rho_1\vartheta^a \vee (d\vartheta_m \wedge *d\vartheta^m)} \quad (9.7)$$

Variation of the second Lagrangian (9.3) takes the form

$$
\begin{aligned}
\delta\left({}^{(2)}\mathcal{L}\right) &= 2\delta(d\vartheta_a \wedge \vartheta^a) \wedge *(d\vartheta_b \wedge \vartheta^b) - \delta(\vartheta_m) \wedge {}^{(2)}J^m \\
&= 2d(\delta\vartheta_a) \wedge \vartheta^a \wedge *(d\vartheta_b \wedge \vartheta^b) + 2d\vartheta_a \wedge \delta\vartheta^a \wedge *(d\vartheta_b \wedge \vartheta^b) - \\
&\quad \delta(\vartheta_m) \wedge {}^{(2)}J^m \\
&= 2d\left(\delta\vartheta_a \wedge \vartheta^a \wedge *(d\vartheta_b \wedge \vartheta^b)\right) + 2\delta\vartheta_a \wedge d\left(\vartheta^a \wedge *(d\vartheta_b \wedge \vartheta^b)\right) \\
&\quad +2\delta\vartheta^a \wedge d\vartheta_a \wedge *(d\vartheta_b \wedge \vartheta^b) - \delta(\vartheta_m) \wedge {}^{(2)}J^m \\
&= 2d\left(\delta\vartheta_a \wedge \vartheta^a \wedge *(d\vartheta_b \wedge \vartheta^b)\right) - 2\delta\vartheta_a \wedge \vartheta^a \wedge d*(d\vartheta_b \wedge \vartheta^b) + \\
&\quad 4\delta\vartheta^a \wedge d\vartheta_a \wedge *(d\vartheta_b \wedge \vartheta^b) - \delta(\vartheta_m) \wedge {}^{(2)}J^m, & (9.8)
\end{aligned}
$$

where the current term is

$$^{(2)}J^m = \left(\vartheta^m \vee (d\vartheta_b \wedge \vartheta^b)\right) \wedge *(d\vartheta_a \wedge \vartheta^a) + (d\vartheta_a \wedge \vartheta^a) \wedge \left(\vartheta^m \vee *(d\vartheta_b \wedge \vartheta^b)\right)$$

$$= -\vartheta^m \vee \left((d\vartheta^a \wedge \vartheta_a) \wedge *(d\vartheta^b \wedge \vartheta_b)\right) + 2\vartheta^m \vee (d\vartheta^a \wedge \vartheta_a) \wedge *(d\vartheta^b \wedge \vartheta_b)$$

$$= -\vartheta^m \vee \left((d\vartheta^a \wedge \vartheta_a) \wedge *(d\vartheta^b \wedge \vartheta_b)\right) + 2(\vartheta^m \vee d\vartheta^a) \wedge \vartheta_a \wedge *(d\vartheta^b \wedge \vartheta_b)$$

$$+ 2d\vartheta^m \wedge *(d\vartheta^b \wedge \vartheta_b) \tag{9.9}$$

Thus the contribution of the Lagrangian $^{(2)}\mathcal{L}$ in the field equation is

$$\boxed{\begin{aligned} &-2\rho_2\vartheta^a \wedge d*(d\vartheta_b \wedge \vartheta^b) + \rho_2\vartheta^a \vee \left((d\vartheta^m \wedge \vartheta_m) \wedge *(d\vartheta^n \wedge \vartheta_n)\right) + \\ &2\rho_2 d\vartheta_a \wedge *(d\vartheta_b \wedge \vartheta^b) + -2\rho_2(\vartheta^a \vee d\vartheta^m) \wedge \vartheta_m \wedge *(d\vartheta^n \wedge \vartheta_n) \end{aligned}} \tag{9.10}$$

For the third Lagrangian (9.4), we obtain by proposition (5.1)

$$\delta\left(^{(3)}\mathcal{L}\right) = 2\delta(d\vartheta_a \wedge \vartheta^b) \wedge *(d\vartheta_b \wedge \vartheta^a) - \delta\vartheta_m \wedge {}^{(3)}J^m = -\delta\vartheta_m \wedge {}^{(3)}J^m +$$

$$2d(\delta\vartheta_a) \wedge \vartheta^b \wedge *(d\vartheta_b \wedge \vartheta^a) + 2\delta\vartheta^b \wedge d\vartheta_a \wedge *(d\vartheta_b \wedge \vartheta^a)$$

$$= 2d\left(\delta\vartheta_a \wedge \vartheta^b \wedge *(d\vartheta_b \wedge \vartheta^a)\right) - 2\delta\vartheta_a \wedge \vartheta^b \wedge d*(d\vartheta_b \wedge \vartheta^a) +$$

$$4\delta\vartheta^b \wedge d\vartheta_a \wedge *(d\vartheta_b \wedge \vartheta^a) - \delta\vartheta_m \wedge {}^{(3)}J^m. \tag{9.11}$$

The correspondent current term is

$$^{(3)}J^m = \left(\vartheta^m \vee (d\vartheta_a \wedge \vartheta^b)\right) \wedge *(d\vartheta_b \wedge \vartheta^a) + (d\vartheta_a \wedge \vartheta^b) \wedge \left(\vartheta^m \vee *(d\vartheta_b \wedge \vartheta^a)\right)$$

$$= -\vartheta^m \vee \left((d\vartheta_a \wedge \vartheta^b)\right) \wedge *(d\vartheta_b \wedge \vartheta^a) + 2\vartheta^m \vee (d\vartheta_a \wedge \vartheta^b) \wedge *(d\vartheta_b \wedge \vartheta^a)$$

$$= -\vartheta^m \vee \left((d\vartheta_a \wedge \vartheta^b)\right) \wedge *(d\vartheta_b \wedge \vartheta^a) + 2(\vartheta^m \vee d\vartheta_a) \wedge \vartheta^b \wedge *(d\vartheta_b \wedge \vartheta^a) +$$

$$= 2d\vartheta_a \wedge *(d\vartheta^m \wedge \vartheta^a) \tag{9.12}$$

Hence the contribution of the Lagrangian $^{(3)}\mathcal{L}$ to the field equation is

$$\boxed{\begin{aligned} &-2\rho_3\vartheta^b \wedge d*(d\vartheta_b \wedge \vartheta^a) + \rho_3\vartheta^a \vee \left((d\vartheta_m \wedge \vartheta^n)\right) \wedge *(d\vartheta_n \wedge \vartheta^m) \\ &-2\rho_3(\vartheta^a \vee d\vartheta_m) \wedge \vartheta^n \wedge *(d\vartheta_n \wedge \vartheta^m) + 2\rho_3 d\vartheta_b \wedge *(d\vartheta^a \wedge \vartheta^b) \end{aligned}} \tag{9.13}$$

The relations (9.7), (9.10) and (9.13) yield the field equation generated by free variations of the total Lagrangian

$$\begin{aligned} -2\ell^2\Sigma_a =\ & 2\rho_1 d*d\vartheta_a - 2\rho_2\vartheta_a \wedge d*(d\vartheta^b \wedge \vartheta_b) - 2\rho_3\vartheta_b \wedge d*(\vartheta_a \wedge d\vartheta^b) \\ &+ \rho_1\left[e_a \vee (d\vartheta^b \wedge *d\vartheta_b) - 2(e_a \vee d\vartheta^b) \wedge *d\vartheta_b\right] \\ &+ \rho_2\left[2d\vartheta_a \wedge *(d\vartheta^b \wedge \vartheta_b) + e_a \vee (d\vartheta^c \wedge \vartheta_c \wedge *(d\vartheta^b \wedge \vartheta_b))\right. \\ &\left. -2(e_a \vee d\vartheta^b) \wedge \vartheta_b \wedge *(d\vartheta^c \wedge \vartheta_c)\right] \\ &+ \rho_3\left[2d\vartheta_b \wedge *(\vartheta_a \wedge d\vartheta^b) + e_a \vee (\vartheta_c \wedge d\vartheta^b \wedge *(d\vartheta^c \wedge \vartheta_b))\right. \\ &\left. -2(e_a \vee d\vartheta^b) \wedge \vartheta_c \wedge *(d\vartheta^c \wedge \vartheta_b)\right], \end{aligned} \tag{9.14}$$

where $\Sigma_a$ depends on matter field energy-momentum current. This equation is equivalent to the field equation given by Kopczyński [7], see also [15].

## 9.8  Conclusion

We discussed the variational procedure on a teleparallel manifold. This case is rather differ from the variation on a pseudo-Riemannian manifold. The commutativity and anti-commutativity of the variation with the Hodge dual map are shown to be related to different covariant types of the constrained variations. We derive a general relation for variation of the quadratic type Lagrangians and apply it to the viable cases of the electro-magnetic Lagrangian and to the translation invariant of Rumpf. The derived formulas can be applied for study the various material fields on teleparallel background.

**Acknowledgments**: My gratitude to Friedrich W. Hehl for various discussions of the teleparallel models. I thank the organizers of First Hermann Minkowski Meeting on the Foundations of Spacetime Physics for a nice conference.

## References

[1] J. H. Schwarz and N. Seiberg, Rev. Mod. Phys. **71**, S112 (1999) doi:10.1103/RevModPhys.71.S112 [hep-th/9803179].

[2] Cartan, E., *Geometry of Riemannian spaces*, Math. Sci. Press, New York, 1983.

[3] R. Hermann *Differential geometry and the calculus of variations.*,Academic Press, N.Y., 1968

[4] A. Einstein:"Auf die Riemann-Metric und den Fern-Parallelismus gegründete einheitliche Field-Theorie", Math. Ann. **102**, (1930), 658-697.

[5] A. Einstein, W.R.P. Mayer,: "Zwei strenge statische Lösungen der Feldgleichungen der einheitlichen Feldtheorie", *Sitzungsber. preuss, Acad. Wiss.,phys.-math.* **K 1** (1930) 110–120.

[6] R. Weitzenböck: *Invariantentheorie*, Noordhoff, Groningen, 1923.

[7] W. Kopczyński, "Problems with metric-teleparallel theories of gravitation", *J. Phys.* **A15**, (1982), 493.

[8] F. Müller-Hoissen, J.Nitsch: "Teleparallelism - A viable theory of gravity", *Phys.Rev.*, **D28**, (1983) 718.

[9] F. W. Hehl, J. D. McCrea, E. W. Mielke, and Y. Ne'eman: "Metric-affine gauge theory of gravity: Field equations, Noether identities, world spinors, and breaking of dilation invariance", *Physics Reports* **258**, (1995) 1–171.

[10] F. Gronwald and F. W. Hehl: "On the gauge aspects of gravity". In: *International School of Cosmology and Gravitation:* 14 Course: Quantum Gravity, held May 1995 in Erice, Italy. Proceedings. P.G. Bergmann et al.(eds.) (World Scientific, Singapore 1996) pp. 148–198, gr-qc/9602013.

[11] J. W. Maluf, "Sparling two-forms, the conformal factor and the gravitational energy density of the teleparallel equivalent of general relativity ", *Gen.Rel.Grav.* **30**, (1998) 413-423

[12] J. W. Maluf and J. F. da Rocha-Neto, "General relativity on a null surface: Hamiltonian formulation in the teleparallel geometry," Gen. Rel. Grav. **31**, 173 (1999)

[13] R. S. Tung, J. M. Nester: "The quadratic spinor Lagrangian is equivalent to the teleparallel theory", gr-qc/9809030

[14] R. S. Tung and J. M. Nester, "The Quadratic spinor Lagrangian is equivalent to the teleparallel theory," Phys. Rev. D **60**, 021501 (1999) doi:10.1103/PhysRevD.60.021501 [gr-qc/9809030].

[15] U. Muench, F. Gronwald, F. W. Hehl: " A small guide to variations in teleparallel gauge theories of gravity and the Kaniel-Itin model", *Gen.Rel.Grav.*, **30**, (1998) 933-961.

[16] O. V. Babourova, B. N. Frolov and E. A. Klimova, "Plane torsion waves in quadratic gravitational theories," Class. Quant. Grav. **16**, 1149 (1999) doi:10.1088/0264-9381/16/4/005 [gr-qc/9805005].

[17] F. W. Hehl: "Four lectures on Poincaré gauge theory", in Proceedings of the 6-th Course on Spin, Torsion and Supergravity, held at Erice, Italy , 1979, eds. P.G. Bergmann, V. de Sabbata (Plenum, N.Y.) p.5

[18] H. Rumpf: "On the translational part of the Lagrangian in the Poincaré gauge theory of gravitation", *Z. Naturf.* **33a** (1978) 1224–1225.

[19] Y. Itin and S. Kaniel, "On a class of invariant coframe operators with application to gravity," J. Math. Phys. **41**, 6318 (2000) doi:10.1063/1.1287434 [gr-qc/9907023].

[20] Y. Itin, "Energy momentum current for coframe gravity," Class. Quant. Grav. **19**, 173 (2002) doi:10.1088/0264-9381/19/1/311

[21] Y. Itin, "Coframe energy momentum current: Algebraic properties," Gen. Rel. Grav. **34**, 1819 (2002) doi:10.1023/A:1020759923382 [gr-qc/0111087].

[22] Y. Itin, "Conserved currents for general teleparallel models," Int. J. Mod. Phys. A **17**, 2765 (2002) doi:10.1142/S0217751X02011928 [gr-qc/0103017].

[23] Y. Itin, "Noether currents and charges for Maxwell - like Lagrangians," J. Phys. A **36**, 8867 (2003) doi:10.1088/0305-4470/36/33/310 [math-ph/0307003].

[24] Y. Itin, "Weak field reduction in teleparallel coframe gravity: Vacuum case," J. Math. Phys. **46**, 012501 (2005) doi:10.1063/1.1819523 [gr-qc/0409021].

[25] Y. Itin, "Coframe geometry, gravity and electromagnetism," J. Phys. Conf. Ser. **437**, 012003 (2013). doi:10.1088/1742-6596/437/1/012003

[26] R. Aldrovandi and J. G. Pereira, *Teleparallel Gravity: An Introduction* (Springer, Dordrecht, The Netherlands, 2013).

[27] Y. Itin, F. W. Hehl and Y. N. Obukhov, "Premetric equivalent of general relativity: Teleparallelism," Phys. Rev. D **95**, no. 8, 084020 (2017) doi:10.1103/PhysRevD.95.084020 [arXiv:1611.05759 [gr-qc]].

# 10 EINSTEIN THEORY OF GRAVITATION IS UNIVERSAL

JERZY KIJOWSKI

**Abstract** Generalizations of general relativity theory, based on a generic Lagrangian scalar density $L = L(g_{\mu\nu}, R^\lambda_{\mu\nu\kappa})$ depending upon the metric $g_{\mu\nu}$ and the full curvature tensor $R^\lambda_{\mu\nu\kappa}$ (not only scalar curvature), are equivalent to the conventional Einstein theory for a (possibly) different metric tensor $\tilde{g}_{\mu\nu}$ and a (possibly) different set of matter fields. This is a mathematical result based on a new approach to variational problems containing metric and connection.

## 10.1 Introduction

Einstein theory of gravity is usually derived from the "metric" variational principle:

$$L(g, \partial g, \partial^2 g, \varphi, \partial\varphi) = L_{Hilbert}(g, \partial g, \partial^2 g) + L_{Matter}(g, \partial g, \varphi, \partial\varphi) , \quad (10.1)$$

with universal Hilbert Lagrangian:

$$L_{Hilbert} = \frac{1}{16\pi}\sqrt{|g|}R = \pi^{\mu\nu}R_{\mu\nu} . \quad (10.2)$$

Here, $R_{\mu\nu}$ is the Ricci tensor, $R$ is the scalar curvature. Moreover, by

$$\pi^{\mu\nu} := \frac{1}{16\pi}\sqrt{|g|}g^{\mu\nu} , \quad (10.3)$$

we have denoted the contravariant density of metric tensor.

Usually, one considers the so called "minimal coupling" of various matter fields (denoted symbolically by $\varphi$) with gravity. This statement means that we begin with a special-relativistic version of the matter Lagrangian $L_{Matter}(g, \varphi, \partial\varphi)$ where $g$ is a flat Minkowski metric. Then, we allow $g$ to be curved and replace partial derivatives $\partial\varphi$ by appropriate "covariant" derivatives $\nabla\varphi = \partial\varphi + "\Gamma \cdot \varphi"$, where $\Gamma$ is the Levi-Civita connection of $g$. The above "definition" of a covariant derivative is merely symbolical. It makes a precise

C. Duston, M. Holman (Eds), *Spacetime Physics 1907 - 2017. Selected peer-reviewed papers presented at the First Hermann Minkowski Meeting on the Foundations of Spacetime Physics, 15-18 May 2017, Albena, Bulgaria* (Minkowski Institute Press, Montreal 2019). ISBN 978-1-927763-48-3 (softcover), ISBN 978-1-927763-49-0 (ebook).

sense only if $\varphi$ is a tensor field. For an arbitrary matter field we only assume that $L_{Matter}$ is an invariant scalar density built from the field, the metric and their first derivatives.

But, we have:

$$\frac{\delta L_{Hilbert}}{\delta g_{\mu\nu}} = -\frac{1}{16\pi}\sqrt{|g|}G^{\mu\nu} , \qquad (10.4)$$

where $G$ denotes the Einstein tensor. Defining the "matter energy-momentum tensor":

$$T^{\mu\nu} := \frac{2}{\sqrt{|g|}}\frac{\delta L_{Matter}}{\delta g_{\mu\nu}} = \frac{2}{\sqrt{|g|}}\left(\frac{\partial L_{Matter}}{\partial g_{\mu\nu}} - \partial_\kappa\frac{\partial L_{Matter}}{\partial g_{\mu\nu,\kappa}}\right) , \qquad (10.5)$$

where $g_{\mu\nu,\kappa} := \partial_\kappa g_{\mu\nu}$, we obtain field equations of the theory:

$$0 = \frac{\delta L}{\delta g_{\mu\nu}} = -\frac{1}{16\pi}\sqrt{|g|}\left(G^{\mu\nu} - 8\pi T^{\mu\nu}\right) ,$$

$$0 = \frac{\delta L}{\delta\varphi} = \frac{\delta L_{Matter}}{\delta\varphi} .$$

Replacing Hilbert Lagrangian by an arbitrary scalar density $L$ depending upon $g_{\mu\nu}$ and $R^\lambda_{\mu\nu\kappa}$, but no longer linear in the curvature, changes substantially the character of our theory. In a generic case, field equations are no longer of the second differential order with respect to the metric, but are fourth order PDE's. Recently, such "generalizations of General Relativity Theory" have been proposed as a possible description of "black matter" or "black energy" (see [1]).

Consider, therefore, a „generalized" theory of gravity, based on an invariant Lagrangian:

$$L = L(g_{\mu\nu}, R^\lambda_{\mu\nu\kappa}, \Gamma^\lambda_{\mu\nu}, \varphi, \partial\varphi) , \qquad (10.6)$$

where $\Gamma^\lambda_{\mu\nu}$ is the Levi-Civita connection of the metric $g_{\mu\nu}$ and $R^\lambda_{\mu\nu\kappa}$ denotes its Riemann tensor. The following mathematical statement has been proved in [2].

**Theorem 1:** There exists a one-to-one change of variables:

$$(g,\varphi) \Longleftrightarrow (\tilde{g},\varphi,\phi) , \qquad (10.7)$$

and a new matter Lagrangian:

$$\tilde{L}_{Matter} = \tilde{L}_{Matter}(\tilde{g},\partial\tilde{g},\varphi,\phi,\partial\varphi,\partial\phi) , \qquad (10.8)$$

such that $(g,\varphi)$ satisfy field equations derived from the Lagrangian (10.6) if and only if the corresponding fields $(\tilde{g},\varphi,\phi)$ satisfy the conventional „Einstein + matter" equations, derived from the conventional variational principle:

$$\tilde{L} := L_{Hilbert}(\tilde{g}) + \tilde{L}_{Matter} . \qquad (10.9)$$

In particular, equations for the new metric $\tilde{g}$ are of the second differential order: $G^{\mu\nu}(\tilde{g}) = 8\pi\tilde{T}^{\mu\nu}$, where

$$\tilde{T}^{\mu\nu} := \frac{2}{\sqrt{|\tilde{g}|}}\frac{\delta\tilde{L}_{Matter}}{\delta\tilde{g}_{\mu\nu}} . \qquad (10.10)$$

Also matter field equations are of the second differential order because $\tilde{L}_{Matter}$ depends upon first derivatives only.

To define new metric $\tilde{g}$ and new matter fields $\phi$ we decompose the curvature tensor into a sum of two irreducible components: 1) Ricci tensor $R_{\mu\nu}$ and 2) Weyl tensor $W^\lambda_{\mu\nu\kappa}$ (see (10.62) and (10.93)). The new metric (or, rather, its contravariant density, cf. formula (10.3)) is defined as the "momentum canonically conjugate" to the Ricci tensor:

$$\tilde{\pi}^{\mu\nu} := \frac{\partial L}{\partial R_{\mu\nu}} \, , \tag{10.11}$$

whereas new matter fields $\phi$ describe: 1) the old metric $g$ (downgraded now to the level of matter fields) and: 2) the field $p^{\mu\nu\kappa}_\lambda$ defined as the "momentum canonically conjugate" to the Weyl tensor:

$$p^{\mu\nu\kappa}_\lambda := \frac{\partial L}{\partial W^\lambda_{\mu\nu\kappa}} \, . \tag{10.12}$$

This means that the new fields $(\tilde{g}, \phi)$ are defined as combinations of the old fields $(g, \varphi)$ and their derivatives up to the second order. The new matter Lagrangian can be uniquely derived from the old one.

Observe that, in case of the conventional Hilbert Lagrangian, equation (10.11) implies: $\tilde{\pi}^{\mu\nu} = \pi^{\mu\nu}$ (both new and old metric tensors coincide) and (10.12) implies $p^{\mu\nu\kappa}_\lambda = 0$, i.e. nothing changes.

The complete proof of the above statement was published in [2]. It is highly technical, but its idea is very simple. The purpose of the present paper is to convince the reader that this idea clarifies considerably the mathematical structure of General Relativity Theory.

The particular case of a Lagrangian $L$ which depends non-linearly upon the Ricci tensor, but does not depend upon the Weyl tensor, was considered by many authors (cf. [3]). Mathematical structure of such theories was thoroughly analyzed already long ago (see e.g. [4]). In particular, equation (10.12) implies that the field $p^{\mu\nu\kappa}_\lambda$ vanishes identically for such theories and there is only one "new matter field", namely the old metric. Probably the first, physically well motivated, proposal of such a theory was the Sacharov's non-linear Lagrangian containing the $R^2$ term (see [5]). In this case, and also for any Lagrangian density depending exclusively upon the scalar curvature $R$, i.e. for $L = \sqrt{|g|} f(R)$, equation (10.11) implies that the old metric $\pi$ is proportional to the new metric $\tilde{\pi}$. Indeed, (10.11) reads:

$$\tilde{\pi}^{\mu\nu} := \frac{\partial L}{\partial R_{\mu\nu}} = f' \sqrt{|g|} g^{\mu\nu} = \pi^{\mu\nu} e^{-\phi} \, . \tag{10.13}$$

Consequently, the new matter field $\pi^{\mu\nu}$ can be encoded by a single scalar field $\phi$ (see [6] and also [4]). Sacharov theory is, therefore, equivalent to the standard Einstein general relativity theory interacting with a non-linear scalar field[1]:

$$\tilde{L} := L_{Hilbert}(\tilde{g}) + \tilde{L}_{Matter}(\phi, \partial\phi, \tilde{g}) \, . \tag{10.14}$$

---

[1] Also Brans-Dicke theory can be mentioned in this context. It is, however, much simpler to handle because its Lagrangian is linear in the curvature.

Special examples of the Lagrangians depending upon the Ricci tensor: $L = L(g_{\mu\nu}, R_{\mu\nu})$, were analyzed also by Stephenson and Higgs (see [7]). These results are, however, purely algebraic and do not apply to a generic Lagrangian of this type. Moreover, the theories considered in [7] belong to a (much simpler) class of "purely affine" theories.

Our result can be summarized as follows: non-conventional theories of gravity, based on non-conventional Lagrangians, are equivalent to conventional gravity interacting with non-conventional matter fields. We stress, that the result is mathematically rigorous, and does not depend upon different "ideologies" which are often attached to specific, technical formulations of the gravity theory. It follows in a straightforward way from the symplectic interpretation of variational principles in hyperbolic field theories, presented below. We illustrate this approach in Section 10.4 on a simple "toy model". Finally, we go back to General Relativity Theory and give the reader more information about canonical structure of gravitational field.

## 10.2  Volume term versus boundary term in variational principles

Calculus of variations was invented by Johann Bernoulli in 1696 for purposes of his "brachistochrone problem". It is based on the following identity:

$$\delta L(x^\mu, \varphi^K, \varphi^K{}_\mu) = \left( \frac{\partial L}{\partial \varphi^K} - \partial_\mu \frac{\partial L}{\partial \varphi^K{}_\mu} \right) \delta \varphi^K + \partial_\mu \left( \frac{\partial L}{\partial \varphi^K{}_\mu} \delta \varphi^K \right). \quad (10.15)$$

Here, $(x^\mu) \in M$ are independent variables (e.g. coordinates on spacetime $M$), $(\varphi^K)$ are dependent variables (e.g. components of gravitational or electromagnetic field), $\varphi^K{}_\mu := \partial_\mu \varphi^K$. Operator of "variation" is denoted by "$\delta$". Naively, it can be understood as a derivative of a one-parameter family of field configurations $\varphi^K = \varphi^K(x^\mu, \epsilon)$ with respect to this extra parameter, i.e. $\delta = \frac{\partial}{\partial \epsilon}$. On a more sophisticated level, one can treat $\delta$ as an external derivative defined on an appropriate jet space of field configurations attached at a fixed point $(x^\mu) \in M$ (cf. [8] and [9]).

In classical optimization problems (like brachistochrone, ortodrome, minimal surface, various logistic or economy problems etc.) identity (10.15) is used in the following way: imposing boudary conditions $\delta \varphi^K \big|_{\partial V} = 0$ on the boundary $\partial V$, we obtain:

$$\delta \int_V L(x^\mu, \varphi^K, \varphi^K{}_\mu) = \int_V \left( \frac{\partial L}{\partial \varphi^K} - \partial_\mu \frac{\partial L}{\partial \varphi^K{}_\mu} \right) \delta \varphi^K \quad (10.16)$$

which implies Euler-Lagrange equations

$$\frac{\partial L}{\partial \varphi^K} - \partial_\mu \frac{\partial L}{\partial \varphi^K{}_\mu} = 0 \quad (10.17)$$

as a necessary condition for an extremum of the functional $\int_V L$. One can say that boundary condition are used to kill *a priori* boundary term in (10.15). The remaining equation is no longer an identity but an equation to solve.

This philosophy (which I like to call a "brachistochrone philosophy") is absolutely illegal in dynamical theories, like classical mechanics or field theories, governed by hyperbolic partial differential equations. Here, dynamically admissible configurations never minimize action! Formulation "least action principle" is simply false. This is, however, a minor problem which is often correctly discussed in good textbooks ("stationary action" instead of "least action"). But the main difficulty is the impossibility to control the field configuration *via* the boundary data. Indeed: boundary value problem is not well posed in a generic situation. Moreover, those data which allow for a solution of field equation constitute typically a small, singular (non-closed in any reasonable topology) subspace of all boundary data. In other words: choosing at random boundary data, the probability that there exists a solution of field equations matching these data is exactly zero! (An analogous statement: choosing at random a real number, the probability that it is rational is exactly zero!) Any "derivation" of field equations which uses the "brachistochrone philosophy" is simply false. Fortunately enough, what really matters are field equations and not the fact that something has been (or has not been) minimized. Below, I present an approach which relates correctly field equations with their generating function, i.e. a Lagrangian.

An appropriate way to use identity (10.15) in dynamical (hyperbolic) theories consists not in killing boundary term and keeping volume term ("brachistochrone philosophy") but in killing the volume term (i.e. taking into accout only those field configurations which are physically admissible) and keeping the boundary term ("on shell" philosophy). The remaining equation is no longer an identity but a system of first order equations to solve. Indeed, denoting canonical field momenta by $p_K^\mu$, i.e.:

$$p_K^\mu := \frac{\partial L}{\partial \varphi^K_\mu} \tag{10.18}$$

we see, that dynamically admissible ("on shell") field configurations, i.e. those which obey field equations (10.17), must fulfil the remaining equation:

$$\delta L(x^\mu, \varphi^K, \varphi^K_\mu) = \partial_\mu \left( p_K^\mu \delta\varphi^K \right) = (\partial_\mu p_K^\mu) \delta\varphi^K + p_K^\mu \delta\varphi^K_\mu \,. \tag{10.19}$$

It contains the entire information carried by the variational principle: 1) definition of momenta (10.18) and 2) field equations (10.17). This way, field equations become a symplectic control theory, analogous to the 1-law of thermodynamics:

$$\mathrm{d}U(V,S) = -p\mathrm{d}V + T\mathrm{d}S\,. \tag{10.20}$$

The correct interpretation of this formula is following: Thermodynamics of *an arbitrary* simple body uses the universal, four dimensional phase space $\mathcal{P}$, which can be parameterized by the pressure $p$, volume $V$, temperature $T$ and the entropy $S$, and equipped with the canonical symplectic form

$$\Omega = \mathrm{d}V \wedge \mathrm{d}p + \mathrm{d}T \wedge \mathrm{d}S\,. \tag{10.21}$$

A specific, physical body is defined by its 2-dimensional Lagrangian subspace $\mathcal{D} \subset \mathcal{P}$ of physically admissible ("on shell") states

$$\mathcal{D} := \left\{ (p, V, T, S) \mid p = -\frac{\partial U}{\partial V}, T = \frac{\partial U}{\partial S} \right\}. \tag{10.22}$$

Classical termodynamics reduces, this way, to a symplectic geometry of this space.

Formula (10.20) is uniquely implied by the above canonical structure as soon as we choose variables $(V, S)$ as "control parameters" and $(p, T)$ as "response parameters". Consequently, we treat $(-p, T)$ as momenta canonically conjugate to $(V, S)$. Denoting the space of control parameters by $Q$, the phase space becomes isomorphic with its cotangent bundle: $\mathcal{P} \cong T^*Q$ and the dynamics $\mathcal{D}$ becomes a graph of the differential $dU$ of the generating function $U$ defined on $Q$ (cf. [8]).

Exchanging some control parameters with some response parameters, we may pass to another "control mode". This way the same dynamics $\mathcal{D}$ can be described by a different generating function. Such a transition between different control modes is called a Legendre transformation. For example, rewriting generating formula (10.20) in the following way:

$$dU(V, S) = -pdV + d(TS) - SdT, \tag{10.23}$$

we obtain the Helmholz control mode:

$$d(U(V, T) - TS) = -pdV - SdT, \tag{10.24}$$

where $F := U - TS$ is the free energy of the system, available during isotermic processes. Also transition from the Lagrangian to the Hamiltonian description of mechanics is an example of such a Legendre transformation. Indeed, in case of a single independent variable $t$, the variational identity (10.15) reduces *on shell* to the boundary term

$$\delta L(t, q^K, \dot{q}^K) = \frac{d}{dt} \left( p_K \delta q^K \right) = \dot{p}_K \delta q^K + p_K \delta \dot{q}^K, \tag{10.25}$$

analogous to (10.19) in field theory or to (10.20) in thermodynamics, and containing the complete Lagrangian description of mechanics:

$$\mathcal{D} := \left\{ (p_K, \dot{p}_K, q^K, \dot{q}^K) \mid p_K = \frac{\partial L}{\partial \dot{q}^K}, \dot{p}_K = \frac{\partial L}{\partial q^K} \right\}, \tag{10.26}$$

whereas exchanging velocities $\dot{q}^K$ with momenta $p_K$ in the role of control parameters, we obtain, in full analogy with (10.24), the Hamiltonian control mode:

$$\delta \left( L - p_K \dot{q}^K \right) = \dot{p}_K \delta q^K - \dot{q}^K \delta p_K. \tag{10.27}$$

Denoting $L - p_K \dot{q}^K =: -H$, we see that Lagrangian dynamics (10.26) may be equivalently described in the Hamiltonian control mode:

$$\mathcal{D} := \left\{ (p_K, \dot{p}_K, q^K, \dot{q}^K) \mid \dot{q}^K = \frac{\partial H}{\partial p_K}, \dot{p}_K = -\frac{\partial H}{\partial q^K} \right\}, \tag{10.28}$$

in the same 4N-dimensional phase space equipped with the canonical symplectic form

$$\Omega = \mathrm{d}\dot{p}_K \wedge \mathrm{d}q^K + \mathrm{d}p_K \wedge \mathrm{d}\dot{q}^K . \tag{10.29}$$

The profit we get when using the above "on shell" description of field theory is not merely moral: we are not obliged to cheat when deriving field equations from the Lagrangian! (to cheat that boundary condition can be imposed in hiperbolic field theories) but also technical: powerful symplectic techniques considerably simplify the structure of the theory. As we shall see in the sequel, equivalence of gravity based on a non-linear Lanrangian, with the conventional, Einsteinian gravity coupled to additional matter fields follows from an appropriate Legendre transformation which relates both theories. (In Appendix I the reader can find a mathematically more comprehensible description of this formalism together with a discussion of gauge and constraints.)

## 10.3 Boundary term in the Hilbert variational principle

Following the "brachistochrone philosophy", most texts concerning variational formulation of gravity consider only the volume term and neglect the boundary term of identity (10.15). For example, C.Misner, K.Thorne and J.A.Wheeler in their monograph, otherwise excellent as an introduction to General Relativity Theory, derive formula (10.4) and then provide the following statement (see e.g. [10], page 520, above formula 21.86): *Variation of the geometry interior to the boundary make no difference in the value of the surface term. Therefore, it has no influence on the equations of motion to drop the term* (21.85). The term which is dropped is precisely the surface term, which was never calculated in this monograph. Moreover, the author of the present paper has never seen in the literature any attempt to calculate properly this term and, therefore, has a tendency to consider the following simple identity as his private discovery (see [11]):

$$\begin{aligned}
\delta L_{Hilbert} &= \delta\left(R_{\mu\nu}\pi^{\mu\nu}\right) = -\frac{1}{16\pi}\sqrt{|g|}G^{\mu\nu}\delta g_{\mu\nu} + \partial_\kappa\left\{\left(\delta_\lambda^\kappa\pi^{\mu\nu} - \delta_\lambda^\mu\pi^{\nu\kappa}\right)\delta\Gamma_{\mu\nu}^\lambda\right\} \\
&= R_{\mu\nu}\delta\pi^{\mu\nu} + \partial_\kappa\left\{\pi_\lambda^{\mu\nu\kappa}\delta\Gamma_{\mu\nu}^\lambda\right\} , \tag{10.30}
\end{aligned}$$

where the following tensor density with four indices:

$$\pi_\lambda^{\mu\nu\kappa} := \delta_\lambda^\kappa\pi^{\mu\nu} - \delta_\lambda^{(\mu}\pi^{\nu)\kappa} = \delta_\lambda^\kappa\pi^{\mu\nu} - \frac{1}{2}\delta_\lambda^\mu\pi^{\nu\kappa} - \frac{1}{2}\delta_\lambda^\nu\pi^{\mu\kappa}, \tag{10.31}$$

encoding the metric tensor, arises in a natural way as a "momentum" canonically conjugate to the connection $\Gamma_{\mu\nu}^\lambda$. Formula (10.30) shows that canonical structure of General Relativity simplifies considerably if we represent the metric structure of spacetime by its contravariant density (10.3), instead of the usual covariant tensor $g_{\mu\nu}$. This observation can already be found in the Fock's monograph [12], formula (60.14) (my modest contribution was to include in momenta "$\pi$" also the gravitational constant, equal to $16\pi$ in appropriate units).

The structure of the boundary term in (10.30) shows, that among the two geometric spacetime structures, *a priori* independent: the metric and the connection $\Gamma$, the second one plays role of the field configuration, whereas the first one should be rather treated as a canonical momentum. Field equations relate the two by the two sets of first order equations: 1) metricity condition for the connection and: 2) Einstein equations. Of course, Legendre transformation consisting in exchanging these roles, are possible. To illustrate these possibilities I will use a toy model, technically much simpler than gravity, and show this way main ideas of the proof of our Theorem.

## 10.4 Toy model

Consider dynamics of a mechanical degree of freedom $\pi$ which, in our toy model, plays role of the metric. There is only one independent variable $t$ and the corresponding derivative which we denote by the "dot". Dynamics of the model is derived from the following Lagrangian function, being of second differential order:

$$L_{Hilbert} = \pi R, \qquad (10.32)$$

where $R := \dot{\Gamma} + \frac{1}{2}\Gamma\Gamma$ and, finally,

$$\Gamma := \frac{\dot{\pi}}{\pi}. \qquad (10.33)$$

Here, $\Gamma$ (derivatives of the metric, multiplied by its inverse) and $R$ (derivatives of $\Gamma$ plus a term quadratic in $\Gamma$) mimic, respectively, the role of connection and curvature. Finally, (10.32) mimics the Hilbert Lagrangian (10.2) in General Relativity theory.

To derive equation of motion we can use one of the following tricks, proposed usually in literature:

1. Even if $L_{Hilbert}$ is second order, we may subtract a complete time derivative and obtain an equivalent, first order Lagrangian. For this purpose we write:

$$L_{Hilbert} = \pi\left(\dot{\Gamma} + \frac{1}{2}\Gamma\Gamma\right) = \frac{\mathrm{d}}{\mathrm{d}t}(\pi\Gamma) - \dot{\pi}\Gamma + \frac{1}{2}\pi\Gamma\Gamma = \frac{\mathrm{d}}{\mathrm{d}t}(\pi\Gamma) - \frac{1}{2}\pi\Gamma\Gamma, \qquad (10.34)$$

and obtain this way an equivalent Lagrangian which we call the Einstein Lagrangian by analogy with General Relativity Theory:

$$L_{Einstein} = -\frac{1}{2}\pi\Gamma\Gamma = -\frac{\dot{\pi}^2}{2\pi}. \qquad (10.35)$$

The reader can easily check that variation of $L_{Einstein}$ gives us "Einstein equations": $R = 0$.

2. We can use the so called "Palatini method" and treat both metric and connection as two independent variables:

$$L_{Palatini} = \pi R = \pi \left( \dot{\Gamma} + \tfrac{1}{2}\Gamma\Gamma \right) . \tag{10.36}$$

The reader can easily check that variation with respect to $\Gamma$:

$$\frac{d}{dt} \frac{\partial L_{Palatini}}{\partial \dot{\Gamma}} = \dot{\pi} = \pi\Gamma = \frac{\partial L_{Palatini}}{\partial \Gamma} ,$$

restores "metricity condition" (10.33) for $\Gamma$, whereas "Einstein equations" are derived form variation with respect to the metric $\pi$.

Above tricks are based on the "brachistochrone philosophy". Let us, however, use the "on shell method". The following identity, analogous to (10.30) in gravity, follows immediately

$$
\begin{aligned}
\delta L_{Hilbert} &= R\delta\pi + \pi\delta\left(\dot{\Gamma} + \tfrac{1}{2}\Gamma\Gamma\right) = R\delta\pi + \frac{d}{dt}(\pi\delta\Gamma) - \dot{\pi}\delta\Gamma + \pi\Gamma\delta\Gamma \\
&= R\delta\pi + \frac{d}{dt}(\pi\delta\Gamma) .
\end{aligned}
\tag{10.37}
$$

Suppose now, that another quantity, denoted by $\varphi$, which plays role of a "matter field", interacts with our "metric" $\pi$, and its dynamics is governed by the "matter Lagrangian" $L_{matter} = L_{matter}(\varphi, \dot{\varphi}; \pi)$. According to (10.15) we have:

$$\delta L_{matter} = \left(\frac{\partial L_{matter}}{\partial \varphi} - \dot{p}\right)\delta\varphi + \frac{\partial L_{matter}}{\partial \pi}\delta\pi + \frac{d}{dt}(p\delta\varphi) , \tag{10.38}$$

where by $p$ we have denoted the momentum canonically conjugate with $\varphi$:

$$p := \frac{\partial L_{matter}}{\partial \dot{\varphi}} .$$

The complete theory: "gravity + matter" is derived from the total Lagrangian, analogous with (10.1):

$$L = L_{Hilbert} + L_{matter} = \pi R + L_{matter}(\varphi, \dot{\varphi}; \pi) . \tag{10.39}$$

According to (10.37) and (10.38) we have

$$\delta L = (R - T)\,\delta\pi + \left(\frac{\partial L_{matter}}{\partial \varphi} - \dot{p}\right)\delta\varphi + \frac{d}{dt}(\pi\delta\Gamma + p\delta\varphi) , \tag{10.40}$$

where we have denoted the "energy-momentum-tensor":

$$T := -\frac{\partial L_{matter}}{\partial \pi} . \tag{10.41}$$

Volume terms give us "field equations": "$R = T$" (Einstein) and "$\dot{p} = \frac{\partial L_{matter}}{\partial \varphi}$" (matter). Hence, volume term vanishes identically on shell and the same dynamics follows from the boundary term:

$$\delta L = \frac{d}{dt}(\pi\delta\Gamma + p\delta\varphi) = \dot{\pi}\delta\Gamma + \pi\delta\dot{\Gamma} + \dot{p}\delta\varphi + p\delta\dot{\varphi} , \tag{10.42}$$

provided we express $L$ as a function of control parameters: $(\Gamma, \dot{\Gamma}, \varphi, \dot{\varphi})$, whereas $(\pi, \dot{\pi}, p, \dot{p})$ are response parameters. Hence, in this control mode the metric $\pi$ belongs to response parameters and has to be calculated "on shell" as a function of the former. For this purpose we must solve algebraically Einstein equations with respect to the metric $\pi$ and plug this solution, i.e. $\pi = \pi(R, \varphi, \dot{\varphi})$, into (10.39), like in case of formula

$$L = p\dot{q} - H(p, q),$$

which gives us correct value of the Lagrangian in classical mechanics, provided we eliminate $p$ and express it in terms of $q$ and $\dot{q}$. Here, the Hilbert Lagrangian "$\pi R$" plays role of "$p\dot{q}$" in mechanics and (10.39) can be viewed as merely the Legendre transformation from the "metric control mode" to the "affine control mode". This way we obtain the so called "affine Lagrangian", depending upon geometry via curvature (connection and its derivative) only: $L_{affine} = L_{affine}(R, \varphi, \dot{\varphi})$ . The complete dynamics is given by equation (10.42). In particular, metric is given by

$$\pi := \frac{\partial L_{affine}}{\partial \dot{\Gamma}} = \frac{\partial L_{affine}}{\partial R}, \tag{10.43}$$

whereas metricity condition (10.33) is derived as one of the Euler-Lagrange equations:

$$\dot{\pi} := \frac{\partial L_{affine}}{\partial \Gamma} = \frac{\partial L_{affine}}{\partial R} \frac{\partial R}{\partial \Gamma} = \pi \Gamma. \tag{10.44}$$

The above Legendre transformation from the metric picture to affine picture:

$$L_{affine} = \pi R + L_{matter}(\varphi, \dot{\varphi}; \pi), \tag{10.45}$$

where $\pi$ must be eliminated "on shell", can be inverted:

$$L_{matter} = -\pi R + L_{affine}(R, \varphi, \dot{\varphi}), \tag{10.46}$$

but now $R$ must be eliminated "on shell". This procedure works universally and there is no restriction for the shape of the function $L_{affine} = L_{affine}(R, \varphi, \dot{\varphi})$.
As an example take:

$$L_{matter} = \frac{1}{2\pi} \dot{\varphi}^2 - \frac{m^2 \pi}{2} \varphi^2, \tag{10.47}$$

which nicely mimics the matter Lagrangian of a massive Klein-Gordon scalar field in relativity. We have:

$$T = -\frac{\partial L_{matter}}{\partial \pi} = \frac{1}{2\pi^2} \dot{\varphi}^2 + \frac{m^2}{2} \varphi^2 = \pi^{-1} \left( \frac{1}{2\pi} \dot{\varphi}^2 + \frac{m^2 \pi}{2} \varphi^2 \right). \tag{10.48}$$

On the other hand, we have

$$p = \frac{\partial L_{matter}}{\partial \dot{\varphi}} = \frac{1}{\pi} \dot{\varphi}$$

and, therefore, the energy carried by the matter field equals:

$$E = p\dot{\varphi} - L_{matter} = \frac{1}{2\pi}\dot{\varphi}^2 + \frac{m^2\pi}{2}\varphi^2 \, .$$

Consequently, equation (10.48) is an analog of the Rosenfeld-Belinfante theorem. It says that the "canonical energy" $E$ is equal to the "symmetric energy" $T$ (given as a derivative of the matter Lagrangian with respect to the metric) modulo "lowering an index" with help of the metric tensor. More precisely, we have

$$E = \pi T \, .$$

(We stress that in field theory canonical energy momentum tensor has one covariant and one contravariant index, whereas symmetric tensor – derivative of Lagrangian with respect to the metric – has both contravariant indices. To compare them, metric tensor is necessary.) To calculate metric in terms of remaining quantities, we must solve Einstein equations:

$$R = T = \frac{1}{2\pi^2}\dot{\varphi}^2 + \frac{m^2}{2}\varphi^2 \, ,$$

and, consequently,

$$\pi = \frac{\dot{\varphi}}{\sqrt{2R - m^2\varphi^2}} \, . \tag{10.49}$$

Inserting this value into (10.45), we finally obtain

$$L_{affine} = \dot{\varphi}\sqrt{2R - m^2\varphi^2} \, . \tag{10.50}$$

The reader may easily check that the inverse Legendre transformation (i.e. formulae (10.43) and (10.46)) reproduce (10.49) and the original matter Lagrangian (10.47).

The procedure described above is universal: it works for an arbitrary affine Lagrangian:

$$L_{affine} = L_{affine}(R, \varphi, \dot{\varphi}) \, , \tag{10.51}$$

which proves that an arbitrary "affine" theory is equivalent to the conventional "metric" theory with an appropriate matter Lagrangian, uniquely defined by (10.43) and (10.46).

Finally, in analogy with (10.6), consider a generalized "metric" theory, where Lagrangian not linear with respect to the $R$, i.e.:

$$L_{nonlinear} = L_{nonlinear}(\pi, \Gamma, R, \varphi, \dot{\varphi}) \, . \tag{10.52}$$

To prove its equivalence with the conventional theory (10.39), we: 1) "downgrade" the metric $\pi$, (together with its derivatives contained in $\Gamma$) to the level of matter fields and: 2) treat (10.52) as an affine version of the new theory:

$$L_{nonlinear} = L_{nonlinear}(R, \varphi, \dot{\varphi}, \pi, \dot{\pi}) \, . \tag{10.53}$$

Using Legendre transformation (10.43) and (10.46) we obtain the conventional theory for a new "metric"

$$\widetilde{\pi} := \frac{\partial L_{nonlinear}}{\partial R}, \qquad (10.54)$$

and the two "matter fields": the original one $\varphi$ and the new one $\pi$. Adding the "Hilbert Lagrangian" $\widetilde{\pi}\widetilde{R}$ of the new metric to the new matter Lagrangian we obtain the conventional, "metric" formulation (10.39) of this theory.

We see that Hilbert Lagrangian plays merely role of the Legendre term, like $p\dot{q}$ in mechanics, and does not carry any specific physical information on the dynamics of the theory.

## 10.5 Back to General Relativity Theory: Lagrangians depending upon Ricci

Also in General Relativity the Hilbert Lagrangian $\pi^{\mu\nu}R_{\mu\nu}$ (see (10.2)) plays role of the Legendre term, like in our toy model. Formula (10.30) suggests to treat the connection $\Gamma$ as the gravitational field configuration, whereas the metric $\pi^{\mu\nu}$ arises as the momentum canonically conjugate. Hence, formula

$$L = \pi^{\mu\nu}R_{\mu\nu} + L_{matter}(\pi, \varphi, \partial\varphi) \qquad (10.55)$$

for the total "gravity + matter" Lagrangian, can be interpreted as a Legendre transformation, like $L = p\dot{q} - H(p, q)$ in mechanics. To complete this transformation, we have to express (on shell) momenta $\pi^{\mu\nu}$ by "positions" $\Gamma^{\lambda}_{\mu\nu}$ and "velocities", i.e. derivatives of $\Gamma$ contained in the Ricci tensor $R_{\mu\nu}$. As a result, we obtain the alternative, affine formulation of general relativity:

$$L_{affine} = L_{affine}(R_{\mu\nu}, \varphi, \partial\varphi). \qquad (10.56)$$

The inverse Legendre transformation, back to the metric formulation, is obtained if we exchange $\pi^{\mu\nu}$ and $R_{\mu\nu}$ in the role of control and response parameters. Consequently, the metric is restored as:

$$\pi^{\mu\nu} := \frac{\partial L_{affine}}{\partial R_{\mu\nu}}. \qquad (10.57)$$

Affine formulation of General Relativity was discovered in [13] (see also [8] and [14]). It implies that a theory based on a generic, non linear Lagrangian (10.6) depending upon the Ricci tensor is equivalent to the conventional relativity theory with additional matter fields. The proof is fully analogous with what we have done in case of our toy model: 1) downgrade the metric $\pi^{\mu\nu}$ to the level of matter fields, 2) define new metric as momentum canonically conjugate to the Ricci tensor:

$$\widetilde{\pi}^{\mu\nu} := \frac{\partial L_{nonlinear}}{\partial R_{\mu\nu}}, \qquad (10.58)$$

3) treat (10.6) as an affine Lagrangian:

$$L_{affine} = L_{affine}(R_{\mu\nu}, \varphi, \partial\varphi, \pi, \partial\pi),$$

and perform an inverse Legendre transformation

$$L_{matter} = -\tilde{\pi}^{\mu\nu} R + L_{affine} \qquad (10.59)$$

where $R_{\mu\nu}$ must be expressed in terms of metric variables $\tilde{\pi}^{\mu\nu}$; 4) finally:

$$L = \tilde{\pi}^{\mu\nu} \tilde{R}_{\mu\nu} + L_{matter} \qquad (10.60)$$

provides the conventional, matter formulation of (always the same) field theory. I stress that all these manipulations are performed "on shell": field equations remain always the same but we express them in terms of different variables. At the beginning, we have fourth order equation for the original metric and for matter fields, at the end we have second order Einstein equations for the new metric plus second order equations (matter equations) for both the matter fields, namely: original matter fields and the old metric downgraded now to the level of matter fields. We see again that Hilbert Lagrangian is merely a mathematical device (Legendre term in (10.60)), which carries no specific physical information about dynamics of gravitational field. In fact, the dynamics can be chosen at will by the choice of the original Lagrangian (10.6).

Different versions of the above theorem were proved in 1988-89 (see [4]). However, they do not cover the case of a generic Lagrangian density (10.6), which can depend upon the entire curvature tensor $R^\lambda_{\mu\nu\kappa}$.

## 10.6 Weyl tensor included

Proof of the complete version of our theorem, covering also the case of Lagrangians depending upon the Weyl tensor, was given recently in [2]. It is based on the following observations.

Riemann tensor $R^\lambda_{\mu\nu\kappa}$ of a generic, symmetric connection fulfills algebraic identities

$$R^\lambda_{\mu\nu\kappa} = -R^\lambda_{\mu\kappa\nu} \quad ; \quad R^\lambda_{[\mu\nu\kappa]} = 0, \qquad (10.61)$$

(the last one is called Bianchi 1-st type) and decomposes into the sum of three irreducible parts:

$$R^\lambda_{\mu\nu\kappa} = \frac{1}{3}\left(\delta^\lambda_\nu K_{\mu\kappa} - \delta^\lambda_\kappa K_{\mu\nu}\right) + \frac{1}{5}\left(2\delta^\lambda_\mu F_{\nu\kappa} + \delta^\lambda_\nu F_{\mu\kappa} - \delta^\lambda_\kappa F_{\mu\nu}\right) + W^\lambda_{\mu\nu\kappa}, \qquad (10.62)$$

where the Weyl tensor $W$ fulfills the same identities (10.61) and is traceless, whereas $K$ and $F$ describe the symmetric and the antisymmetric parts of the Ricci tensor, respectively:

$$R_{\mu\nu} := R^\lambda_{\mu\lambda\nu} = K_{\mu\nu} + F_{\mu\nu}. \qquad (10.63)$$

We have also:

$$R^\lambda_{\lambda\mu\nu} = 2F_{\mu\nu}. \qquad (10.64)$$

If $\Gamma$ is a metric connection, the second part of (10.62) vanishes because Ricci is symmetric: $F_{\mu\nu} = 0$.

In case of a generic affine theory, the boundary term (10.30) of the variational formula reduces to:

$$
\begin{aligned}
\delta L_{affine}(R^\lambda_{\mu\nu\kappa}, \varphi, \partial\varphi) &= \partial_\kappa \left\{ P^{\mu\nu\kappa}_\lambda \delta\Gamma^\lambda_{\mu\nu} + p^\kappa \delta\varphi \right\} \\
&= (\partial_\kappa P^{\mu\nu\kappa}_\lambda) \delta\Gamma^\lambda_{\mu\nu} + P^{\mu\nu\kappa}_\lambda \delta\Gamma^\lambda_{\mu\nu\kappa} \quad (10.65) \\
&+ (\partial_\kappa p^\kappa) \delta\varphi + p^\kappa \delta\varphi_\kappa,
\end{aligned}
$$

where we denote: $\Gamma^\lambda_{\mu\nu\kappa} := \partial_\kappa \Gamma^\lambda_{\mu\nu}$ and $\varphi_\kappa = \partial_\kappa \varphi$. Hence, momentum $P^{\mu\nu\kappa}_\lambda$ is defined as

$$
P^{\mu\nu\kappa}_\lambda = \frac{\partial L_{affine}}{\partial \Gamma^\lambda_{\mu\nu\kappa}}. \tag{10.66}
$$

Derivatives of the connection are contained in the Riemann tensor only. Unfortunately, the quantity

$$
Q^{\mu\nu\kappa}_\lambda := \frac{\partial L_{affine}}{\partial R^\lambda_{\mu\nu\kappa}},
$$

although carrying exactly the same information as $P^{\mu\nu\kappa}_\lambda$ does, has different symmetries. This obscures heavily the structure of the proof. A simple remedy for that is to use a different representation of the curvature. It is based on a novel theory of spacetime connection, well adapted to Canonical Relativity, which is presented in the Appendix. Below, I omit all these computational difficulties and briefly sketch main ideas of the proof.

Using decomposition (10.62) we obtain:

$$
P^{\mu\nu\kappa}_\lambda = \left( \delta^\kappa_\lambda \widetilde{\pi}^{\mu\nu} - \delta^{(\mu}_\lambda \widetilde{\pi}^{\nu)\kappa} \right) - \frac{1}{2} (\delta^\mu_\lambda \mathcal{F}^{\nu\kappa} + \delta^\nu_\lambda \mathcal{F}^{\mu\kappa}) + p^{\mu\nu\kappa}_\lambda, \tag{10.67}
$$

where $\widetilde{\pi}$, $\mathcal{F}$ and $p^{\mu\nu\kappa}_\lambda$ are defined by derivatives of the affine Lagrangian with respect to the irreducible parts of Riemann: the symmetric part of Ricci, the antisymmetric part of Ricci and Weyl, respectively. In particular:

$$
\widetilde{\pi}^{\mu\nu} = \frac{\partial L_{affine}}{\partial R_{\mu\nu}} \quad ; \quad p^{\mu\nu\kappa}_\lambda = \frac{\partial L_{affine}}{\partial W^\lambda_{\mu\nu\kappa}} \tag{10.68}
$$

In case of a generic Lagrangian (10.6), the connection is metric *a priori* and, therefore, $F$ and $\mathcal{F}$ vanish identically. Hence, formula (10.65) reduces to

$$
\delta L_{affine}(R^\lambda_{\mu\nu\kappa}, \varphi, \partial\varphi) = \partial_\kappa \left\{ (\pi^{\mu\nu\kappa}_\lambda + p^{\mu\nu\kappa}_\lambda) \delta\Gamma^\lambda_{\mu\nu} + p^\kappa \delta\varphi \right\}. \tag{10.69}
$$

At this point we first use Legendre transformation in the Weyl part:

$$
\partial_\kappa \left\{ p^{\mu\nu\kappa}_\lambda \delta\Gamma^\lambda_{\mu\nu} \right\} = \delta \left\{ \partial_\kappa \left( p^{\mu\nu\kappa}_\lambda \Gamma^\lambda_{\mu\nu} \right) \right\} - \partial_\kappa \left\{ \Gamma^\lambda_{\mu\nu} \delta p^{\mu\nu\kappa}_\lambda \right\}.
$$

Putting now:

$$
\Lambda_{affine} := L_{affine} - \left\{ \partial_\kappa \left( p^{\mu\nu\kappa}_\lambda \Gamma^\lambda_{\mu\nu} \right) \right\} \tag{10.70}
$$

and replacing *on shell* the Weyl part $W^\lambda_{\mu\nu\kappa}$ by the corresponding momentum $p^{\mu\nu\kappa}_\lambda$ we enlarge the set of "control parameters" by the latter but reduce the dependence of the Lagrangian upon the curvature to the Ricci tensor only. This way we reduce our problem to the one discussed in the previous Section.

# 10.7 Conclusions

The main advantage of the reformulation presented above is the applicability of the standard Hamiltonian formalism developed for purposes of general relativity theory. In particular, the "positive energy" theorem applies here if and only if the matter energy is positive. This is probably the simplest way to analyze stability of different models of this type (cf. also [15]).

There might be doubts about which one of the two metric tensors is "the true one". This question was already considered by Higgs, who had the following remark: *it seems likely that more direct physical significance may be attached to the new metric than to the original dynamic variables g, which enter into the action principle* (see [7]).

In this context I want to stress that already in the absolutely standard, Einsteinian formulation, gravitational waves can propagate along different "light cones" than the ones defined by the metric tensor. Indeed, if the matter Lagrangian depends also upon connection coefficients (which are necessary, e.g., for the covariant derivatives of matter fields) then the energy momentum tensor contains also second derivatives of the metric, contained in the last term of equation (10.5). Hence, expression $G_{\mu\nu} - 8\pi T_{\mu\nu}$ contains second derivatives of the metric, multiplied not only by the metric tensor (coming from $G_{\mu\nu}$) but also by functions depending upon matter fields (coming from $T_{\mu\nu}$). The effective light cone is, therefore, different from the one defined by the metric. The existence of two metric tensors is, therefore, not so controversial as one could feel at the beginning.

Having already accepted the existence of different metric tensors in the theory, the question: "which one among them is more physical than the remaining ones" is irrelevant as far as the dynamical properties of the field evolution are considered. Indeed, these properties do not depend upon the set of equivalent variables in (10.7), which we use to parameterize field configurations. But the very mathematical structure of the theory distinguishes our metric $\widetilde{\pi}^{\mu\nu}$, arising as a momentum canonically conjugate to the symmetric part $K_{\mu\nu}$ of the Ricci tensor $R_{\mu\nu}$. Using it, the gravitational part of field equations will always be written in the universal form of Einstein equations $G^{\mu\nu}(\tilde{g}) = 8\pi \tilde{T}^{\mu\nu}$, no matter how exotic and complicated is the Lagrangian (10.6) of the theory.

Our theory shows that, instead of "generalizing" general relativity theory, one can concentrate on inventing new matter fields describing phenomena which we want to model (e.g. black energy). In this context our theorem can be a good starting point. In particular, "purely affine" theories, which do not contain any "primary metric" $\pi$, are especially interesting. Here, there is a unique metric tensor $\widetilde{\pi}$, arising dynamically as a momentum canonically conjugate to the Ricci part of the curvature.

Using ideas presented in [9], one can easily generalize our Theorem to the case of Lagrangians depending not only upon curvature tensor, but also upon its (covariant) derivatives.

152

# Acknowledgments

This research was supported by Narodowe Centrum Nauki, Poland (grant DEC-2016/21/B/ST1/00940).

# Appendix I: Constraints *versus* gauge

Once a control mode has been fixed, there is one-to-one correspondence between the Lagrangian submanifold $\mathcal{D}$ describing dynamics of the system and its generating function (e.g.: $L$ in case of the Lagrangian mode (10.26) or $H$ in case of the Hamiltonian mode (10.28)). The correspondence is valid *up to an additive constant* and may be formalized as follows (see [8] or [9] for more details).

Choice of a control mode is equivalent to the choice of a one-form $\Theta$, primary with respect to the symplectic form $\Omega$: $d\Theta = \Omega$. For example: $\Theta_L = \dot{p}_K dq^K + p_K d\dot{q}^K$ in the Lagrangian mode and $\Theta_H = \dot{p}_K dq^K - \dot{q}^K dp_K$ in the Hamiltonian mode. Such a primitive form is closed when restricted to the dynamics $\mathcal{D}$ because it is a Lagrangian submanifold:

$$d(i_D^* \Theta) = i_D^*(d\Theta) = i_D^* \Omega = 0 \,,$$

where by $i_D^*$ we denote restriction (pull-back) of the differential form to a submanifold $\mathcal{D}$. Assuming trivial topology of $\mathcal{D}$ we conclude that this form is exact, i.e. it is a differential of a function:

$$i_D^* \Theta = d\tilde{F} \,. \tag{10.71}$$

This function (called a generating function of $\mathcal{D}$ in the control mode $\Theta$ is given on $\mathcal{D}$ uniquely up to an additive constant. With a little abuse of notation we can thus write

$$\dot{p}_K dq^K + p_K d\dot{q}^K = dL \tag{10.72}$$

for the control mode and

$$\dot{p}_K dq^K - \dot{q}^K dp_K = d(-H) \tag{10.73}$$

for the Hamiltonian mode. This abuse of notation is very useful in practical calculation. It is based on identification of a function defined on $\mathcal{D}$ and its projection on the space of control parameters. Such an identification is obvious when $D$ is a graph of a mapping, i.e. when: 1) every point of the control space $Q$ is admissible, and when 2) there is a unique response to every control.

But there are important physical theories which do not belong to this class. For example: it may happen that not all points of the control space are admissible but only those which belong to a *constraint manifold* $\mathcal{C} \subset Q$. This means that $\mathcal{D}$ covers only a subset $\mathcal{C}$ of $Q$. Denote by $k$ the co-dimension of $\mathcal{C}$. Because dimension of $\mathcal{D}$ is equal to the number of control (or response) parameters, we conclude that to every control there must be a $k$-dimensional collection of admissible responses of the system. But, due to the fact that $\mathcal{D}$

is a Lagrangian submanifold (i.e. $\Omega$ vanishes when restricted to $\mathcal{D}$) a miracle occurs: the function $\tilde{F}$ defined by formula (10.71) is constant on every such a $k$-dimensional, vertical submanifold of $\mathcal{D}$. Hence, $\tilde{F}$ projects uniquely from $\mathcal{D}$ to a uniquely defined function $F$ defined on the space of control parameters.

Conversely: given a function $F$ defined on a constraint submanifold $\mathcal{C} \subset Q$, it defines uniquely a Lagrangian submanifold $\mathcal{D}$ by formula

$$\mathcal{D} := \operatorname{graph}(\mathrm{d}F),$$

if by $\mathrm{d}F$ we understand the collection of all the responses (i.e. covectors on the control space $Q$) which, when restricted to $\mathcal{C}$, agree with the differential of $F$. This way we obtain a $k$-dimensional collection of admissible responses for every control belonging to the constraint submanifold $\mathcal{D}$. One can summarize this construction shortly: a constraint imposed on a control parameter $q$ induces gauge in the conjugate (dual) response variable $p$, i.e. there is no mapping (control) $\rightarrow$ (response). But the Legendre transformation between the two modes *is perfectly invertible*. The generating function in the first mode is defined on the constraint manifold only. Replacing the constrained parameter $q$ by its dual $p$ we obtain a new control mode where the generating function does not depend upon this particular $p$. We see that the transition from the $p$ mode back to the $q$ mode does not fulfill the (very often required in classical textbooks) condition of "invertibility of the Hessian matrix". Indeed, the mapping $p \rightarrow q$ is not invertible because its value is always $q_0$: the value imposed by constraints. Nevertheless, Legendre transformation between these two modes goes smoothly in both directions.

More complicated situations are, of course, possible, when $\mathcal{D}$ does not project on a submanifold but is not a graph of a mapping (control) $\rightarrow$ (response). A bit more sophisticated tools (Morse function, catastrophe theory) are necessary here. Situations of this type occur in control theory, logistics etc., but not in field theory. Here, we restrict ourselves that the case when $\mathcal{D}$ is a bundle of $k$-dimensional "vertical" subspaces over a constraint submanifold $\mathcal{C}$ having codimension $k$.

**Example:** Classical electrodynamics (Maxwellian, but also non-linear, like Born-Infeld) is based on a Lagrangian which depends upon the electromagnetic field $f = \mathrm{d}A$. It generates the field equations according the formula (10.19), which reads:

$$\delta L(A_\nu, A_{\nu\mu}) = \partial_\mu(\mathcal{F}^{\nu\mu}\delta A_\nu) = (\partial_\mu \mathcal{F}^{\nu\mu})\delta A_\nu + \mathcal{F}^{\nu\mu}\delta A_{\nu\mu}. \tag{10.74}$$

For a free field, the Lagrangian does not depend upon the potentials $A_\nu$. Moreover, it uses only $f_{\mu\nu} = \partial_\mu A_\nu - \partial_\nu A_\mu = A_{\nu\mu} - A_{\mu\nu}$, i.e. the antisymmetric part of its derivatives. Hence, formula (10.74) implies following field equations:

$$\partial_\mu \mathcal{F}^{\nu\mu} = \frac{\partial L}{\partial A_\nu} = 0 \quad , \qquad \mathcal{F}^{\nu\mu} = \frac{\partial L}{\partial A_{\nu\mu}} = 2\frac{\partial L}{\partial f_{\mu\nu}}. \tag{10.75}$$

In this formulation we insert "manually" the first pair of Maxwell equations $\mathrm{d}f = 0$ (which is implied by our assumption $f = \mathrm{d}A$), whereas the second pair (first equation above) and the constitutive relation between electromagnetic tensor $f$ and the induction tensor $\mathcal{F}$ is derived from variational principle. The complete Legendre transformation is also possible:

$$\delta\left(L - \mathcal{F}^{\nu\mu}A_{\nu\mu}\right) = -A_\nu\delta(\partial_\mu\mathcal{F}^{\nu\mu}) - A_{\nu\mu}\delta\mathcal{F}^{\nu\mu}. \qquad (10.76)$$

The new Lagrangian

$$\tilde{L} := L - \mathcal{F}^{\nu\mu}A_{\nu\mu} = L + 2\mathcal{F}^{\nu\mu}f_{\nu\mu}$$

is defined on the constraints submanifold: $\partial_\mu\mathcal{F}^{\nu\mu} = 0$ and $\mathcal{F}^{\nu\mu}$-antisymmetric, and must be expressed with respect to the new control parameters $\mathcal{F}^{\nu\mu}$. This way, imposing the second pair of Maxwell equations in the form of Lagrangian constraints, we derive the remaining equations of the form the (constrained) variational principle

$$\delta\tilde{L} = -A_\nu\delta(\partial_\mu\mathcal{F}^{\nu\mu}) + 2f_{\mu\nu}\delta\mathcal{F}^{\nu\mu}.$$

In particular, parameters dual to constraints (the potential $A_\nu$ and the symmetric part of its derivatives: $A_{(\nu\mu)}$) may assume an arbitrary value.

# Appendix II: An alternative description of the curvature tensor: gravity is a field of local reference frames

Variation with respect to a connection field $\Gamma$ leads to an ugly algebra, which obscures considerably description of the corresponding canonical (Hamiltonian) structure of the theory. This is due to the fact that the momentum canonically conjugate to $\Gamma$:

$$P_\lambda^{\mu\nu\kappa} := \frac{\partial L}{\partial\Gamma_{\mu\nu\kappa}^\lambda} \qquad (10.77)$$

(where $\Gamma_{\mu\nu\kappa}^\lambda = \partial_\kappa\Gamma_{\mu\nu}^\lambda$), and the derivative of the Lagrangian with respect to the Riemann tensor:

$$Q_\lambda^{\mu\nu\kappa} := \frac{\partial L}{\partial R_{\mu\nu\kappa}^\lambda}, \qquad (10.78)$$

although related by a one-to-one correspondence, have different symmetries (symmetry *versus* antisymmetry; cf. also [10], formula (21.20) on p. 500). Also Bianchi I-st type identities are implemented in a completely different way on $P$ and $Q$. Below, we propose an alternative description of fundamental geometric structures which we need to formulate gravity theory: the connection and the curvature. This approach, although perfectly equivalent to the standard one, simplifies enormously all the canonical and Hamiltonian considerations in general relativity theory.

Newton' first law: "there is a global inertial system" says that there are spacetime coordinates $(y^\alpha)$ such that, in absence of any force, a test body moves along straight lines with constant velocity, i.e. its trajectory fulfills equation

$$\ddot{y}^\alpha = 0 \,. \tag{10.79}$$

Once we have one such coordinate system, any coordinate system $(x^\mu)$ is equally good if and only if it is "equivalent" with the previous one with respect to the following *global* equivalence relation:

$$\{(y^\alpha) \sim (x^\lambda)\} \iff \left\{\frac{\partial^2 y^\alpha}{\partial x^\mu x^\nu} = 0\right\} \,. \tag{10.80}$$

We conclude that an "inertial frame" is not a coordinate system but rather an equivalence class of such systems.

If $(x^\mu)$ is not inertial, we can recalculate equations of motion (10.79) in the following way:

$$\dot{y}^\alpha = \frac{\partial y^\alpha}{\partial x^\nu}\dot{x}^\nu$$

$$\ddot{y}^\alpha = \frac{\partial y^\alpha}{\partial x^\nu}\ddot{x}^\nu + \frac{\partial^2 y^\alpha}{\partial x^\mu \partial x^\nu}\dot{x}^\mu\dot{x}^\nu = 0$$

$$\frac{\partial x^\lambda}{\partial y^\alpha}\ddot{y}^\alpha = \delta^\lambda_\nu\ddot{x}^\nu + \frac{\partial x^\lambda}{\partial y^\alpha}\frac{\partial^2 y^\alpha}{\partial x^\mu \partial x^\nu}\dot{x}^\mu\dot{x}^\nu = \ddot{x}^\lambda + \Gamma^\lambda_{\mu\nu}\dot{x}^\mu\dot{x}^\nu = 0\,, \tag{10.81}$$

where the coefficients:

$$\Gamma^\lambda_{\mu\nu}(\mathbf{x}) := \frac{\partial x^\lambda}{\partial y^\alpha}\frac{\partial^2 y^\alpha}{\partial x^\mu \partial x^\nu}(\mathbf{x})\,, \tag{10.82}$$

measure at each spacetime point $\mathbf{x} \in M$ how much non-inertial is the system $(x^\lambda)$.

Equation of motion (10.81) can be written in a "Newtonian" form:

$$m\ddot{x}^\lambda = -m\Gamma^\lambda_{\mu\nu}\dot{x}^\mu\dot{x}^\nu =: F^\lambda \tag{10.83}$$

where $m$ is the mass and $F^\lambda$ are fictitious forces due to non-inertiality of the system. They can be *globally* eliminated if we use an inertial frame for which $\Gamma^\lambda_{\mu\nu} = 0$.

Einstein says: gravity is nothing but the above fictitious force. It can *locally* be eliminated (freely falling elevator!). Maybe, there is no *global* inertial frame: it is not necessary! But, Einstein says, at each spacetime point $\mathbf{x} \in M$ separately there is a *local* inertial frame, i.e. coordinates $(y^\alpha)$ such that a freely falling test bodies fulfill (10.79) at this particular point.

Define a "reference frame at a point $\mathbf{x} \in M$" of a manifold $M$ as an equivalence class of coordinate charts with respect to the following *local* relation: "$\sim_{\mathbf{x}}$": given two charts in a neighbourhood of $\mathbf{x}$, we declare them to be equivalent if the second derivatives of any coordinate from one chart with respect to coordinates of the other chart vanish at $\mathbf{x}$:

$$((x^\mu) \sim_{\mathbf{x}} (y^\alpha)) \iff \left(\frac{\partial^2 y^\alpha}{\partial x^\mu x^\nu}(\mathbf{x}) = 0\right) \,. \tag{10.84}$$

156

It is easy to check that, indeed, it is an equivalence relation.

Given a reference frame $\Upsilon_0$ at $\mathbf{x}$, we may parameterize any other reference frame $\Upsilon$ at $\mathbf{x}$ by the following table of numbers:

$$\Gamma^\lambda_{\mu\nu}(\mathbf{x}) := \frac{\partial x^\lambda}{\partial y^\alpha} \frac{\partial^2 y^\alpha}{\partial x^\mu x^\nu}(\mathbf{x}) \, , \tag{10.85}$$

where $(y^\alpha)$ is a representative of $\Upsilon_0$ and $(x^\mu)$ a representative of $\Upsilon$. It is easy to check that $\Gamma^\lambda_{\mu\nu}(\mathbf{x})$ does not depend upon the choice of these representatives. This way the set of all reference frames acquires a structure of an affine fiber bundle over $M$.

Symmetric connection on a manifold $M$ is a "field of local reference frames" $M \ni \mathbf{x} \to \Upsilon(\mathbf{x})$, i.e. a section of this bundle. The "privileged" reference frame $\Upsilon(\mathbf{x})$ at $\mathbf{x} \in M$ can be called a "*local inertial frame* at $\mathbf{x}$". Its coordinate description with respect to any coordinate chart $(x^\mu)$ is provided by the set of functions $\Gamma^\lambda_{\mu\nu} = \Gamma^\lambda_{\mu\nu}(\mathbf{x})$. If $(x^\mu)$ belongs to this privileged class: $(x^\mu) \in \Upsilon(\mathbf{x})$, i.e. if $\Gamma^\lambda_{\mu\nu}(\mathbf{x}) = 0$, then $x^\mu$ will be called "inertial coordinates at $\mathbf{x}$". Gravity is nothing but a symmetric spacetime connection i.e. a field of inertial frames.

Gravity vanishes if this connection is *flat*, i.e. if there exists a *global* inertial frame, represented by a *global* coordinate chart which is inertial not just at a single point, but everywhere. Given a connection $\Upsilon$, how to check whether or not it is flat? First, we can choose coordinates $(x^\mu)$ which are inertial at $\mathbf{x}$, i.e. such that $\Gamma^\lambda_{\mu\nu}$ vanish at $\mathbf{x}$. Without any loss of generality we can assume that $\mathbf{x} = (0, 0, \ldots, 0)$. Is it possible to "improve" these coordinates in such a way that $\Gamma^\lambda_{\mu\nu}$ vanish also outside of $\mathbf{x}$? As a first step to answer this question let us try to kill derivatives $\Gamma^\lambda_{\mu\nu\kappa}(\mathbf{x}) = \partial_\kappa \Gamma^\lambda_{\mu\nu}(\mathbf{x})$. Is it possible?

Consider such an improved system of coordinates:

$$y^\lambda := x^\lambda + \frac{1}{6} Q^\lambda_{\mu\nu\kappa} x^\mu x^\nu x^\kappa + \text{term of order higher than 3} \, , \tag{10.86}$$

where coefficients $Q$ are symmetric: $Q^\lambda_{\mu\nu\kappa} = Q^\lambda_{(\mu\nu\kappa)}$. Only such coordinate transformations are interesting because:

1. terms of order 0 vanish under differentiation (10.85), i.e. do not influence the connection coefficients $\Gamma^\lambda_{\mu\nu}$;

2. terms of order 1 produce only a linear (with constant coefficients) transformation of $\Gamma^\lambda_{\mu\nu}$ and, whence, a linear homogeneous (tensorial type) transformation of the coefficients $\Gamma^\lambda_{\mu\nu\kappa}(\mathbf{x})$: if they do not vanish before, they will not vanish after such a transformation;

3. non-vanishing terms of order 2 would change, due to (10.85), the value of $\Gamma$ at $\mathbf{x}$. We try to avoid such a change because we have already $\Gamma^\lambda_{\mu\nu}(\mathbf{x}) = 0$ and we do not want to spoil this!

4. a possible non-symmetric part of $Q$ vanishes when contracted with the totally symmetric expression $x^\mu x^\nu x^\kappa$;

5. 4th and higher order terms produce 2nd and higher order term in $\Gamma^\lambda_{\mu\nu}$ and, whence, do not change the value of derivatives $\Gamma^\lambda_{\mu\nu\kappa}(\mathbf{x})$.

Using (10.85) we calculate the new connection coefficients $\tilde{\Gamma}$. They contain an extra linear term proportional to $Q$. Finally, after differentiation, we obtain:

$$\tilde{\Gamma}^\lambda_{\mu\nu\kappa}(\mathbf{x}) = \Gamma^\lambda_{\mu\nu\kappa}(\mathbf{x}) + Q^\lambda_{\mu\nu\kappa} \ . \tag{10.87}$$

Using an arbitrary (but symmetric!) tensor $Q^\lambda_{\mu\nu\kappa}$ we are able to kill the totally symmetric part $\Gamma^\lambda_{(\mu\nu\kappa)}$ of $\Gamma^\lambda_{\mu\nu\kappa}$. The remaining part, if any:

$$K^\lambda_{\mu\nu\kappa} := \Gamma^\lambda_{\mu\nu\kappa} - \Gamma^\lambda_{(\mu\nu\kappa)} \ , \tag{10.88}$$

constitutes an obstruction against a possibility to kill derivatives of $\Gamma$, i.e. against its flatness. It measures, therefore, how *non-flat*, i.e. how *curved*, is the connection. We call it the *curvature tensor*.

Formula (10.88) is valid only in inertial coordinates. In a generic coordinate system we calculate the value of the curvature tensor (10.88) at a point $\mathbf{x}$ in three steps: 1) recalculate $\Gamma$ to any inertial frame at $\mathbf{x}$, then: 2) calculate curvature tensor $K$ according to (10.88) and, finally: 3) recalculate components of the tensor $K$ back to original coordinate system. It is easy to prove that this way we obtain the following, universal formula, valid in an arbitrary coordinate system:

$$
\begin{aligned}
K^\lambda_{\mu\nu\kappa} &= \Gamma^\lambda_{\mu\nu\kappa} - \Gamma^\lambda_{(\mu\nu\kappa)} + \left( \Gamma^\lambda_{\sigma\kappa}\Gamma^\sigma_{\mu\nu} - \Gamma^\lambda_{\sigma(\kappa}\Gamma^\sigma_{\mu\nu)} \right) \\
&= \Gamma^\lambda_{\mu\nu\kappa} + \Gamma^\lambda_{\sigma\kappa}\Gamma^\sigma_{\mu\nu} - \left( \Gamma^\lambda_{(\mu\nu\kappa)} + \Gamma^\lambda_{\sigma(\kappa}\Gamma^\sigma_{\mu\nu)} \right) \ .
\end{aligned}
\tag{10.89}
$$

Due to the definition, curvature tensor $K$ is symmetric in first two indices and its totally symmetric part vanishes:

$$K^\lambda_{\mu\nu\kappa} = K^\lambda_{\nu\mu\kappa} \quad ; \quad K^\lambda_{(\mu\nu\kappa)} = 0 \ . \tag{10.90}$$

The last identity can be called Bianchi I-st type identity.

The above curvature tensor carries the same information as the Riemann tensor $R^\lambda_{\mu\nu\kappa}$: antisymmetrization of $K$ in last two indices produces $R$ and symmetrization of $R$ in first two indices produces $K$. More precisely, the following relations are obvious:

$$R^\lambda_{\mu\nu\kappa} = -2K^\lambda_{\mu[\nu\kappa]} \quad ; \quad K^\lambda_{\mu\nu\kappa} = -\frac{2}{3}R^\lambda_{(\mu\nu)\kappa} \ , \tag{10.91}$$

and identities (10.90) for $K$ are equivalent to the analogous identities for $R$:

$$R^\lambda_{\mu\nu\kappa} = -R^\lambda_{\mu\kappa\nu} \quad ; \quad R^\lambda_{[\mu\nu\kappa]} = 0 \ . \tag{10.92}$$

The curvature tensor can be decomposed into three irreducible parts: the symmetric and antisymmetric part of the Ricci tensor and the traceless (Weyl) tensor. More precisely, we have:

$$K^\lambda_{\mu\nu\kappa} = -\frac{1}{9}\left( \delta^\lambda_\mu K_{\nu\kappa} + \delta^\lambda_\nu K_{\mu\kappa} - 2\delta^\lambda_\kappa K_{\mu\nu} \right) - \frac{1}{5}\left( \delta^\lambda_\mu F_{\nu\kappa} + \delta^\lambda_\nu F_{\nu\kappa} \right) + U^\lambda_{\mu\nu\kappa} \ , \tag{10.93}$$

where $K_{\mu\nu} = K_{\nu\mu}$ and $F_{\mu\nu} = -F_{\nu\mu}$. All the three terms on the right hand side of (10.93) satisfy the same symmetries (10.90). Moreover, the last term is traceless: $U^\lambda_{\mu\nu\lambda} = U^\lambda_{\lambda\nu\kappa} = 0$. The coefficients have been chosen in such a way that $K_{\mu\nu}$ and $F_{\mu\nu}$ are respectively the symmetric and the antisymmetric part of the Ricci tensor:

$$R_{\mu\nu} := R^\lambda_{\mu\lambda\nu} = K_{\mu\nu} + F_{\mu\nu} \ . \tag{10.94}$$

The traces of the curvature $K^\lambda_{\mu\nu\kappa}$ can be obtained from (10.93):

$$K^\lambda_{\lambda\nu\kappa} = -\frac{1}{3}K_{\nu\kappa} - F_{\nu\kappa} \ ; \quad K^\lambda_{\mu\nu\lambda} = \frac{2}{3}K_{\mu\nu} \ . \tag{10.95}$$

For the sake of completeness let us mention that the corresponding decomposition of the Riemann tensor can be obtained directly from (10.93) and (10.91):

$$R^\lambda_{\mu\nu\kappa} = \frac{1}{3}\left(\delta^\lambda_\nu K_{\mu\kappa} - \delta^\lambda_\kappa K_{\mu\nu}\right) + \frac{1}{5}\left(2\delta^\lambda_\mu F_{\nu\kappa} + \delta^\lambda_\nu F_{\mu\kappa} - \delta^\lambda_\kappa F_{\mu\nu}\right) + W^\lambda_{\mu\nu\kappa} \ , \tag{10.96}$$

where the Weyl tensor $W$ fulfills identities (10.92) and is traceless. We have also:

$$R^\lambda_{\lambda\mu\nu} = 2F_{\mu\nu} \ . \tag{10.97}$$

If $\Gamma$ is a metric connection, the second part of both (10.93) and (10.96) vanishes because we have $F_{\mu\nu} = 0$ in this case. Finally, observe that $U$ contains the complete information about the Weyl tensor $W$ because we have:

$$W^\lambda_{\mu\nu\kappa} = -2U^\lambda_{\mu[\nu\kappa]} \ ; \quad U^\lambda_{\mu\nu\kappa} = -\frac{2}{3}W^\lambda_{(\mu\nu)\kappa} \ . \tag{10.98}$$

The $K^\lambda_{\mu\nu\kappa}$ representation (instead of $R^\lambda_{\mu\nu\kappa}$ representation) of the curvature simplifies considerably variational considerations because of the following identity:

$$P^{\mu\nu\kappa}_\lambda := \frac{\partial L}{\partial \Gamma^\lambda_{\mu\nu\kappa}} = \frac{\partial L}{\partial K^\lambda_{\mu\nu\kappa}} \tag{10.99}$$

Riemann representation $R^\lambda_{\mu\nu\kappa}$ gives us:

$$Q^{\mu\nu\kappa}_\lambda := \frac{\partial L}{\partial R^\lambda_{\mu\nu\kappa}} = -\frac{2}{3}P^{(\mu\nu)\kappa}_\lambda \ , \tag{10.100}$$

which highly obscures all the formulae of canonical and Hamiltonian relativity.

# References

[1] T. P. Sotiriou, S. Liberati, J. Phys. Conf. Ser. 68 (2007) 012022,
S. Capozziello, M. De Laurentis, M. Francaviglia, S. Mercadante, Found. Phys. 39 (2009) 1161.
S. Capozziello, M. De Laurentis, *Extended Theories of Gravity*, Physics Reports (2011) Elsevier

[2] J. Kijowski, *Universality of the Einstein theory of gravitation*, Int. J. of Geometric Methods in Modern Physics, **13** (2016), p. 1640008

[3] P. Havas; General Relativity and Gravitation, 8 (1977) 631;
G.T.Horowitz and R.M.Wald; Phys. Rev. D 17 (1978) 414;
K.S.Stelle; Gen. Relativ. Gravit. 9 (1978) 353;
K.I.Macrae and R.J.Rieger; Phys. Rev. D24 (1981) 2555;
A.Frenkel and K.Brecher,ibid. 26 (1982) 368;
V.Müller and H.-J.Schmidt, Gen. Relativ. Gravit. 17 (1985) 769 and 971.

[4] A. Jakubiec and J. Kijowski, On the universality of Einstein equations, Gen. Relat. Grav. Journal, 19 (1987) 719
A. Jakubiec and J. Kijowski: On theories of gravitation with nonlinear Lagrangians, Phys. Rev. D. 37 (1988) 1406; A. Jakubiec and J. Kijowski, On the Cauchy problem for the theory of gravitation with nonlinear Lagrangian Journ. Math. Phys. 30 (1989) 2923; A. Jakubiec and J. Kijowski, On the universality of linear Lagrangians for gravitational field Journ. Math. Phys. 30 (1989) 1073.

[5] A.D.Sakharov, Dok. Akad. Nauk SSSR 177 (1967) 70;

[6] M.Ferraris, in *Atti del VI Convegno Nazionale di Relativita' Generale e Fisica della Gravitazione*, Firenze, 1984 (Pitagora, Bologna, Italy, 1986), p. 127;
R. Kerner, Gen. Relativ. Gravit. 14 (1982) 453.

[7] G. Stephenson; Il Nuovo Cimento, 9 (1958) 263; P. W. Higgs; Il Nuovo Cimento, 11 (1959) 817.

[8] J. Kijowski and W.M. Tulczyjew, *A Symplectic Framework for Field Theories*, Lecture Notes in Physics No.107 (Springer-Verlag, Berlin, 1979)

[9] J. Kijowski and G. Moreno, *Symplectic structures related with higher order variational problems*, Int. Journ. Geom. Meth. Modern Phys., 12 (2015) 1550084

[10] C.W. Misner, K.S. Thorne, J.A. Wheeler, *Gravitation*, N.H. Freeman and Co, San Francisco, Cal. (1973).

[11] J. Kijowski, *A simple derivation of canonical structure and quasi-local Hamiltonians in general gelativity*, Gen. Relat. Grav. **29** (1997) 307.

[12] V. Fock The theory of space time and gravitation (translated from the Russian), Pergamon Press, London (1959)

[13] J. Kijowski, *On a new variational principle in general relativity and the energy of the graviatational field*, Gen. Relat. Grav. **9** (1978) 857;

[14] J. Kijowski, R. Werpachowski, Universality of affine formulation in General Relativity Rep. Math. Phys. 59 (2007) 1.

[15] A. Jakubiec, J. Kijowski , On the Cauchy problem for the theory of gravitation with nonlinear Lagrangian Journ. Math. Phys. 30 (1989) p. 2923 – 2924.

# Part III

# Relativistic Cosmology and Astrophysics

C. Duston, M. Holman (Eds), *Spacetime Physics 1907 - 2017. Selected peer-reviewed papers presented at the First Hermann Minkowski Meeting on the Foundations of Spacetime Physics, 15-18 May 2017, Albena, Bulgaria* (Minkowski Institute Press, Montreal 2019). ISBN 978-1-927763-48-3 (softcover), ISBN 978-1-927763-49-0 (ebook).

# 11 WARPED 5D COSMIC STRINGS, CONFORMAL INVARIANCE AND THE QUASAR LINK

REINOUD JAN SLAGTER

**Abstract**     Topological defects formed in the early stages of our universe can play a crucial role in understanding anisotropic deviations of the FLRW model we observe today. The most interesting defects are cosmic strings, vortex-like structures in the famous gauged U(1) abelian Higgs model. This local gauge model is the basis of the standard model of particle physics, where the Higgs-mechanism provides elementary particles with mass. It cannot be a coincidence that this model also explains the theory of superconductivity. The decay of the high multiplicity ($n$) super-conducting vortex into a lattice of $n$ vortices of unit magnetic flux is energetically favourable and is experimentally confirmed. This process could play an essential role by the entanglement of cosmic strings just after the symmetry breaking. The questions is how the imprint of the cosmic strings could be observed at present time. Up to now, no evidence is found. The recently found alignment of the spinning axes of quasars in large quasar groups on Mpc scales, could be a first indication of the existence of these cosmic strings. The temporarily broken axial symmetry will leave an imprint of a preferred azimuthal-angle on the lattice. This effect is only viable when a scaling factor is introduced. This can be realized in a warped five dimensional model. The warp factor plays the role of a dilaton field on an equal footing with the Higgs field. The resulting field equations can be obtained from a conformal invariant model. Conformal invariance, the missing symmetry in general relativity, will then spontaneously be broken, just as the Higgs field. It is conjectured that the dilaton field has a dual meaning. At very early times, when the dilaton field approaches zero, it describes the small-distance limit of the model, while at later times it is a warp (or scale) factor that determines the dynamical evolution of the universe.

C. Duston, M. Holman (Eds), *Spacetime Physics 1907 - 2017. Selected peer-reviewed papers presented at the First Hermann Minkowski Meeting on the Foundations of Spacetime Physics, 15-18 May 2017, Albena, Bulgaria* (Minkowski Institute Press, Montreal 2019). ISBN 978-1-927763-48-3 (softcover), ISBN 978-1-927763-49-0 (ebook).

# 11.1 Introduction

The standard model (SM) of the electroweak and strong interactions is a successful framework in which one studies elementary particles and includes the principles of quantum mechanics (QM). On the other hand, general relativity (GR) is also a very impressive theory constructed by theoretical physicists. It describes large scale structures in our universe and one can construct solutions which are related to real physical objects, for example the Kerr solution, the end stage of a collapsing spinning star. A legitimate question is if there are other axially symmetric solutions in GR. It came as a big surprise that there exist vortex-like solutions in Einstein's theory. These vortex solutions occur as topological defects at the symmetry breaking scale in the Einstein-abelian U(1) scalar-gauge model, where the gauge field is coupled to a complex charged scalar field[1, 2, 3, 4]. The solution shows a surprising resemblance with type II superconductivity of the Ginzburg-Landau(GL) theory[5, 6], where the electromagnetic(EM) gauge invariance is broken and the well-known Meissner effect occurs[7, 8]. One says that the phase symmetry is spontaneously broken and the EM field acquires a length scale, which introduces a penetration depth of the gauge field $A_\mu$ in the superconductor and a coherence length of $\Phi$. In the relativistic case one says that the photon acquires mass. Because we have three space dimensions, these solutions of the GL theory behave like magnetic flux vortices (or Nielsen-Olesen (NO) strings[3]) extended to tubes and carry a quantized magnetic flux $2\pi n$, with $n$ an integer, the topological charge or winding number of the field. It was discovered by Abrikosov[8] that these vortices can form a lattice. These localized vortices (or solitons) in the GL-theory are observed in experiments. The phenomenon of magnetic flux quantization in the theory of superconductivity is characteristic for so-called ordered media. The vortex solution possesses mass, so it will couple to gravity. The resulting self-gravitating cosmic strings (CS) still show all the features of superconductivity, but the stability conditions complicate considerably. The stability of the formed lattices depends critically on the parameters of the model, certainly when gravity comes into play. The force between the gauged vortices depends on the the strength of the self-interaction potential of the Higgs field, the gauge-coupling constant, the energy scale at which the phase transition takes place and the spacetime structure. When the mass of the Higgs field is greater than the mass of the gauge field, vortices will repel each other. So gravity could balance the vortices. The energy of the vortex grows by increasing multiplicity $n$, so configurations with $n > 1$ can be seen as multi-soliton states and it is energetically favourable for these to decay into $n$ well separated $n = 1$ solitons. Vortices with high multiplicity can be formed during the symmetry breaking. The total vortex number $n$ is the sum of multiplicities $n_1, n_2, ..$ of isolated points (zero's of $\Phi$)[6].

Our universe, described by a spatially homogeneous and isotropic Friedmann Lemaître Robertson Walker (FLRW) spacetime, shows significant large-scale inhomogeneous structures, for example, the cosmic web of voids with galaxies and clusters in sheets, filaments and knots, the angular distribution in the cosmic microwave background (CMB) radiation and the recently found alignment of polarization axes of quasars in large quasar groups(LQG's) on Mpc-scales[9, 10]. The question is if these complex nonlinear structures of deviation from isotropy and homogeneity have a cosmological origin at a moment in the early stage of the universe. One possibility of this origin could be a CS-network formed by the self-gravitating Einstein-scalar-gauge model. A pleasant fact is that this model has very few parameters and hence more appealing than other models such as inflationary models. It is believed that the mass per unit length of the CS is of the order of the GUT scale, $G\mu \approx 10^{-7}$. Observational bounds, however, predict a negligible contribution of CS's to initial density perturbation from which galaxies and clusters grew. Besides the inconsistencies with the power spectrum of the CMB, radiative effects of the CS embedded in a FLRW spacetime are rapidly damped in any physical regime[11]. Further, the lensing effect of these CS's are not found yet.

There is, however, another possibility to detect the presence of CS's. On a warped spacetime, the fields can become temporarily super-massive by the warp factor in the framework of string theory (or M-theory). Naively one expect that gravity will play a subordinate role compared with the other fields. In 4D counterpart models this is true, but not in warped spacetimes. The super-massive CS's can be formed at a symmetry breaking scale much higher than the GUT scale, i.e., $G\mu >> 1$. So their gravitational impact increases considerably, because the CS builds up a huge mass in the bulk space. Here we consider the warped brane world model of Randall-Sundrum (RS)[12, 13], with one large extra dimension. The result is that effective 4D Kaluza-Klein(KK) modes are obtained from the perturbative 5D graviton. These KK modes will be massive from the brane viewpoint. The modified Einstein equations on the brane and scalar gauge field equations will now contain contributions from the 5D Weyl tensor[14, 15, 16, 17]. In order to explore these effective field equations, we apply an approximation scheme, i.e., a multiple scale method(MSM). In this method one can handle the decay of the $n$-vortex in a perturbative way. The MSM or high-frequency method is an approved tool to handle nonlinearities and secular terms arising in the partial differential equations(PDE) in GR. When there is a high curvature situation, a linear approximation of the Einstein equations is not suitable[18, 19, 20].

Other issue related to our 5D warped spacetime is the behavior at small scales, i.e., when the warp factor or scale factor of the spacetime becomes very small. We conjecture that our warp factor becomes the dilaton field which is needed to make the Lagrangian conformal invariant. Breaking of the conformal symmetry (which will occur when other fields come into play; after all we experience today a huge discrepancy in scales), can be compared with the Brout-Englert-Higgs Mechanism (BEH) in the standard model of particle physics.

In section 2 we will outline the model under consideration. This section is a review of a former study[21, 22, 23, 24]. In section 3 we will explain the connection with conformal invariance.

## 11.2 The superconducting string model in warped spacetime

### 11.2.1 Outline of the model

Our model will be based on a warped five-dimensional FLRW spacetime

$$ds^2 = \mathcal{W}(t,r,y)^2 \left[ e^{2(\gamma(t,r)-\psi(t,r))}(-dt^2 + dr^2) + e^{2\psi(t,r)}dz^2 \right. $$
$$\left. + r^2 e^{-2\psi(t,r)}d\varphi^2 \right] + dy^2, \qquad (11.1)$$

with $\mathcal{W} = W_1(t,r)W_2(y)$ is the warp factor. All standard model fields reside on the brane, while gravity can propagate into the bulk. An extended treatment of the warped five-dimensional spacetime and the effective gravitational equations can be found in Shiromizu et al.[17] and Slagter[21]. The 5D Einstein equations are

$$^{(5)}G_{\mu\nu} = -\Lambda_5 {}^{(5)}g_{\mu\nu} + \kappa_5^2 \delta(y)\left(-\Lambda_4 {}^{(4)}g_{\mu\nu} + {}^{(4)}T_{\mu\nu}\right), \qquad (11.2)$$

with $\kappa_5 = 8\pi^{(5)}G = 8\pi/{}^{(5)}M_{pl}^3$, $\Lambda_4$ the brane tension, $^{(4)}g_{\mu\nu} = {}^{(5)}g_{\mu\nu} - n_\mu n_\nu$, and $n^\mu$ the unit normal to the brane. The effective 4D Einstein-Higgs-gauge field equations are

$$^{(4)}G_{\mu\nu} = -\Lambda_{eff} {}^{(4)}g_{\mu\nu} + \kappa_4^{2(4)}T_{\mu\nu} + \kappa_5^4 \mathcal{S}_{\mu\nu} - \mathcal{E}_{\mu\nu}, \qquad (11.3)$$

$$D^\mu D_\mu \Phi = 2\frac{dV}{d\Phi^*}, \qquad {}^{(4)}\nabla^\mu F_{\nu\mu} = \frac{1}{2}ie\left(\Phi(D_\nu\Phi)^* - \Phi^* D_\nu\Phi\right), \qquad (11.4)$$

with $D_\mu\Phi \equiv {}^{(4)}\nabla_\mu\Phi + ieA_\mu\Phi$, $^{(4)}\nabla_\mu$ the covariant derivative with respect to $^{(4)}g_{\mu\nu}$, $V(\Phi) = \frac{1}{8}\beta(\Phi\Phi^* - \eta^2)^2$ the potential of the Abelian Higgs model and $\eta$ the symmetry breaking scale. $F_{\mu\nu}$ is the Maxwell tensor. The righthand side of the Einstein equations contains a contribution $\mathcal{E}_{\mu\nu}$ from the 5D Weyl tensor and carries information of the gravitational field outside the brane. The quadratic term in the energy-momentum tensor, $\mathcal{S}_{\mu\nu}$, arising from the extrinsic curvature terms in the projected Einstein tensor. $^{(4)}T_{\mu\nu}$ represents the matter content on the brane, in our case the scalar and gauge fields[17, 21]. These terms are responsible for the effective 4D modes, i.e., KK-modes, of the perturbative 5D graviton.

It is clear that the general solution for the vortex will be cylindrical symmetric (in polar coordinates $(r, z, \varphi)$ and in the notation of Nielsen and Olesen), so we parameterize the self-gravitating scalar gauge field as

$$\Phi = \eta X(t,r)e^{in\varphi}, \qquad A_\mu = \frac{n}{e}\left[P(t,r) - 1\right]\nabla_\mu\varphi, \qquad (11.5)$$

with n the topological charge or winding number of the scalar field. The two real functions $P(t,r)$ and $X(t,r)$ are determined by the field equations. The resulting rotationally symmetric configuration was the first example of a string-like solution in classical field theory and is equivalent to the Ginzberg-Landau theory of superconductivity.

For a detailed treatment of the issue, we refer to chapter III.1 of Jaffe and Taubes[5]. The warp factor can be solved from the 5D Einstein equations:

$$\mathcal{W}^2 = W_2(y)^2 W_1(t,r)^2 = \frac{e^{2\sqrt{-\frac{1}{6}\Lambda_5}(y-y_0)}}{\tau r}$$
$$\cdot \left(d_1 e^{(\sqrt{2\tau})t} - d_2 e^{-(\sqrt{2\tau})t}\right)\left(d_3 e^{(\sqrt{2\tau})r} - d_4 e^{-(\sqrt{2\tau})r}\right), \tag{11.6}$$

On the one hand, the constants $(d_i, \tau)$ will be fixed at late time by the features of the FLRW-model. They will appear in the modified second Friedmann equation. On the other hand, on the very small scale, the warp factor equals the dilaton field in conformal invariant models and these constant will be determined by constraint equations. We will return to this issue in section 3 and is currently under study by the author. In figure 1 we plotted several possible solutions. Note that the warp factor depends on $r$ and $t$, so the contribution to the spacetime evolutions will be different for different stages in time. In the

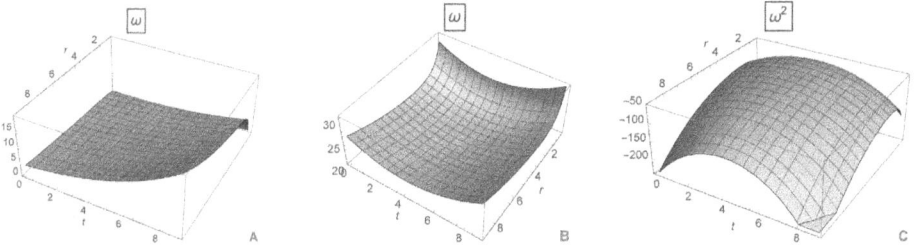

Figure 11.1: **Three different plots of the warp factor for some values of the constants $d_i$ and $\tau$.**

early universe the warp factor represents a dilaton field, conformally coupled to gravity. In section 3 we will return to this issue. The model under consideration is invariant under the group U(1) of local gauge transformations (of the second kind)

$$\Phi(\mathbf{x}) \rightarrow e^{i\chi(x)}\Phi(x), \qquad A_a(\mathbf{x}) \rightarrow A_a(x) + \frac{1}{e}\partial_a \chi(\mathbf{x}), \tag{11.7}$$

The conserved electromagnetic current becomes now $J^a = \frac{ie}{2}(\Phi D^a \Phi^* - \Phi^* D^a \Phi)$. Since the minima of $V(\Phi)$ are at $|\Phi| = \eta$, this symmetry is spontaneously broken and the field acquires a non-zero vacuum expectation value. Let us now take a closer look at the potential (figure 2). Suppose we take a closed loop L in physical space in polar coordinates$(r, z, \varphi)$ around the string, where z is a kind of "dummy" coordinate: $\delta\varphi = 2\pi$. The far field will then take the form $\Phi \approx \eta e^{i\varphi}$. The position of the string is located by taking the closed loop $\Gamma$. If we shrink $\Gamma$ to a point, then the phase jump of $2\pi$ is no longer defined. This phase jump can only be resolved continuously if the field rises to the top of the potential $\Phi = 0$. We say that the energy of the "false" vacuum is trapped. This is a *topological defect*. The vacuum manifold $\mathcal{M}$ is not simply connected. $\mathcal{M}$ contains enclosed holes about which loops can be trapped. In a non-cylindrical

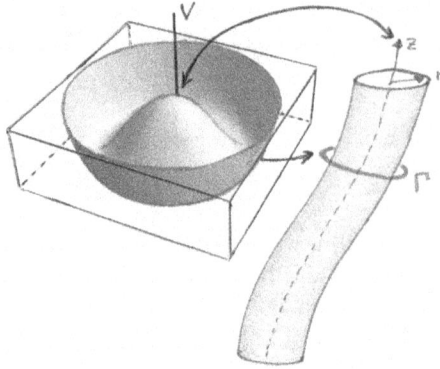

**Figure 11.2: Mapping the degenerated minima of the potential to position space.**

symmetric configurations the curve $\Gamma$ would shrink to a point and we produce a discontinuity in the phase factor, contradicting the smoothness of the Higgs field. But if there is a string within the circle, Stoke's theorem implies that the winding number is non-zero due to the flux through the surface. One says that the abelian Higgs model is topological stable by the cylindrical symmetry. In this physical cylindrical symmetric space we can have again similar non-trivial windings n about a degenerated circle of minima. By calculating the magnetic flux ($\frac{q}{\hbar} = e$)

$$\Theta = \oint_\Gamma \vec{A}.d\vec{r} = \frac{\hbar}{q} \oint_\Gamma \vec{\nabla}\phi.d\vec{r}, \tag{11.8}$$

and using Stokes's theorem, we obtain that the magnetic flux lines are quantized by $\frac{2\pi n}{e}$.

One could wonder what happens when quantum fluctuations excites the vortex. This will be treated in the next section. The energy of the string in flat spacetime is given by

$$E = \frac{1}{2}\frac{n^2}{e^2 r^2}(\partial_r P)^2 + \frac{1}{2}\eta^2(\partial_r X)^2 + \frac{1}{2}n^2\eta^2\frac{P^2 X^2}{r^2} + \frac{1}{8}\beta\eta^4(X^2 - 1)^2 \tag{11.9}$$

The energy is proportional with $n^2$, so there can be no exact ground state for the string carrying multiple flux quanta (the expression changes when gravity comes into play and new features will emerge). There are some characteristic parameters:

$$penetration\ and\ coherence\ lenght : \nu = \frac{1}{e\eta}, \quad \zeta = \frac{\sqrt{2}}{\eta\sqrt{\lambda}}$$

$$masses : m_\Phi = \eta\sqrt{\lambda}, \quad m_A = e\eta$$

$$widths : \delta_\Phi \sim \frac{1}{m_\Phi}, \quad \delta_A \sim \frac{1}{m_A}$$

$$Bogomol'nyi\ parameter : \alpha_b \equiv \frac{m_A^2}{m_\Phi^2} = \frac{e^2}{\lambda}$$

$$Ginsburg - Landau\ parameter : \kappa = \frac{\nu}{\zeta} = \frac{\sqrt{\lambda}}{\sqrt{2}e}$$

The two parameters $\alpha_b$ and $\eta$ play an important role by the calculation of the forces between the vortices. For small r, $E$ can be approximated by $E \approx n^2 \frac{\ln \kappa}{e^2 \hbar^2 \nu^2}$ The field equations become

$$\partial_{rr} P = \frac{\partial_r P}{r} + e^2 \eta^2 P X^2, \quad \partial_{rr} X = -\frac{\partial_r X}{r} + n^2 \frac{X P^2}{r^2} + \frac{1}{2}\lambda\eta^2 X(X^2-1) \quad (11.10)$$

In figure 3 a typical solution of the NO vortex string is visualized.

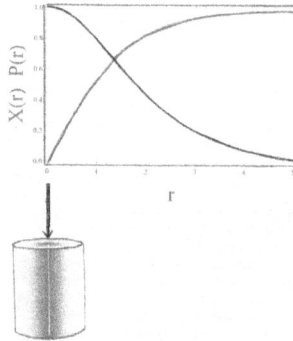

**Figure 11.3: Typical numerical solution of the Higgs (P) and gauge field (X).**

A particular feature of these particle-like solutions is their topological structure characterized by an integer n, the topological charge, or winding number of the field. The topological charge can also be identified as the net number of the new type of particle. As can be seen from Eq. (11.9), the energy increases with $n^2$. For $n = 1$ we have the minimal energy situation, which is stable as it cannot decay into a topological trivial field. These field configurations are also called solitons.

## 11.2.2 The approximation

In order to study perturbations in the model, one can apply the multiple-scale approximation, very suited at high curvature situations[18, 19, 20]. The method is called a "two-timing" method, because one considers the relevant fields $V_i$ in point $\mathbf{x}$ on a manifold M dependent on different scales $(\mathbf{x}, \xi, \chi, ...)$:

$$V_i = \sum_{n=0}^{\infty} \frac{1}{\omega^n} F_i^{(n)}(\mathbf{x}, \xi, \chi, ...). \tag{11.11}$$

Here $\omega$ represents a dimensionless parameter, which will be large (the "frequency", $\omega \gg 1$). So $\frac{1}{\omega}$ is a small expansion parameter. Further, $\xi = \omega\Theta(\mathbf{x})$, $\chi = \omega\Pi(\mathbf{x}), ...$ and $\Theta, \Pi, ...$ scalar (phase) functions on M. The parameter $\frac{1}{\omega}$ can be the ratio of the characteristic wavelength of the perturbation to the characteristic dimension of the background. On warped spacetimes it could also be the ratio of the extra dimension y to the background dimension or even both. When one substitutes the expansions of the field variables

$$g_{\mu\nu} = \bar{g}_{\mu\nu}(\mathbf{x}) + \frac{1}{\omega}h_{\mu\nu}(\mathbf{x}, \xi) + \frac{1}{\omega^2}k_{\mu\nu}(\mathbf{x}, \xi) + ...,$$
$$A_\mu = \bar{A}_\mu(\mathbf{x}) + \frac{1}{\omega}B_\mu(\mathbf{x}, \xi) + \frac{1}{\omega^2}C_\mu(\mathbf{x}, \xi) + ...,$$
$$\Phi = \bar{\Phi}(\mathbf{x}) + \frac{1}{\omega}\Psi(\mathbf{x}, \xi) + \frac{1}{\omega^2}\Xi(\mathbf{x}, \xi) + ..., \tag{11.12}$$

into the equations, one obtains first order equations in $u = t - r$ for the first and second order perturbations[22, 23, 24]. They are of the form

$$\partial_u \dot{\mathbf{U}}_1 = \bar{\mathbf{A}}, \qquad \partial_u \dot{\mathbf{U}}_2 = \bar{D}_1 \dot{\mathbf{U}}_2 + D_2 \dot{\mathbf{U}}_1 + D_3 \tag{11.13}$$

where $\bar{A}$ and $\bar{D}_1$ depends solely on the background fields, while $D_2, D_3$ depend on the first order perturbations and background fields. The dot represents differentiation with respect to the rapid scale $\xi$. In principle one could push the approximation to higher orders. In this way one obtains a wavelike approximation which is asymptotically finite[18]. To highest order in $\omega$, one obtains the equations $l^\alpha(\ddot{h}_{\alpha\nu} - \frac{1}{2}\bar{g}_{\alpha\nu}\ddot{h}) = 0, l_\alpha l^\alpha \ddot{\Psi} = 0$ and $l^\alpha \ddot{B}_\alpha = 0$, where $l_\alpha \equiv \frac{\partial\Theta}{\partial x^\mu}$ is the wave vector perpendicular to the hyper-surface. These equations deliver constraints on $h, B$ and $C$. In other approximations, they are a priori used as gauge conditions. Moreover, the original symmetry on the gauge field will be broken by the appearance of $B_t$ besides $B_\varphi$.

In our approximation scheme we can gain a lot of insight in the behavior of the clustering of vortices when gravity is present. The equations (Eq. (11.13)) are hard to solve. However, the energy-momentum tensor components can tell us a lot about the behavior of the model.

## 11.2.3 Excitation of the vortices and the quasar link

In the expansion of Eq. (11.12) we parameterized the scalar field in subsequent orders as

$$\bar{\Phi} = \eta\bar{X}(t, r)e^{in_1\varphi}, \qquad \Psi = Y(t, r, \xi)e^{in_2\varphi}, \qquad \Xi = Z(t, r, \xi)e^{in_3\varphi}. \tag{11.14}$$

It turns out that the solution to second order is no longer axially symmetric. There appear terms like $\sin(n_2 - n_1)\varphi$ in the field equations. The most interesting information can be found in the energy-momentum tensor components

$$^4T_{t\varphi}^{(0)} = \bar{X}\bar{P}\dot{Y}n_1\sin[(n_2 - n_1)\varphi], \tag{11.15}$$

$$^4T_{tt}^{(0)} = \dot{Y}^2 + \dot{Y}(\partial_t\bar{X} + \partial_r\bar{X})\cos[(n_2 - n_1)\varphi]$$
$$+ \frac{e^{2\bar{\psi}}}{\bar{W}_1^2 r^2 e}\left(e\dot{B}^2 + n_1\dot{B}(\partial_r\bar{P} + \partial_t\bar{P})\right), \tag{11.16}$$

While $^4\bar{T}_{t\varphi} = 0$, we conclude from Eq. (16) that the axial symmetry is broken already to first order. The energy $^4T_{tt}^{(0)}$ contribution to first order contains

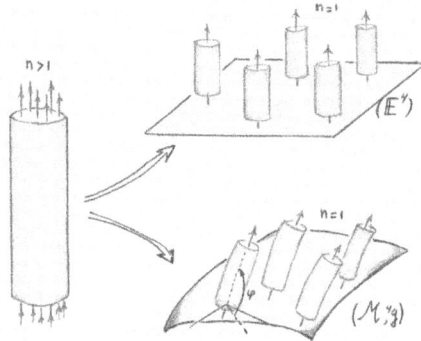

Figure 11.4: **Excitation and decay of a high multiplicity vortex into correlated vortices of unit flux $n = 1$. Top: the Abrikosov lattice in Euclidean space. Bottom: correlated vortices with preferred azimuthal angle $\varphi$ in curved spacetime after the symmetry breaking.**

the warp factor in the denominator. So the energy depends crucially on the age of the universe. Terms in the energy can dominate at early times and are negligible at late times in the evolution of the universe. In the second order contributions there appear terms like $\cos(n_3 - n_1)\varphi$[24]. The azimuthal-angle dependency are expressed in trigonometrical functions with extrema which differ $mod(\frac{\pi}{k})$. After the excitation of the vortex with multiplicity n, it will decay into n vortices of unit flux in a regular lattice (figure 4). The Abrikosov vortices form a hexagonal lattice such that the energy is minimal. This process depends on the Bogomol'nyi parameter and is observed in laboratory experiments. In the special case of $\alpha_b = 1$ ( mass of scalar and gauge field are equal) are the forces between the vortices easier to understand. It was a great achievement of Bogomol'nyi[25] to find the decoupled equations

$$\partial_{rr}X = -\frac{1}{r}\partial_r X + \frac{1}{X}\partial_r X^2 + \frac{1}{2}e^2\eta^2 X(X^2 - 1), \qquad P = \frac{r}{n\eta X}\partial_r X \tag{11.17}$$

Without the Bogomol'nyi equations it is difficult to understand the cancellation of the forces. The movement of the gauged vortices are even harder to understand[6].

There is another characterization of the winding number. It is the total vortex number, i.e., the number of points in the plane with multiplicity taken into count where $\Phi = 0$. The zero's of $\Phi$ are then a set of n isolated points $z_i, i = 0..n$ in $C$ such that $\Phi(z, z^*) \sim c_j(z - z_j)^{n_j}$ with $n_j$ the multiplicity of $z_j$ and $n = \sum_{z_j} n_j$. This n-vortex solution represents a finite energy configuration with n flux quanta, provided $\Phi$ and $A$ satisfy the boundary conditions

$$\lim_{r \to 0} X = 0, \quad \lim_{r \to 0} P = 1, \quad \lim_{r \to \infty} X = 1, \quad \lim_{r \to \infty} P = 0 \qquad (11.18)$$

The collection of n vortices of unit flux is energetically more appealing than a n-flux vortex. The energy density is peaked around the zero's of $\Phi$. Hence they can be identified by the location of the vortices. In general, one must solve the time-dependent GL equations in order to get insight in the stability issues. In this case there is a gradient flow, which makes the analysis very complicated. The PDE's are badly nonlinear and one relies often on numerical simulation. The temporarily broken axial symmetry will be the onset of emission of electromagnetic and gravitational waves. From Eq. (11.15) we see that the angular momentum will fade away when $n_2$ approaches $n_1$ and the axial symmetry is restored. The first and second order perturbations of the scalar and gauge fields in higher winding number decay into NO strings of n=1. In order to understand the azimuthal-angle ($\varphi$) preference, one must consider the terms of ${}^4T_{\varphi\varphi}$, for example

$$\begin{aligned}
{}^4T_{\varphi\varphi}^{(0)} &= e^{-2\gamma}r^2\dot{Y}(\partial_t\bar{X} - \partial_r\bar{X})\cos[(n_2 - n_1)\varphi] \\
&+ \frac{n_1 e^{2\bar{\psi} - 2\bar{\gamma}}}{\bar{W}_1^2 e}\dot{B}(\partial_r\bar{P} - \partial_t\bar{P}),
\end{aligned} \qquad (11.19)$$

$$\begin{aligned}
{}^4T_{\varphi\varphi}^{(1)} &= e^{-2\gamma}r^2\dot{Z}(\partial_t\bar{X} - \partial_r\bar{X})\cos[(n_3 - n_1)\varphi] + \frac{e^{2\bar{\psi} - 2\bar{\gamma}}}{\bar{W}_1^2 e}n_1\dot{C}(\partial_r\bar{P} - \partial_t\bar{P}) \\
&+ e^{-2\bar{\gamma}}r^2\dot{Y}(\partial_t Y - \partial_r Y) + \bar{X}^2 n_1\bar{P}eB \\
&+ \left[\frac{e^{2\bar{\psi} - 2\bar{\gamma}}}{\bar{W}_1^2}\dot{Y}(\partial_t\bar{X} - \partial_r\bar{X})(h_{44} + e^{-2\bar{\gamma}}r^2 h_{11})\right. \\
&+ n_1\bar{X}\bar{P}Y(n_2 - n_1 + n_1\bar{P}) + \frac{1}{2}\beta e^{-2\bar{\psi}}\bar{W}_1^2 r^2\bar{X}Y(\eta^2 - \bar{X}^2) \\
&\left.+ e^{-2\bar{\gamma}}r^2(\partial_t\bar{X}\partial_t Y - \partial_r\bar{X}\partial_r Y)\right]\cos[(n_2 - n_1)\varphi] \\
&+ \frac{e^{4\bar{\psi} - 4\bar{\gamma}}}{\bar{W}_1^4 r^2 e^2}\left[r^2 e\dot{B}n_1(\partial_r\bar{P} - \partial_t\bar{P}) + \frac{1}{2}r^2 n_1^2(\partial_r\bar{P}^2 - \partial_t\bar{P}^2)\right. \\
&\left.+ \frac{1}{2}\bar{W}_1^2 e^2 e^{2\bar{\psi}}(\partial_t\bar{X}^2 - \partial_r\bar{X}^2)\right]h_{11} + \left[\frac{1}{2\bar{W}_1^2}e^{2\bar{\psi} - 2\bar{\gamma}}(\partial_t\bar{X}^2 - \partial_r\bar{X}^2)\right. \\
&\left.- \frac{1}{8}\beta(\bar{X}^2 - \eta^2)^2\right]h_{44}.
\end{aligned} \qquad (11.20)$$

$T_{\varphi\varphi}$ plays an important role in the interaction of the strings. Positive terms in the expression indicate "pressure" in the direction of the Killing vector field $(\frac{\partial}{\partial\varphi})^i$ (and negative "tension"). The result is that the interaction contribution can change sign dynamically (dependent of the warp factor). This Killing vector must be normalized such that, along a closed integral curve, the parameter $\varphi$ varies van 0 to $2\pi$ with $\varphi = 0$ and $\varphi = 2\pi$ identified. This will provide boundary conditions for the metric fields close to the axis of the string, such as $\partial_r(r^2 e^{-2\psi})(0) = 1$. We observe in the expressions of ${}^4T_{\varphi\varphi}^{(0)}$ and ${}^4T_{t\varphi}^{(0)}$ that when $sin(n_2 - n_1)\varphi$ becomes zero, $cos(n_2 - n_1)\varphi$ has its maximum. So there is an emergent imprint of a preferred azimuthal angle $\varphi$ on the lattice of vortices when the ground state is reached ( n=1). This effect can also be seen in the ${}^4T_{zz}^{(0)}$ component which is not equal to $-{}^4T_{tt}^{(0)}$ as is the case in static models. The second order contribution ${}^4T_{\varphi\varphi}^{(1)}$ contains terms like $cos(n_3 - n_1)\varphi$ and produces a complicated extrema[24].

The recently observed alignment of the spinning axes of quasars in LQG's on Mpc-scales can be explained by our model. The observations were carried out at the European Southern Observatory, Paranal with the Very Large Telescope equipped with the FORS2 instrument. There was a confirmation of the alignment for radio galaxies by the Giant Metrewave Radio Telescope in the ELAIS-N1 field. This curious effect cannot be the result of statistical fluctuations[26]. The origin must be found in the early universe just after the symmetry breaking, as described in our model. Specially, the two preferred orientations perpendicular to each other in quasar groups of less richness could be the second order effect in our model by the appearance of the trigonometrical terms with periodicity difference of $\frac{\pi}{2}$. The correlated $n = 1$ vortices with preferred azimuthal angle, emerged on a correlation length smaller than the horizon on that moment and took place at the Ginzburg temperature $\sim \frac{1}{\beta\eta}$. These correlated regions will survive to later times, because at this moment the gravity contribution from the 5D bulk comes into play. The warp factor ( see Figure 2) will have different contributions to the field equations for different times. The mass per unit length will contain the warp factor. Just after the symmetry breaking, the vortex will acquire a huge mass $G\mu > 1$ and will initiate the perturbations of high-frequency and justifies our high-frequency approximation. This is the reason that the regions with $(n = 1, \varphi = \varphi_0)$ will stick together and are observed in LQG's with aligned polarization axes[9, 10]. The most striking observation is the alignment of the spin axes of the quasars parallel to the major axes of their host LQG, while the spin axes can also become perpendicular to the the LQG major axes when the richness decreases. This can be explained in our model as second order effect: the higher multiplicity terms like $cos(n_3 - n_1)\varphi$ enters the energy-momentum tensor component ${}^4T_{\varphi\varphi}^{(1)}$ and is out phase with the term $cos(n_2 - n_1)\varphi$ (see Eq. (11.2.3)). In figure 5 we sketched a global distribution of the polarization axes as found by the observations. Some specific features of this alignment which must be confirmed by more observations on quasars and radio sources at high redshift. Alignment at high redshifts would confirm that the mechanism took place indeed in the early universe.

174

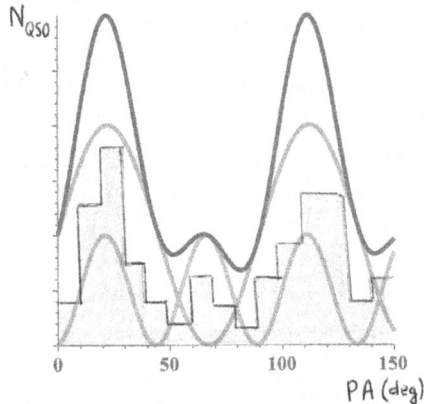

Figure 11.5: Possible fit of the theoretical predicted curve (black) on the distribution of a number of observed polarization angles. This curve can be the sum of the first (red) and second (green) contribution.

## 11.2.4 Breaking of the axial symmetry from a different viewpoint

Self-gravitating objects in equilibrium exhibit a striking analogue with the mathematical model of the Maclaurin-Jacobi sequences and its bifurcation points[27]. Bifurcation points that are of particular interest to us here are those marking the onset secular instability, i.e., the dynamical breaking of axially symmetry ( or better formulated: the spacetime possesses 2 in stead of 3 Killing vectors). This means the appearance of an off-diagonal metric function. In our model it is the transition from the Weyl metric (after the substitution $t \rightarrow iz, z \rightarrow it$) to the Papapetrou metric, expressed by the appearance of the $T_{t\varphi}$ components in first and second order. It is remarkable that this symmetry breaking can be compared with the second order phase transition in type II superconductivity[28, 29, 30], which is the basis of our model (see also Slagter[24] for more details). An initial axially symmetric configuration, as is the case in our perturbative model, can dynamically spontaneously be broken, where equatorial eccentricity plays the role of order-parameter. The equatorial eccentricity $\varepsilon \equiv \frac{b}{a}$, with b and a the two equatorial axes, can be expressed through the azimuthal-angle $\varphi(t)$. The particular orientation of the ellipsoid in the frame $(r, \varphi, z)$ expressed through $\varphi_0 \equiv \varphi(t_0)$, will be at $t > t_0$ determined by the transformation $\varphi \rightarrow \varphi_0 - Jt$, where $J$ is the rotation frequency (circulation or "angular momentum") of the coordinate system. The angle $\varphi_0$ is fixed arbitrarily at the onset of symmetry breaking. This arbitrariness of $\varphi_0$, i.e., the orientation of the ellipsoid at $t = t_0$ can be compared with the massless Goldstone-boson modes of the spontaneously broken symmetry of continuous groups. The phase transition take place on the same time scale that the vorticity is destroyed by dissipative mechanism and $J$ is lost. The end point is a lower energy state that belongs to the Jacobi or Dedekind sequence of equilibrium ellipsoids[31]. In the original paper of Chandrasekhar and Lebovitz[28], in the Newtonian case, the deformations of the axisymmetric configuration by an infinitesimal non-axisymmetric deformation is described in terms of a Lagrangian displacement $\varsigma^a(r, z, \varphi) = \bar{\varsigma}^a(r, z)e^{in\varphi}$, with $n$ an integer. However, the real part of the $e^{in\varphi}$ must be put in by hand, in contrast to our result: it appears in a perturbative way as a first and second order effect. The temporarily broken axial symmetry will be the onset of emission of electro-magnetic and gravitational waves, while the string relaxes to the NO configuration. It is a consequence of the coupled system of PDE's that a high-frequency scalar field can create through an electro-magnetic field, a high frequency gravitational field and conversely. It is the appearance of the term $sin(n_2 - n_1)\varphi$ in the first order term $^4T_{t\varphi}^{(0)}$ (Eq.(11.15)) and explained in section 2.3, which triggers this angular momentum and the axially symmetry will be restored when $n_2$ becomes equal to $n_1$ again. The second order contribution $^4T_{tt}^{(1)}$ shows terms like $\dot{B}^2 h_{44}$[24], indicating the interaction between the high-frequency EM and gravitational waves. It contains the warp factor in the denominator. In the early stages of the universe $W_1$ is still small and the term is significant. As time increases, it will fade away.

## 11.3   Relation with conformal invariance

In the preceding sections we found that the warp factor $\mathcal{W}$ plays the role of a "scaling" factor, different at different epochs in time. There is, however, another interpretation, related to the dilaton field in conformal invariant gravity theory. The brane-part $W_1(t, r)$ could be solved from the 5D Einstein equations and was given in Eq.(11.6). The differential equation could be separated and reads ( we rename, for historical reason, $W_1$ in $\omega$, not to confuse with the expansion parameter in Eq.(11.11))

$$\partial_{tt}\omega = \partial_{rr}\omega + \frac{1}{\omega}\left((\partial_r\omega)^2 - (\partial_t\omega)^2\right) + \frac{2}{r}\partial_r\omega. \qquad (11.21)$$

One can then write the spacetime[32]

$$g_{\mu\nu} = \omega^2 W_2^2 \tilde{g}_{\mu\nu} + n_\mu n_\nu \qquad (11.22)$$

where the dilaton is conformally coupled to gravity and embedded in a smooth $M_4 \otimes R$ manifold by the action

$$\mathcal{I} = \int d^4x\sqrt{-\tilde{g}}\Big\{-\frac{1}{12}\left(\tilde{\Phi}\tilde{\Phi}^* + \bar{\omega}^2\right)\tilde{R} - \frac{1}{2}\left(\mathcal{D}_\alpha\tilde{\Phi}(\mathcal{D}^\alpha\tilde{\Phi})^* + \partial_\alpha\bar{\omega}\partial^\alpha\bar{\omega}\right)$$
$$-\frac{1}{4}F_{\alpha\beta}F^{\alpha\beta} - V(\tilde{\Phi}, \bar{\omega}) - \frac{1}{36}\kappa_4^2\Lambda_4\bar{\omega}^4\Big\} \quad (11.23)$$

We wrote $\Phi = \frac{1}{\omega}\tilde{\Phi}$ and Newton's constant is absorbed in a redefinition of $\bar{\omega}$. We redefined $\bar{\omega}^2 \equiv -6\omega^2/\kappa^2$, in order to make the dilaton field comparable with the scalar field ( see solution C in figure 1). A term $\sim \omega^4$ can be added to the action.

Such a term could play a role in the generation of a cosmological constant. This action is local (Weyl-) conformal invariant by the transformation

$$\tilde{g}_{\mu\nu} \to \Omega^2 \tilde{g}_{\mu\nu}, \qquad \bar{\omega} \to \frac{1}{\Omega}\bar{\omega}, \qquad \tilde{\Phi} \to \frac{1}{\Omega}\tilde{\Phi}. \tag{11.24}$$

when there is no mass term in $V$, because a mass term spoils the tracelessness of the energy momentum tensor. One could say that the conformal symmetry is spontaneously broken, just as the gauge symmetry of in the Brout-Englert-Higgs mechanism is spontaneously broken.

However, there will be no singular behavior when the former "scale"-function $\omega$ approaches zero (the small distance limit), because the Einstein field equations will contain in the dominator the term $\omega^2 + |\Phi|^2$ (the scalar field and dilaton field are treated here on equal footing). So we could have a regular description of gravity at very small distances[33, 34, 35, 36].

The action appears to be entirely renormalizable for the dilaton field. After integrating over $\omega$ but not yet over $\tilde{g}_{\mu\nu}$, the resulting action stays local conformal invariant. The problem is that $\tilde{g}_{\mu\nu}$ in Eq.11.22 is not flat. One could consider an additional gauge transformation on $\tilde{g}_{\mu\nu}$ to make it Ricci-flat. If we write $\tilde{g}_{\mu\nu} = \Omega^2 \hat{g}_{\mu\nu}$, one can try to find the additional $\Omega$ for Ricci flat $\hat{g}_{\mu\nu}$[32]. Calculations on renormalizability and the appearance of anomalies would then be simplified[34].

On a flat background the choice of $\omega$ is unique. In curved spacetime it is fixed only if we know the evolution of spacetime and after choosing a coordinate frame. We conjecture that in our warped 5D model is it fixed by the evolution of the Einstein equations: the dilaton field plays the role of the warp factor.

It is a tantalizing idea to connect the mass generation in the Higgs-mechanism with the tracelessness of $T_{\mu\nu}$ ( see for example the treatment of Mannheim[36]) and the cosmological constant problem. If one omits a kinematic mass term ( $\sim \beta\eta^2 X^2\omega^2$) in the action (so $T_{\mu\nu}$ is traceless) and include a fermion field in the action, then one can dynamically generate massive particles without breaking the tracelessness of $T_{\mu\nu}$. Moreover, the cosmological constant can naturally arise in these dynamical mass theories and its value is constrained.

There is another possible way out for the breaking of the tracelessness of $T_{\mu\nu}$. In our 5D warped spacetime, the trace of the energy momentum tensor[32]

$$\frac{1}{\bar{\omega}^2 + X^2}\left[16\kappa_4^2\beta\eta^2 X^2\bar{\omega}^2 - \kappa_5^4 n^4 \left(\frac{(\partial_r P)^2 - (\partial_t P)^2}{r^2 e^2}\right)^2 e^{8\tilde{\psi} - 4\tilde{\gamma}}\right] \tag{11.25}$$

will contain contributions from $\mathcal{S}_{\mu\nu}$ ( see Eq.(11.3)). The demand of tracelessness will deliver constraints on the parameters of the model in a dynamical way. Newton's constant, for example, reappears by the conformal breaking ( note that this constant is hidden in the effective quartic interaction term for the $\Phi$ field). It is conjectured that constraints on the small distance behavior, all the physical constants, including the masses and cosmological constant, are constrained to values that are computable in terms of the Planck unit. For example , all $\beta$-coefficients of the renormalization group must vanish by the adjustment of the coupling constants (note that the zeros of the beta functions are isolated stationary points in quantum field theory).

In the conformal model there are still many problems unsolved, for example, the anomalies, which must be constrained to cancel out. Further, the black hole complementarity in conformal gravity is not yet well-understood[33].

## 11.4 Conclusion

By considering an axially symmetric warped five-dimensional warped space-time, were the standard model fields are confined to the brane, we find in a nonlinear approximation, an emergent azimuthal-angle dependency of Nielsen-Olesen vortices just after the symmetry breaking at GUT scale. Using a approximation scheme, the azimuthal-angle dependency appears in the first and second order field equations as trigonometrical functions $\sin(n_i - n_j)\varphi$ and $\sin(n_i - n_j)\varphi(i > j)$, with $n_i$ the multiplicities of subsequent perturbation terms of the scalar field. Vortices with high multiplicity decay into a lattice with entangled Abrikosov vortices. The stability of this lattice of correlated flux $n = 1$ vortices with preferred azimuthal-angle is guaranteed by the contribution from the bulk spacetime by means of the warp factor: the cosmic string becomes super-massive for some time during the evolution. These so-called super-massive cosmic strings arise in a natural way in string theory or M-theory., i.e., brane-world models and are produced when the universe underwent phase transitions at energies much higher than the GUT scale, $G\mu > 1$. The imprint on the effective 4D brane spacetime can be caused by wavelike disturbances triggered by the huge mass of the cosmic strings in the bulk. One conjectures that these disturbances could act as an effective dark energy field. Another possible effect can be the formation of massive KK-modes as the imprint of the 5D gravitational field on the 4D brane.

We used the azimuthal-angle correlation for the explanation of the recently observed alignment of polarization axes of quasars in large quasar groups. The detailed behavior of this alignment can be explained with our model. The two different orientations perpendicular to each other in quasars groups of less richness could be a second order effect in our model.

When gravity is coupled to standard model fields and one demands the validity on all distance scales, one runs into problems. These are: the dark energy problem, the cosmological constant problem, the hierarchy problem and the problem how probe the small distance structure of our spacetime. Conformal invariance could be the solution for at least some of these problems. It could be the missing symmetry of nature. In our model, by identifying the warp factor as a dilaton field, one will not encounter singular behavior when the dilaton field becomes very small. At present time, the warp-like manifestation of the dilaton field describes the exponential expansion of our universe. Moreover, the exceptional smallness of the cosmological constant, $\Lambda \asymp\sim 10^{-120}$ compared to the calculated vacuum energy could be explained in the warped 5D spacetime by the warp factor.

More data of high-redshift quasars will be needed in order to test the second order effect predicted in our model.

# References

[1] Vilenkin, A. and Shellard, E.P.S. (1994) Cosmic Strings and Other Topological Defects. Cambridge University press, Cambrigde, UK.

[2] Anderson, M.R. (2003) The Mathematical Theory of Cosmic Strings. IoP publishing, Bistol, UK.

[3] Nielsen, H.B. and Olesen, P. (1973) *Nucl. Phys.* **B61**, 45.

[4] Felsager, B. (1987) Geometry, particles and fields. Odense University press: Odense, Denmark.

[5] Jaffe, A. and Taubes, C. (1981) page 53, Vortices and monopoles. Birkhauser press: Boston, USA.

[6] Manton, N. and Sutcliffe, P. (2007) Topological solitons. Cambridge University press, Cambrigde, UK.

[7] Ginzburg, V. L. and Landau, L. D. (1950) *Zh. Eksp. Teor. Fiz.* **20**, 1064.

[8] Abrikosov, A. A. (1957) *Soviet Physics JETP* **5**, 1774.

[9] Hutsemekers, D., Braibant, L., Pelgrims, V. and Sluse, D. (2014) *Astron. Astrophys.* **572**, A18.

[10] Taylor, A.R. and Jagannathan, P. (2016) *Mon. Not. Roy. Astr. Soc.* **459**, *459*, L36.

[11] Gregory, R. (1989) *Phys. Rev. D* **39**, 2108.

[12] Randall, L. and Sundrum, R. (1999) *Phys. Rev. Lett.*, **83**,3370.

[13] Randall, L. and Sundrum, R. (1999) *Phys. Rev. Lett.* **83**, 4690.

[14] Maartens, R. (2007) *J. Phys. Conf. Ser.* **68**, 012046.

[15] Maartens, R. (20076) *Lect. Notes. Phys.* **720**, 323.

[16] Maartens, R. and Koyama, K. (2010) *Living Rev. Relativity* **13**, 5

[17] Shiromizu, T., Maeda, K. and Sasaki, M. (2000) *Phys. Rev. D* **62**, 024012.

[18] Choquet-Bruhat, Y. (1969) *Commun. Math. Phys.* **12**, 16.

[19] Choquet-Bruhat, Y. (1977) *Gen. Rel. Grav.* **8**, 561.

[20] Slagter, R. J. (1986) *Astroph. J.* **307**, 20.

[21] Slagter, R. J. and Pan, S. (2016) *Found. of Phys.* **46**, 1075.

[22] Slagter, R. J. (2016) *J. of Mod.Phys.* **7**, 501.

[23] Slagter, R. J. (2017) *J. of Mod.Phys.* **8**, 163 .

[24] Slagter, R. J. (2018) *Int. J. of Mod. Phys. D* doi.org/10.1142/S0218271818500943. ArXiv: gr-qc/1609.05068

[25] Bogomol'nyi, E. (1976) *Sov. J. Nucl. Phys.* **24**, 449.

[26] Pelgrims, V. Thesis: ArXiv: astr-ph/160405141v1

[27] Lebovitz, N. R. (1967) Bifurcation and stability problems in astrophysicsm, in Applications of bifurcation theory. Academic Press, New York, U

[28] Chandrasekhar, S. and Lebovitz, N.R. (1973) *Astroph. J.* **185**, 19.

[29] Chandrasekhar, S.and Friedman, J. L. (1973) *Astrophys. J* **185**, 1.

[30] Bertin, G. and Radicati, L. A. (1976) *Astrophys. J* **206**, 815.

[31] Christodoulou, D. M., Kazanas, D. Shlosman, I and Tohline, J. E. (1995) *Astrophys. J* **446** , 472

[32] Slagter, R. J. (2018) ArXiv: gr-qc/171108193, gr-qc/181008793.

[33] 't Hooft, G. (2015) ArXiv: gr-qc/151104427v1

[34] 't Hooft, G. (2011) ArXiv: gr-qc/11044543v1

[35] 't Hooft, G. (2010) ArXiv: gr-qc/10110061v1

[36] Mannheim, P. D. (2015) ArXiv: hep-th/161008907v2

# 12 PROPAGATION OF GRAVITATIONAL WAVES THROUGH THE STOCHASTIC BACKGROUND OF GRAVITATIONAL WAVES

CARLOS FRAJUCA, FABIO DA SILVA BORTOLI,
FRANCISCO Y. NAKAMOTO, GIVANILDO ALVES DOS SANTOS

**Abstract**    Given the recent claim that gravitational waves (GW) have finally been detected and the other efforts around the world for GWs detection, it is reasonable to imagine that the relic gravitational wave background could be detected some time in the near future and with such information the scientific community could gather some hints about the origin of the universe. Nevertheless, it must be considered that gravity has self-interaction; under such assumption, it's reasonable to expect that these gravitational waves will interact with the relic or non-relic GW background by scattering, for example. Such interaction should decrease the distance from which such propagating waves could be detected The propagation of gravitational waves is analyzed in an asymptotic de Sitter space by the perturbation expansion around a Minkowski space using a scalar component. Using the case of de Sitter inflationary phase scenario, the perturbation propagates through an FRW background. The GW, using the actual value for the Hubble scale ($H_0$), has a damping factor with a very small valor for the size of the observational universe; the stochastic relic GW background is given by a dimensionless function of the frequency. In this work we analyze this same damping including the gravitational wave background originated by astrophysical sources such background is 3 orders of magnitude bigger in some frequencies and produces a higher damping factor.

C. Duston, M. Holman (Eds), *Spacetime Physics 1907 - 2017. Selected peer-reviewed papers presented at the First Hermann Minkowski Meeting on the Foundations of Spacetime Physics, 15-18 May 2017, Albena, Bulgaria* (Minkowski Institute Press, Montreal 2019). ISBN 978-1-927763-48-3 (softcover), ISBN 978-1-927763-49-0 (ebook).

## 12.1 Introduction

The detection of gravitational waves came after a long road of experiments planned in 2010 [1], and in 2016 the detection was finally made [2, 3], as can be seen in figure 12.1. Gravitational waves got a very strong evidence with the binary **PSRB1913 + 16** (also known as **PSRJ1915 + 1606**, and the Hulse–Taylor binary after its discoverers), which is a pulsar (a radiating neutron star) and, along with another neutron star, form a binary star system. PSR 1913+16 was the first binary pulsar to be discovered and its orbital period has been decreasing over time due to the emission of gravitational waves [4], as can be seen in figure 12.1. The first attempts toward gravitational wave detection started in the early sixties [5], with the resonant mass gravitational wave detectors [6, 7, 8, 9]. The efforts towards the detection of gravitational waves using this kind of detector culminated with the spherical detector, where six sensors are connected to the surface of the sphere, arranged according to a work based on the distribution of Merkowitz and Johnson [10, 11]. These transducers are located as if they were on the center of 6 pentagons connected to a surface that corresponds to a half dodecahedron [12, 13]. Each transducer mechanically amplifies the motion occurring at region of the sphere surface in which it is connected. The already amplified movement excites the membrane of one resonant microwave cavity, these vibration creates sidebands on the microwave signal which amplitude is proportional to the gravitational wave signal. The Brazilian efforts on the field of Gravitational Wave detection started in the 90´s and can be seen in [14, 15, 16, 17, 18, 19, 20, 21, 22, 23, 24, 25, 26, 27].

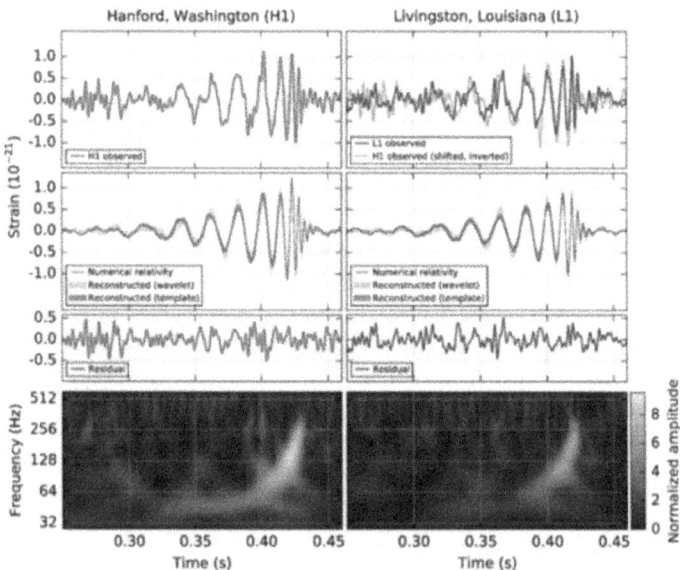

**Figure 12.1:** The signal detected in the first gravitational wave detection. From top to bottom: signal detected on the LIGOs, simulated expected signal, residuals and the frequency amplitude diagram versus time.

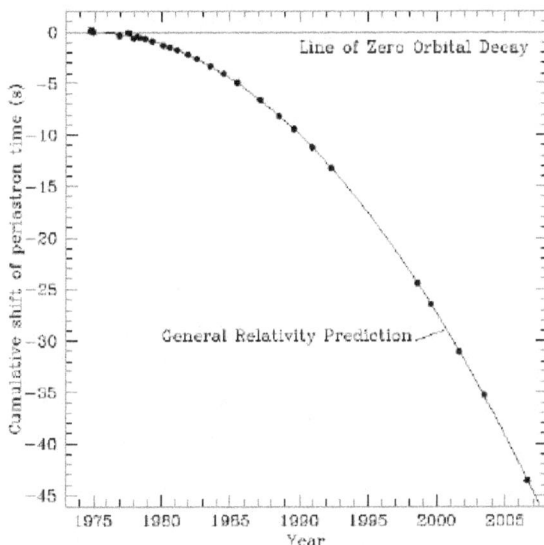

**Figure 12.2:** The gravitational wave emitted by the **PSR J1915+1606, PSR 1913+16** binary carries more momentum than energy, which causes the binary to get closer together and have its periastron time reduced.

## 12.2 The scope of this work

Given the detection of gravitational waves, it is reasonable to imagine that the relic gravitational wave background could be detected some time in the near future, and with such information the scientific community could gather some hints about the origin of the universe [28, 29]. Some calculations on the amplitude of some background have been made, as can be seen in figure 12.2.

Nevertheless, it must be considered that gravity has self-interaction, which means that gravitons interact with gravitons [30, 31], as, for an example, can be seem in figure 12.2. Under such assumption, it's reasonable to expect that these gravitational waves will interact with the relic or non-relic GW background by scattering, and make some of the gravitational waves that are traveling in space lose coherence and deviate in another direction. Such interaction should decrease the distance from which such propagating waves could be detected or at least make the signal weaker, which would makes it seem like the source is far away. It is not the intention of this work to make calculations in quantum gravity, these relation is only a hint that such phenomena should occur.

The propagation of gravitational waves (GWs) is analyzed in an asymptotic de Sitter space by the perturbation expansion around Minkowski space. The stochastic relic GW background is given by a dimensionless function of the frequency [28, 29]. Another way to characterize such background can be seen in [32, 33]. Using the case of de Sitter inflationary phase scenario [28, 29, 34, 35],

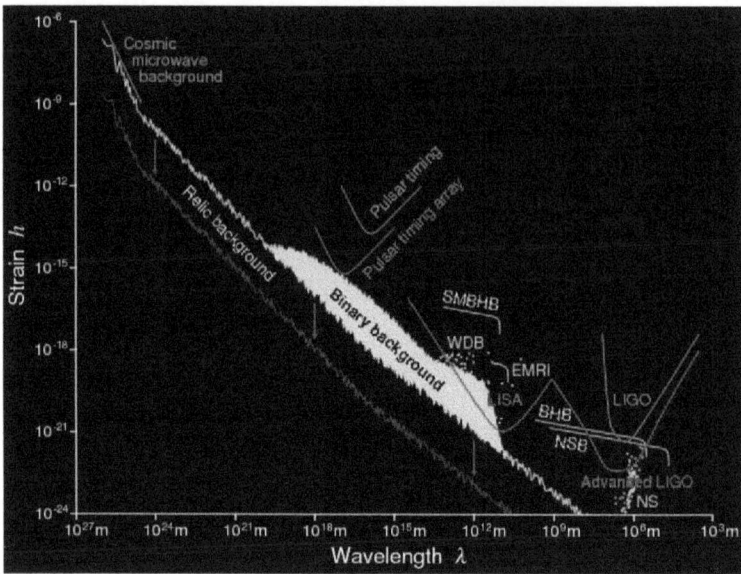

**Figure 12.3:** The gravitational wave background calculated by the group in Caltech and the possible detector for each frequency band as can be found in the web page of Caltech http://www.tapir.caltech.edu/teviet/Waves/gwave_spectrum.html, SMBHB means Supermassive Black Hole Binaries; WDB means White Dwarf Binaries; EMRI means Extreme Mass Ratio Inspirals and NSB means Neutron Star Binaries.

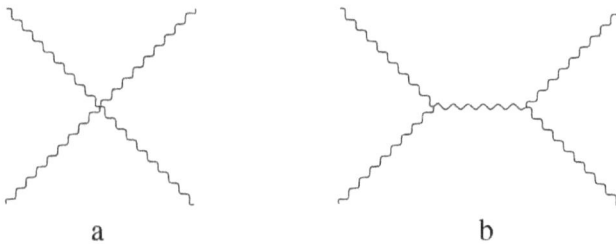

**Figure 12.4:** Four-graviton scattering in the first two orders of the perturbation theory (panels a and b, respectively).

the perturbation propagates through an FRW background with a metric:

$$ds^2 = a^2(\eta)(-d\eta^2 + dx^2) \qquad (12.1)$$

where the negative term is the conformal time.

The GW is represented by a scalar field to make the calculations easier, and can be written as follow:

$$h_{\mu\nu} = e_{\mu\nu}\phi(\eta)e^{ik.x}. \qquad (12.2)$$

The perturbation propagation metric could be expressed as:

$$g_{\mu\nu} = a^2(\eta)(-d\eta^2 + dx^2 + h_{\mu\nu}dx^\mu dx^\nu).$$ (12.3)

The amplitude of the GW must satisfy the following equation:

$$(\frac{d^2}{d\eta^2} + \frac{2}{a}\frac{da}{d\eta} + \frac{d}{d\eta} + |k|)\phi = 0.$$ (12.4)

The relevant result is the one corresponding to the de Sitter inflationary phase, where the scale factor behaves as

$$a \propto e^{H_{ds}t}.$$ (12.5)

In terms of the physical time t. In this scenario the solution becomes:

$$\phi(\eta) = \frac{a(\eta_1)}{a(\eta)}(1 + iH_{ds}\omega^{-1})e^{-ik(\eta-\eta_1)}$$ (12.6)

here $H_{ds}$ is the Hubble factor during the de Sitter phase.

Such solution shows that the amplitude has a damping effect that goes with

$$\phi \propto e^{-H_{ds}t}.$$ (12.7)

This shows that, using the actual value for the Hubble scale $H_0$ , the GW, has a damping factor with a very small value for the size of the observational universe. In this work we analyze this same damping including the gravitational wave background originated by astrophysical sources. Such background is 3 orders of magnitude bigger in some frequencies and produces a considerable damping at shorter distances.

The results presented in [34] gave the authors the idea of developing this work; nevertheless, it has been done by using the cosmological constant and not the GW background, but the approach used in [34] does not apply to the GW background.

## 12.3   Results

This shows that, by using the actual value for the Hubble scale Ho, the amplitude of the GW decreases by a factor of 1/e after 100BLY(Billion Light Year).

What will happen if the gravitational background relic has the values that the GW background has today?

Figure 3 shows that the gravitational wave background has an amplitude bigger than the relic gravitational background by a factor of $10^{1.5}$ in strain h in all of the wavelengths. The stochastic GW background is an energy density distribution $\rho GW$ , which is proportional to the GW strain squared ($h^2$).

The critical density necessary to close the universe is given by:

$$\rho_c = \frac{3H_0^2}{8\pi G}.$$ (12.8)

186

Let´s assume that, as the density distribution ρGW is increase so is the critical power density then $H_0^2$ should increase by the same order of $h^2$.

This implies that, by using the actual value for the Hubble scale $(H_0)$ corrected by a factor of 10, the amplitude of the GW decreases by a factor of $1/e$ after 3BLY.

In figure 3 the amplitude of the background is even stronger in the frequency of LISA detector. In this region the gravitational wave background has a amplitude bigger than the relic gravitational background by a factor of $10^3$ in strain h in all of the wavelengths approximately. This shows that, by using the actual value for the Hubble scale $H_0$, corrected by a factor of $10^3$ ,the amplitude of the GW decreases by a factor of $1/e$ after 100MLY (100 Million Light Year).

## 12.4   Conclusions

Taking into consideration that gravitons should scatter as they interact with other gravitons, gravitational waves should do the same as they interact with the gravitational wave background. This effect is small but as gravitational waves travel very long distances the effect should become important. The results find in [36, 37] shows that interaction of GW with photons may also introduce an attenuation of GW propagating from astrophysical sources, corroborating the results here presented.

Since the gravitational wave measured in [2] was identified to have occurred in a distance of 1BLY without taking this effect into account, it could be located in a distance 30% closer, if the approximations used here are correct.

We also have shown that the LISA detector could have a serious problem of finding sources because it will be able to detect sources in a radius of only 100 MLY.

The next step to this work is to make the same calculations using only General Relativity, without the use of scalar components.

Another follow-up work is to do a formal calculation of the amplitude cross-section using a quantum field theory approach.

## Acknowledgment

Carlos Frajuca acknowledges FAPESP for grant #2013/26258-4 and grant #2006/56041-3 and Valeria M. Frajuca for reviewing this article.

## References

[1] The Gravitational Waves International Committee Roadmap. A Global plan. 2010. Glasglow: University of Glasglow - Kevin Building G28QQ (2010)

[2] Abbott B P: *et al.* "Observation of Gravitational Waves from a Binary Black Hole Merger". *Phys.Rev.Lett.* **116** 061102 (2016)

[3] Castelvecchi D D: "Einstein's gravitational waves found at last". *Nature News* doi: 10.1038/nature.2016.19361 (2016)

[4] Hulse R A, Taylor J H: *et al.*: 'Discovery of a pulsar in a binary system". *Astrophysical Journal* **195** L51-L53 (1975)

[5] Weber J: "Detection and Generation of Gravitational Waves". *Physical Review* **117** 306 (1960)

[6] Richard J P: "Wide-Band Bar Detectors of Gravitational Radiation". *Physical Review* **167** 165 (1984)

[7] Richard J P: "Wide-Band Bar Detectors of Gravitational Radiation ". Proc. Second Marcel Grossmann Meeting on General Relativity, edited R. Rufini, (North-Holland, Amsterdam)(1979)

[8] Frajuca C: "A noise model for the Brazilian gravitational wave detector 'Mario Schenberg'". *Class. Quantum Grav.* **21** 1107 (2004)

[9] Frossati G: "A 100 TON 10mK spherical gravitational wave detector". Proc. First International Workshop for an Omnidirectional Gravitational Radiation Observatory, W.F. Velloso, Jr., O.D. Aguiar and N.S. Magalhaes, editors (Singapore, World Scientific, 1997).

[10] Merkowitz S M, Johnson W W "Techniques for detecting gravitational waves with a spherical antenna". *Phys. Rev.* D **56** 7513 (1997)

[11] Merkowitz S M, Johnson W W "Truncated icosahedral gravitational wave antenna". *Phys. Rev. Lett.* **70** 2367 (1993)

[12] Frajuca, C. "Otimização de transdutores de dois modos mecânicos para detectores de ondas gravitacionais." São Paulo. 97 p. Phd Dissertation - Universidade de São Paulo, (1996).

[13] Frajuca C. et al , "Perspectives on transducers for spherical gravitational wave detectors". Proc. 3rd Edoardo Amaldi Conference on Gravitational Waves (Pasadena, USA, July 1999). AIP Conf. Proc. 523 (New York, AIP) p.417 (2000)

[14] Aguiar O D: *et al.* "Status report of the Schenberg gravitational wave antenna". *Journal of Physics:Conference Series* **363** 012003 (2012)

[15] Aguiar O D: *et al.* "The Brazilian gravitational wave detector Mario Schenberg: progress and plans". *Class. Quantum Grav.* **22** 209 (2005)

[16] Frajuca C: *et al.* "Transducers for the Brazilian gravitational wave detector 'Mario Schenberg'". *Class. Quantum Grav.* **19** 1961 (2002)

[17] Magalhaes N S: *et al.* "A geometric method for location of gravitational wave sources". *Astrophysical Journal* **475** 462 (1997)

[18] Magalhaes N S: *et al.* "Determination of astrophysical parameters from the spherical gravitational wave detector data". *Monthly Notices of Royal Astronomical Society* **274** 670 (1995)

[19] Frajuca C, Bortoli F S, Magalhaes NS: "Resonant transducers for spherical gravitational wave detectors". *Brazilian Journal of Physics* **35** 1201 (2005)

[20] Aguiar O D: *et al.* "The Brazilian gravitational wave detector Mario Schenberg: status report". *Class. Quantum Grav.* **23** 239 (2006)

[21] Frajuca C, Bortoli F S, Magalhaes NS: "Studying a new shape for mechanical impedance matchers in Mario Schenberg transducers". *Journal of Physics: Conference Series* **32** 319 (2006)

[22] Aguiar O D: *et al.* 'The gravitational wave detector" Mario Schenberg": status of the project". *Brazilian Journal of Physics* **32** 866 (2002)

[23] Aguiar O D: *et al.* "The Brazilian spherical detector: progress and plans". *Journal of Class. Quantum Grav.* **21** 457 (2004)

[24] Frajuca C: *et al.* : "Study of six mechanical impedance matchers on a spherical gravitational wave detector" *Journal of Physics:Conference Series* **122** 012029 (2008)

[25] Bortoli F S: *et al.*: "A physical criterion for validating the method used to design mechanical impedance matchers for Mario Schenberg's transducers" *Journal of Physics:Conference Series* **228** 012011 (2010)

[26] Magalhaes N S: *et al.* 'Possible resonator configurations for the spherical gravitational wave antenna". *Gen. Rel. Grav.* **29** 1511 (1997)

[27] Andrade L A *et al.*: "Ultra-low phase noise 10 GHz oscillator to pump the parametric transducers of the Mario Schenberg gravitational wave detector". *Class. Quantum. Grav.* **21** 1215 (2004)

[28] Allen B: "The stochastic gravity-wave background: Sources and Detection, in Relativistic Gravitation and Gravitational Radiation" (Les Houches, 1995), p. 373

[29] Maggiore M: "Gravitational wave experiments and early universe cosmology" *Phys.Rep.* **331** 283 (2000)

[30] Tsvi P: "The quantum interaction of macroscopic objects and gravitons" *Int. J. Mod. Phys. D* **25** 1644020 (2016)

[31] Anirban B: "Non-BPS interactions from the type II one loop four graviton amplitude" *Class. Quantum Grav.* **33** 125028 (2016)

[32] Cappozziello S, Corda C, De Laurentis M: "STOCHASTIC BACKGROUND OF RELIC SCALAR GRAVITATIONAL WAVES FROM SCALAR–TENSOR GRAVITY" *Mod. Phys. Lett. A* **22** 2647 (2007)

[33] Cappozziello S, Corda C, De Laurentis M: 'STOCHASTIC BACK-GROUND OF GRAVITATIONAL WAVES "TUNED" BY f(R) GRAVITY" *Mod. Phys. Lett. A* **22** 1097 (2007)

[34] Nowakowski M, Arraut I: "The Fate of a gravitational wave in de Sitter spacetime" *Acta Physica Polonica B* **41** 911 (2010)

[35] Arraut I: "About the Propagation of the Gravitational Waves in an Asymptotically de Sitter Space: Comparing Two Points of View" *Modern Physics Letters A* **28** 1350019 (2013)

[36] Jones P *et al.*: "Particle production in a gravitational wave background" *Phys. Rev. D* **95** 065010 (2017)

[37] Jones P, Singleton D: "Gravitons to photons — Attenuation of gravitational waves" *Int. J. Mod. Phys. D* **24** 1544017 (2015)

# Part IV

# Quantum Aspects of Space and Time

C. Duston, M. Holman (Eds), *Spacetime Physics 1907 - 2017. Selected peer-reviewed papers presented at the First Hermann Minkowski Meeting on the Foundations of Spacetime Physics, 15-18 May 2017, Albena, Bulgaria* (Minkowski Institute Press, Montreal 2019). ISBN 978-1-927763-48-3 (softcover), ISBN 978-1-927763-49-0 (ebook).

# 13 MODELING SPACETIME AS A BRANCHED COVERING SPACE OVER 2-KNOTS

CHRISTOPHER DUSTON

**Abstract**   In this paper we review a proposal to represent the geometric degrees of freedom of the gravitational field as a branched covering space, and introduce a new application of this in which the branch loci are 1- or 2-knots. This allows one to construct arbitrary smooth, closed 3- and 4-manifolds with enough geometric and topological information to write down a partition function and calculate statistical quantities in the thermodynamic limit. Further, we find clear evidence for a dimensional reduction of the spacetime geometry from four to two. As an example, we choose a family of smooth 4-manifolds presented in this way, and calculate the entropy of the system.

## 13.1   Introduction

Differential geometry and topology are fundamentally important to the analysis of spacetime models. Not just because the physical theory which describes spacetime is built upon them, but they are also naturally associated with the physical concept of space. Like so many close connections in theoretical physics and mathematics, interesting features and results from one inform the explorations of the other. A specific example of this is that of the classification problem of smooth 4-manifolds, in which the key issue is exotic smooth structure. Attempts to understand and model exotic smooth structure have lead to new threads of exploration, such as the paradigm of using branched covering spaces and partition functions to study semiclassical gravity that will be utilized in this paper.

What we present here is a particular reparameterization of the gravitational field. In the traditional approach to gravitational theory, the field variable is the metric, which lives on a smooth 4-manifold. Briefly, a 4-manifold is a mathematical structure which provides a background upon which to precisely

C. Duston, M. Holman (Eds), *Spacetime Physics 1907 - 2017. Selected peer-reviewed papers presented at the First Hermann Minkowski Meeting on the Foundations of Spacetime Physics, 15-18 May 2017, Albena, Bulgaria* (Minkowski Institute Press, Montreal 2019). ISBN 978-1-927763-48-3 (softcover), ISBN 978-1-927763-49-0 (ebook).

define the usual operations of calculus, but on a curved surface. In order for the early relativists to postulate that "spacetime is curved", this mathematical structure was needed so that one could define what it would mean for a test mass to move on such a curved surface. To build a manifold, you begin with points, collect these points together into sets, define coordinate systems on those sets, and then define transition functions on the sets which allows motion between them. The manifold becomes smooth (or topological, or piecewise-linear) when the transition functions are required to be infinitely differentiable (or continuous, or piecewise-continuous).

Once this structure is in place, the metric can be defined on the tangent space at points of the spacetime. Of course, the nature of the Einstein equations is such that one does not build the manifold first; the metric and the matter influence each other and are inexorably linked. This *background independence* marks general relativity as unique from all other field theories. If we find a consistent metric and matter configuration, we are implicitly assuming that a manifold could be constructed which furnishes such a metric. Solving the Einstein equations then amounts to solving a set of second-order coupled partial differential equations. Not only is this quite a difficult task except for only the very simplest of manifolds and matter distributions, but the precise nature of the spacetime manifold is generally hidden. Topological characteristics are difficult to detect, and physical results are necessarily restricted to local regions of the spacetime, where the metric has been determined.

The approach followed in this paper will allow us to recover some of this lost information. The backbone of this construction is a branched covering space, and the geometric and topological information will be encoded in 2-dimensional submanifolds of an auxiliary 4-sphere. In essence, we will describe how to construct any closed, smooth model for a 4-dimensional spacetime as a number of 4-spheres sewn together along surfaces.

By doing this, some information will be retained about the spacetime model that is actually being used. Since the topological structure is recorded, it will be known from the outset. In addition, since this can be done generically, calculations which require a "sum over states" can now be formally complete. The only information which is lost in this approach is the physical intuition for the associated spacetime, but there is no fundamental reason why this is true, only that our intuition originates from the standard parameterization. Since any 4-manifold can be presented in this way, one can alternate between the traditional representation, in which the topology is obscured but the field equations solved, and this alternative representation using a branched covering space.

The rest of this work will be presented as follows. We will begin by introducing branched covering spaces in 3- and 4-dimensions (§13.2), and then discuss using 1- and 2-knots as branch loci (§13.3). We will then relate the singularities in the base to those in the cover (§13.4), and discuss how we will relate these singularities to the knot type of the locus (§13.5). Then we will present our two primary results, dimensional reduction, demonstrated with a partition function approach (§13.6), and a specific example of a 3-fold branched cover over a trefoil (§13.7). We will conclude in §13.8 with a summary and

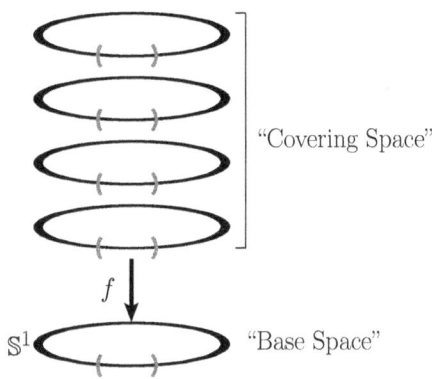

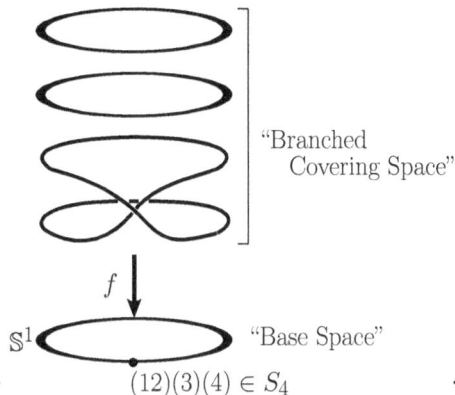

**Figure 13.1:** A 4-fold covering space of the 1-sphere, illustrating that the preimage of every open set in the base is an identical copy of the open set in the cover.

**Figure 13.2:** A 4-fold branched covering space of the 1-sphere. This is a regular covering space everywhere except for a single point, over which there are only three preimages.

outlook for the future.

## 13.2 Branched Covering Spaces as Representations of 3- and 4-Manifolds

The possibility of presenting a spacetime model as a branched covering space has been explored previously by the author and others [26, 12, 13, 14, 15, 16]. In this approach, the base space of a covering space serves as something of a prototype, copies of which are sewn together to create the spacetime model. The sewing is defined by the covering space, which can be described using a representation of a permutation group corresponding to deck transformations. These approaches are generic enough to enable the construction of any closed manifold in 3- or 4-dimensions, which makes them particularly convenient for studying large classes of spacetime models. Depending on the context, statements like "all the closed spacetimes of dimension $d$" with some particular parameters can be made precise.

We begin by carefully defining the mathematical constructions we will be using.

**Definition 1** *A **covering map** $p : M \to B$ is a continuous, surjective map between topological manifolds $M$ and $B$ such that for every open set $U \subset B$ the inverse image $p^{-1}(U)$ can be written as the union of disjoint open sets $V_i \subset M$ such that $p|_{V_i}$ is a homeomorphism of $V_i$ onto $U$. $B$ is called the **base space** and $M$ is called the **covering space**. The number of inverse images $m$ in $p^{-1}(U)$ is called the **order** of the covering.*

An example of a 4-fold covering space of the 1-sphere is shown in figure 13.1. The copies of the base in the cover are referred to the **sheets**, and when these sheets "collide" we get a branched covering space:

**Definition 2** *A map $p : M \to B$ is a **branched covering map** if there is a subset $L \subset B$ such that the restriction of $p$ to $M - p^{-1}(L)$ is a covering map. The set $L$ is called the **branch locus**, the preimage of the branch locus is the **ramification locus**, and $M$ is a **branched covering space**.*

This is illustrated in figure 13.2. It should be noted that although the base is the same, the cover is topologically distinct, being four disjoint copies of the base in the first case as compared to three disjoint copies for the branched. It should also be emphasized that the branched cover spaces are smooth, non-singular 4-manifolds. The singular structure is completely contained in the covering map, which is singular when evaluated on the ramification locus.

Clearly, the exact nature of the branched covering space depends on the base, the branch locus, and how the sheets "collide" over the locus. A convenient (and complete) way to specify this is an element of the permutation group attached to the ramification locus (see figure 13.2). This permutation group provides topological information about how the multiple copies of the base are sewn together to form the cover. Technical details of this construction can be found in the next section.

It turns out that the description of closed manifolds in this manner is complete, at least for dimensions of physical interest. It is a classic result of Alexander[1, 8] that this can be done in any dimension $p$ over $p$-spheres, branched along $(p-2)$ complexes. More recent work has demonstrated that one only needs 3-fold covers in three dimensions [24, 28, 30] to represent any smooth 3-manifold in this way. In dimension four the situation is similar, but depends on the singular character of the branch locus:

**Theorem 1 (Montesinos-Piergallini)** *[29] Any smooth, oriented, closed 4-manifold can be represented as a simple[1] 4-fold covering of the 4-sphere branched over an immersed piecewise linear (PL) surface.*

This result actually applies equally well to embedded surfaces, and a variation of this approach using surfaces instead of knots can be found in [16]. This can also be strengthened by restricting to locally flat PL surfaces at the cost of adding a 5th sheet [25]. However, since our branched loci will actually be non-locally flat, we will only be using the theorem as presented above.

## 13.3  1- and 2-dimensional Knots as the Branch Locus

Clearly, utilizing such representations of closed manifolds as spacetime models will only be productive if there is a straightforward way of keeping track of both

---

[1] *Simple* means that the representations $\rho : G \to \mathcal{S}_n$ discussed in the next section can be chosen to be single transpositions [31].

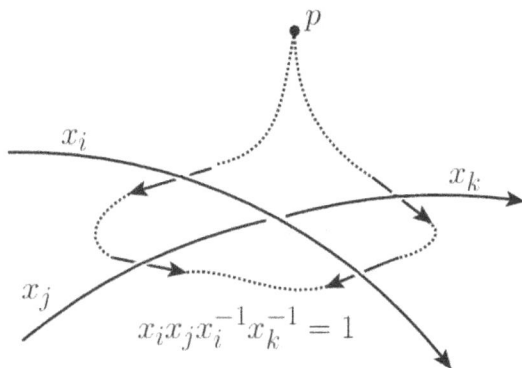

**Figure 13.3:** The definition of a vertex relation for the presentation of a knot group.

the branch locus and the permutation labels presented in §13.2. Fortunately, such techniques are well-known in knot theory, which is concerned with the classical problem of the equivalences of circles $\mathcal{K} \sim \mathbb{S}^1$ embedded in 3-spheres. There are a number of excellent introductions to this field [10, 19, 31, 30], and we will briefly review the key material here.

**Definition 3** *A p-**knot** $\mathcal{K}_p \simeq \mathbb{S}^p$ is an embedding $\mathbb{S}^p \to \mathbb{S}^{p+2}$. A **link** is a disjoint union of knots $\mathcal{K}_p^{(1)} \bigcup \mathcal{K}_p^{(2)} \bigcup ... \bigcup \mathcal{K}_p^{(n)}$. Two knots or links $\mathcal{K}_p$, $\mathcal{K}_p'$ are equivalent if there is a homeomorphism $h : \mathbb{S}^{p+2} \to \mathbb{S}^{p+2}$ such that $h(\mathcal{K}_p') = \mathcal{K}_p$.*

In this paper, we will only be concerned with the classical 1-knots and 2-knots. The equivalence of knots is easiest to (partially) determine by looking at the fundamental group of their complement, $G = \pi_1(\mathbb{S}^{p+2} \setminus \mathcal{K}_p)$, which is called the **knot group** (it is common to refer to both the knot and the image of the knot as $\mathcal{K}$, and we will not break this tradition). Intuitively, the fundamental group is the set of non-contractible paths under the operation of composition, and removing the knot from the sphere produces particular paths which are non-contractible. To determine a presentation of the knot group, a generator $x_i$ is assigned to each arc $i$ in the knot, and a set of relations $r_i$ for each vertex are defined, which correspond to paths traveling around that vertex. For example, see figure 13.3. These generators and relations form the group of a particular knot,

$$G = \pi_1(\mathbb{S}^{p+2} \setminus \mathcal{K}_p) = (x_1, \, x_2, ..., \, x_n | r_1, \, r_2, ..., \, r_m)$$

Once the knot group is found, there are a number of polynomial invariants which can be derived to determine if two knots are equivalent. We will not explore these in detail here, but examples are the Alexander polynomial, the Kauffman polynomial, and the Jones polynomial. What is important to know about these invariants is that they are incomplete - two knots with different sets of polynomial invariants are different, but two knots with equivalent polynomial invariants are not necessarily different.

In addition to the structure of the knot giving insight into the topology of the knot complement, it can also be used to determine the topology of covering spaces of the base sphere when branched over the knot. The essence is the following: as is clear in figure 13.2, we can think of the sheets being labeled with integers, and the "exchanging" of the sheets over the ramification locus to be a permutation label. Technically, to a given knot we assign a representation of the knot group $G$ into the symmetric group of order $n$, which matches the order of the cover:

$$\rho : G \to \mathcal{S}_n.$$

Each arc of the knot has an associated element $g \in G$ in the following way: from any point in $B$ there is a nontrivial loop $fgf^{-1}$, where $f$ is a path from the base to the neighborhood of the arc, and $g$ is the element arbitrarily close to the arc. By associating each of these elements $g_i$ with a cycle in $\mathcal{S}_n$, we can describe what happens to paths which start on some sheet $j$, cross through the region of the preimage of the knot, and end up on sheet $k$. As discussed in [19], the set of admissible representations $\{\rho_i\}$ of a branched covering space can be indexed by the subgroups of $G$.

The fundamental group of the cover will be the pullback of the fundamental group in the base as long as the reconnections of the sheets over the branch points are taken account of. This can be done by applying the elements of the permutation group to the knot relations, as long as few technical details are dealt with. The full algorithm is detailed in [19], and for some examples of using this in practice see [14]. The upshot is that given a knot and an assignment of permutation labels to each arc of the knot, we can reconstruct the branched covering space. In fact, in light of the presentation theorems discussed above, we can do this in full generality to construct any closed 3-manifold branched over 1-knots.

Of course, if we want to study spacetime models we want to be able to do the equivalent construction in dimension 4, and it turns out there are several ways to do that using branched covering spaces. In previous work, this has been done by using algebraic curves of specified degree ([33, 13]), piecewise linear surfaces ([12, 14]), and the Weierstrass representation of surfaces [16] for the branch locus. Here we will represent the surfaces as 2-knots, so we want to extend the construction outlined above to that case.

An easy way to represent 2-knots (also called *knotted spheres*) is to consider the cross-sections of these surfaces. If the surface $\mathcal{K}_2$ is in general position, define the family of hyperplanes $H_t$ for $-\infty < t < \infty$ which intersects the knot for some values $t_a < t < t_b$. Restricted to each hyperplane, the 2-knot will either be an image of a 1-knot or an isolated singular point. A singular point represents either the creation or destruction of a knot, or a reconnection of a knot vertex. Since each hyperplane contains a knot, one can determine the fundamental group of the complement of each level,

$$\pi_1(H_t \setminus \mathcal{K}_1^{(t)}),$$

and then determine the knot group $\pi_1(\mathbb{S}^4 \setminus \mathcal{K}_2)$ of the van Kampen theorem, which essentially amounts to identifying reconnections. This entire process

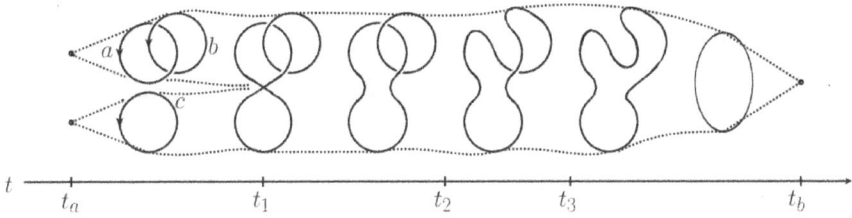

**Figure 13.4**

is presented in more detail in [19]. Here we will just present an illustrative example, shown in figure 13.4.

In the figure is shown a 2-knot with 1-knot (actually 1-link) cross-sections. In the hyperspace at $H_{t_a}$, there are two singularities, one of which is due to an unknot, and another which is due to a Hopf link (we will need to say a little bit more about the nature of this second singularity shortly). In this region, the fundamental group of knot complement in the hyperspace $H_t$, $t_a < t < t_1$ is given by

$$\pi_1(H_t \setminus \mathcal{K}_1^{(t)}) = (a, b, c | a = b^{-1}ab).$$

When we reach $t_1$, two arcs of the link reconnect forming a singular point. Between $t < t_1$ and $t > t_1$, the generator $a$ is being identified with the generator $c$, and the fundamental group becomes $(a, b | a = b^{-1}ab)$. Upon crossing $t_2$, $a$ gets identified with $b$, and we get the final fundamental group of the 2-knot complement,

$$\pi_1(\mathbb{S}^4 \setminus \mathcal{K}_2) = (a) \simeq \mathbb{Z}.$$

It is easy to see that in our example, there is something different about the tips of the cone over the unknot as compared to the tip over the Hopf link. In the case of the unknot, every point on the cone has a region around it which looks like a disk (even the tip itself). This fails on the tip of the cone above the rings; as you take smaller and smaller neighborhoods around the tip, you still get ring cross-sections rather than disks. This is a case of the violation of local flatness:

**Definition 4** *For a submanifold $N^n$ of $M^m$, a point $p \in N$ is **locally flat** if it has closed neighborhood $U \in M$ such that $(U, U \cap N)$ is homeomorphic to the disks $(D^m, D^n)$.*

This definition is from [31]. We can see that this condition is always satisfied for 1-knots; for every point of the knot, a neighborhood in $\mathbb{S}^3$ will be $U = D^3$, while $U \cap \mathcal{K}_1$ will be $D^1$. In the case of our example 2-knot in figure 13.4, aside from the point at the tip of the cone over the ring, every point in $\mathcal{K}_2$ has a neighborhood homeomorphic to $D^2$. This point at the tip of the cone is a **non-locally flat point**, and will be a key feature of our later analysis.

Before moving on to the calculation of the curvature of these manifolds, we would like to point out that modeling spacetime as a branched covering space has been done previously in at least one notable case. It is natural to incorporate it into loop quantum gravity (LQG), because the spin networks used to track the geometry are already 1-dimensional complexes (graphs), and many of the techniques discussed here can be carried over. In [12], they show that for both the spin network and spin foam case, topological labels can be added (creating topspin networks and topspin foams) without modifying the graphs, which allows for the tracking of both geometry and topology in LQG. In [14], it is demonstrated that the algebra of LQG can be altered slightly to include this modification, and some example calculations are presented in [17].

## 13.4 The Curvatures and Singularities of the Branched Cover

Thus far we have shown how to construct 3- and 4-dimensional smooth manifolds as branched covers, with spheres for bases, branched over 1- and 2-knots. These manifolds can serve as spacetime models, and are generic. They are also conceptually simple - they are just copies of the base, sewn together in a way which is given by a knot, labeled with permutation indices. Geometrically they are spheres, with topology being recorded by a set of finite data packaged together in a convenient way.

Our concern here is to calculate the curvatures of the branched covers, so that we can use them to construct partition functions in the next section. Since these are just copies of 4-spheres, we know what the curvature will be everywhere but near the non-locally flat points and the branch locus in the covers. We focus first on the non-locally flat points.

In the base sphere, these regions are cones, and can be assigned a metric of the form

$$ds^2 = dr^2 + r^2 d\phi^2 + (\gamma_{ij}(\theta) + r^2 h_{ij}(\theta))d\theta^i d\theta^j, \qquad (13.1)$$

where the $(r, \phi)$ coordinates are in the cone and $i, j, k, \ldots$ indices denote the coordinates $\theta^1$ and $\theta^2$, which are normal to the cone. Following [20], we will call the singular set $\Sigma$, which will be the $(\theta^1, \theta^2)$ plane that contains the singular point at $r = 0$. The angular coordinate has a range $(0, 2\pi\alpha)$, and it is conventional to speak of the angular deficit $2\pi(1 - \alpha)$.

In [20], the scalar curvature $^{(\alpha)}R$ of the singular manifold $M_\alpha$ is calculated by "rolling off" the singular surface to a regular manifold, calculating the curvature, and removing the regulator. The result is

$$\int_{M_\alpha} {}^{(\alpha)}R = 4\pi(1 - \alpha)A_\Sigma + \int_{M_\alpha \backslash \Sigma} R, \qquad (13.2)$$

where $R$ is now the scalar curvature of the manifold with the singular point removed, and $A_\Sigma$ is the area of the singular set, $\int \delta(\Sigma)$. What this means for us is that since the scalar curvature of the 4-sphere is constant, $\alpha$ and $\Sigma$ are our free parameters. For now we will move on to discussing the curvature near the branch locus, and we will discuss the angle deficit further in the next section.

Calling the neighborhood of the 2-knot $N(\mathcal{K}_2)$, we can write the integral of the scalar curvature as

$$\int_{\mathbb{S}^4} R = \int_{\mathbb{S}^4 \setminus N(\mathcal{K}_2)} R + \int_{N(\mathcal{K}_2)} R.$$

This expression will need to be pulled back to the covers through $p : M \to \mathbb{S}^4$. For a covering space of order $n$, the first term will simply give $n$ copies of the integral of the complement of a neighborhood of the knot. In the second term, on the singular surface $\Sigma$ there will be zero contribution to the integral because it is codimension 2. It's also easy to see that just near the branch locus, integrals of constant curvature will simply be additive in the covering space (see [13] for example). If there are $m$ singularities (non-locally flat points) in the 2-knot there will be $r$ singular surfaces $\Sigma_i$ in the cover with angle deficits $\alpha_i$, for $m \leq r \leq rn$. This $r$ is not unknown, but could be determined with an analysis of the permutation labels. In the end, the integral of the action on the covers will be

$$\int_M R = n \int_{\mathbb{S}^4} R + 4\pi \sum_i (1 - \alpha_i) A_{\Sigma_i} \qquad (13.3)$$

## 13.5 Singularity Type of the Knot and Angle Deficit

The angular deficit, parameterized by $\alpha$, is a measure of "how conical" the spacetime around the cone is. There are several interesting features of geometric structures of this type. For instance, since any surface can be taken as a flat surface with singularities [38], we can associate those singularities with the conical ones and consider these surfaces to be flat. This is part of the origin of the equation (13.2) above.

In addition, conical metrics are used to describe one-dimensional topological structures in physical models of spacetime in the form of cosmic strings, with the tension related to the angular deficit [37]. These are topological defects in the early universe which have not been observed but could contribute to a variety of phenomena, such as structure formation and inflation. The conical geometries have been used in previous studies of exotic smooth structure and topology of the gravitational field [13, 16, 18].

Here we will expand on the previous work by utilizing the 2-knot construction explicitly to determine the angular deficit of the comic string. The key feature which allows us to do this is the inclusion of the non-locally flat points discussed in the previous section. If the 2-knot is locally flat, the metric is flat everywhere and there is no angle deficit. The presence of the non-locally flat point indicates the presence of an angle deficit, the size of which is directly related to the characteristics of the knot cross-section in the region around the conical point.

We will now restrict ourselves to a particular kind of knot, *torus knots*. This is a knot (or link) which lies on a torus in $\mathbb{S}^3$, wrapped around the latitude of the torus $p$ times and the longitude $q$ times. We call these $(p, q)$-torus knots. If $p$ and $q$ are not coprime, the knot forms completely disconnected parts and is actually a link. Torus knots can be described with a 1-parameter action on complex coordinates [27, 35]:

$$t \cdot (z_1, z_2) = (t^{1/p} z_1, t^{1/q} z_2),$$

for a $(p, q)$ torus knot, and the coordinates $(z_1, z_2)$ represent the latitude and longitude of the torus. This set of points is described by the polynomial

$$f : \mathbb{S}^3 \to \mathbb{C}, \qquad f(z_1, z_2) = z_1^p + z_2^q$$

when $f(z_1, z_2) = 0$. The tangent spaces can be described with the derivative,

$$Df : T\mathbb{S}^3 \to T\mathbb{C}, \qquad Df(z_1, z_2) = (pz_1^{p-1}, qz_2^{q-1}).$$

The inverse image of $f$ is a codimension 1 manifold everywhere except at $(z_1, z_2) = 0$, since $Df = 0$ vanishes there. So the origin is a singularity of the submanifold described by points which satisfy

$$z_1^p + z_2^q = 0. \tag{13.4}$$

We can develop a metric near these points in the following way. Begin with a flat metric on $\mathbb{R}^4$,

$$ds^2 = dx_1^2 + dy_1^2 + dx_2^2 + dy_2^2,$$

and pick two sets of polar coordinates with periodicity $p$ and $q$:

$$x_1 = r_1^p \cos(p\phi_1), \qquad\qquad y_1 = r_1^p \sin(p\phi_1)$$
$$x_2 = r_2^q \cos(q\phi_2), \qquad\qquad y_2 = r_2^q \sin(q\phi_2)$$

Notice that this choice satisfies the defining equation (13.4) with $z_1 = x_1 + iy_1$ and $z_2 = x_2 + iy_2$. Pulling these back to the flat metric and enforcing the conditions $r_1^p = r_2^q$, $p\phi_1 = q\phi_2$ and then rescaling $r = r_1\sqrt{2}$ and $\theta = (q/p)\phi_2$ we find

$$ds^2 = dr^2 + \left(\frac{q}{p}\right)^2 r^2 d\theta^2.$$

This is a conical metric with $\alpha = q/p$.

We will pause here to emphasize what we have done. Conical metrics of the form (13.1) are well-studied, and used to represent topological defects with angle deficit $2\pi(1 - \alpha)$. $\alpha$ can be related to the string tension if the metric is modeling a string, but it is otherwise an unknown topological parameter of a 2-dimensional manifold. Here, we are using a fundamental feature of our particular spacetime model (torus knots which are cross-sections of the ramification locus) to characterize them, which gives us information about the topology of the spacetime. The topology is no longer a "free parameter", or something which cannot be kept track of, as is most common when studying solely local geometry.

One would naturally ask if this construction is specific to torus knots. Is it possible to generally write down a polynomial $f = \sum_{ij} z_1^{a_i} z_2^{a_j} = 0$ which represents *any* knot? Recent work has shown that this is indeed the case [5], but the polynomials are complicated enough that we are unlikely to be able to find a simple set of coordinates like we could for the torus knots.

Finally, we must mention that there is a strict balance between the ability to make any surface flat with singularities and the angle deficit of those singularities. This comes from the Gauss-Bonnet theorem, which in the flat case is

$$2\pi\chi(\Sigma) + \sum_i (\theta_i - 2\pi) = 0, \tag{13.5}$$

where $\chi(\Sigma)$ is the Euler characteristic of each surface containing non-locally flat points with angle deficits $\theta_i$ [20]. Since the angle deficits can be written $\theta_i = 2\pi(1 - \alpha_i)$, this condition reduces to

$$\chi(\Sigma) = \sum_i \alpha_i.$$

In other words, although we can always choose the knots to be locally flat, we cannot freely choose what the angle deficits will be; specifically, their sum has to be an integer. This could have already been a problem in our illustrative example (figure 13.4), but the Hopf link is a $(2,2)$-torus link, with $\alpha = 1$. Therefore, as long as the other two singular points are unknots in a region arbitrarily close to them, this is a surface with $\chi(\Sigma) = 1$, *i.e.* a projective plane.

# 13.6 Partition Functions and Dimensional Reduction

The formulation of the general theory of relativity in the language of thermodynamics, specifically with a partition function as a generating functional, is a useful approach when looking for new or interesting behaviors from an alternative representation of the gravitational field. Since it does not require a specific "belief structure" in regards to the quantum nature of the gravitational field (loops, strings, noncommutativity, *etc.*), it is a very flexible tool which allows one to study the transition region between classical and quantum gravity. It really only requires a belief that the transition between the classical and quantum gravitational field resembles that found in the other quantum fields, which is almost automatically true if one accepts the effective, Wilsonian viewpoint. Alternatively, since our experimental sophistication does not actually reach the quantum gravitational transition, these approaches can also be viewed as toy models for studying the abstract behavior of spacetime. For background into these approaches see [23, 22].

The partition function in our case is heuristically

$$Z = \int_{geometries} [dg] e^{-\frac{1}{\hbar} S_{EH}[g]},$$

where $g$ is the metric corresponding to a particular smooth manifold-metric pair $(M, g)$, and the action is the Einstein-Hilbert one,

$$S_{EH} = -\kappa \int_M (R - 2\Lambda)\sqrt{g}d^4x,$$

where $\kappa = c^3/(8\pi G)$ for the case of Einstein-Hilbert gravity. This is the Euclidean action, which can be related to the Lorentzian one through a Wick rotation $t = -i\tau$.

A major problem with this definition is naturally the "integral over geometries", and is a partial motivation for our particular choice of branched covers and knots as representations of the spacetime model (*e.g.* the geometry). The traditional viewpoint is that, at least semiclassically, this integral is a sum over all gauge-inequivalent metrics. In the case of general relativity, the gauge symmetry is encoded by diffeomorphisms, so gauge-equivalence corresponds to metrics which are equivalent under coordinate changes (sometimes called passive diffeomorphisms [36], although we do not favor this categorization). However, the action is invariant under both passive and active diffeomorphisms (that is, all diffeomorphisms $f : M \to N$ between two smooth 4-manifolds), so the sum needs to be over all inequivalent pairs of complete geometries $(M, g)$. This complete geometry must specify both the topological and the smooth structure of the manifold.

A complete determination of the diffeomorphism group $Diff(M)$ for a particular manifold is difficult in general, and is an outstanding problem in dimension 4[2]. A particular feature of this unique situation is the presence of *exotic smooth structure* in dimension four. These are manifolds which are topologically equivalent but nondiffeomorphic, and represent gauge-inequivalent solutions to the Einstein field equations (for more details on exotic smooth structure see [34, 2]). The most dramatic example of this is that there are infinitely many exotic smooth structures on $\mathbb{R}^4$ [21]. The implications of this startling fact are far reaching, and have been studied in the context of dark matter models [7], quantum gravity [13], string theory [3], and inflation [4].

For our purpose, we are going to solve this "classification problem" by using the branched covers and knot presentations to completely enumerate the smooth 4-manifolds. The fact that this works is simply due to the Montesinos-Piergallini theorem; any four-manifold can be presented by a 2-knot $\mathcal{K}_2$ and a covering space $p : M \to \mathbb{S}^4$, which itself is given by a particular representation of the fundamental group of the knot, $\rho_j : \pi_1(\mathbb{S}^4 \setminus \mathcal{K}_2) \to \mathcal{S}_n$, indexed by the number of subgroups of $\pi_1(\mathbb{S}^4 \setminus \mathcal{K}_2)$. So our geometries are collections

---

[2]Two particular manifestations of this: 1) there are simply-connected 4-manifolds with diffeomorphism groups that have infinitely many connected components [32], and 2) there are compact 4-manifolds whose diffeomorphisms group is not Jordan [11].

$(K_2^{(i)}, \rho_j, g) \equiv (i, j, g)$, and the partition function is now presented as

$$Z = \sum_{(i,j,g)} e^{-\frac{1}{\hbar} S_{EH}[g]}, \qquad (13.6)$$

and is formally complete. Of course, this is not everything we could ask for, since the classification problem is not solved; this is just a reparameterization of our ignorance. Rather than making this a sum over all elements of the complicated $Diff(M)$, we are summing over all of the well-defined 2-knots $K_2^{(i)}$, representations $\rho_j$, and metrics $g$. In addition, we are provided with a partial method to determine when a pair $(K_2, \rho, g)$, $(K_2', \rho', g')$ are equivalent using standard techniques in knot theory.

Most importantly, by creating an explicit enumeration we can formally say that such a partition function is complete, whereas previously exotic smooth structures are "missed" because they are generally presented in rather abstract ways, almost always without a metric (for several examples, see Chapter 9 of [21]). Using this approach, we are able to study particular classes of 4-manifolds without having to "worry" about exotic smooth structure. Of course, this is a little bit of a linguistic trick; the partition function (13.6) includes smooth manifolds $M'$ which might normally be classified as "exotic" with respect to a homeomorphic manifold $M$. The point is that rather than being unsure about the existence of $M'$ (because that would require checking all the possible diffeomorphisms of $M$), we know that if $M'$ can be presented as a member of a particular class of $(i, j, g)$, we have included it. And of course, we know that $M'$ can definitely be presented as *some* $(i, j, g)$.

Using the integral of the scalar curvature (13.3), our partition function is now

$$Z = \sum_{(i,j,g)} \exp\left[ \frac{\kappa}{\hbar} \left( n \int_{\mathbb{S}^4} (R - 2\Lambda) + 4\pi \sum_i (1 - \alpha_i) A_{\Sigma_i} \right) \right]. \qquad (13.7)$$

We will proceed to an extended example of this shortly, but we would first like discuss the dimensional reduction implied by the expression. Since the curvature over the 4-sphere is constant, $R = 12/r^2$, all the field dynamics are found in the second term, which is simply an area integral,

$$\int d\zeta_1 d\zeta_2 \sqrt{dg^{(i)}}$$

for a 2-dimensional metric $g^{(i)}$ on each of the singular surfaces $\Sigma_i$. Recall, these surfaces are normal to the 2-knots, and additionally the gravitational coupling constant for these surfaces is topological, $\gamma_i = \frac{4\kappa}{\hbar}\pi(1 - \alpha_i)$. They have no classical dynamics, but thinking of $g$ as a free quantum field we could construct generating functions and calculate transition probabilities between topological states represented as different branched covering spaces.

So rather then having to deal with general sets of smooth 4-manifolds (with both topological and geometrical data), we have reduced the problem to generic 2-manifolds with singular points and finite presentations on $\mathcal{S}_n$. This drastically simplifies the problem, since we are now out of the wild realm of 4-dimensional geometry and into the well-understood dimension 2. This dimensional reduction from four to two is commonly found when studying the quantum nature of the gravitational field (for a recent review, see [9]), and the above partition function is a new, independent example of this behavior.

For a more concrete realization of the dynamics implied by this action which emphasizes the dimensional reduction, we could rewrite the action on the surface by the standard Nambu-Goto string action,

$$\sum_i \frac{T_i}{c} \int d\sigma d\tau \sqrt{(\dot{x}x')^2 - (\dot{x})^2(x')^2},$$

where $T_i$ is the "string tension" $T_i = 4\pi(1 - \alpha_i)c$. Note that this does not need to be a physical string - it may simply represent a topological deficit, indistinguishable from a string except at very short distances. In this way, we can finally assign some physical characteristics to this model. The spatial section is topologically $\mathcal{K}_2 \times \mathbb{S}^1$ (or $\mathcal{K}_2 \times I$ for an open string), and evolves along the proper time $\tau$ because of the form of the string action. The string worldsheet is dual to the 2-knot, and we have corresponding $D$-branes, with $D = 0, 1$, or 2. We can even describe the interactions of strings using our 2-knot cross-sections with non-locally flat points. This can happen in two distinct ways, depending on the physical restrictions one considers. The topological condition coming from the Gauss-Bonnet theorem (13.5) is required in all cases, and these could be considered "topological strings". However, physical strings must conserve energy as well, and these could be considered "geometric strings".

An example of a topological string can be seen in figure 13.5. In this case, we are representing a split string, with surfaces chosen to be tori with $\chi = 0$. Of course, these types of strings are not usually considered in most approaches to either quantum or classical strings, because the strings are no longer submanifolds.

For geometric strings, we enforce energy conservation as well, which depends on the length of the string, $E_i = L_iT_i$. So when a string splits, the tension must change to conserve energy [39]. In our model, this corresponds to the conical deficit changing during the interaction. For an example of this, see figure 13.6. If the initial string length is $L$, and the final string lengths are $fL$ and $(1 - f)L$, we have a condition on the angle deficits coming from the condition of energy conservation,

$$(1 - \alpha_i)L = (1 - \alpha_1)fL + (1 - \alpha_2)(1 - f)L \to f = \frac{\alpha_2 - \alpha_i}{\alpha_2 - \alpha_1}.$$

Since the angle deficit is not changing along the length of the string, each of these surfaces must be tori as well, with $\chi = 0$. For example, picking $\alpha_i = 1/2$, $f = 1/2$, we have $\alpha_1 = 1/3$ and $\alpha_2 = 2/3$ as one possibility.

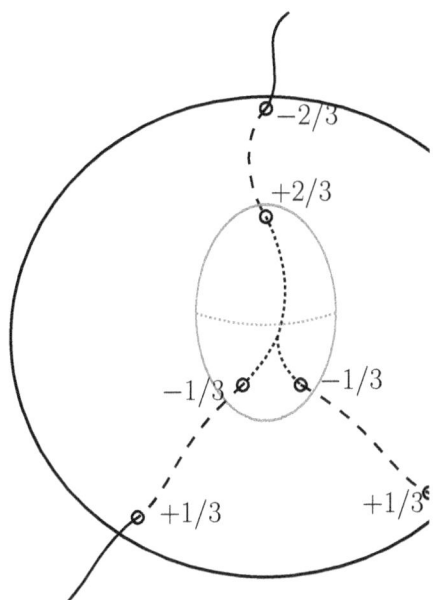

**Figure 13.5:** The splitting of a "topo-logical" string.

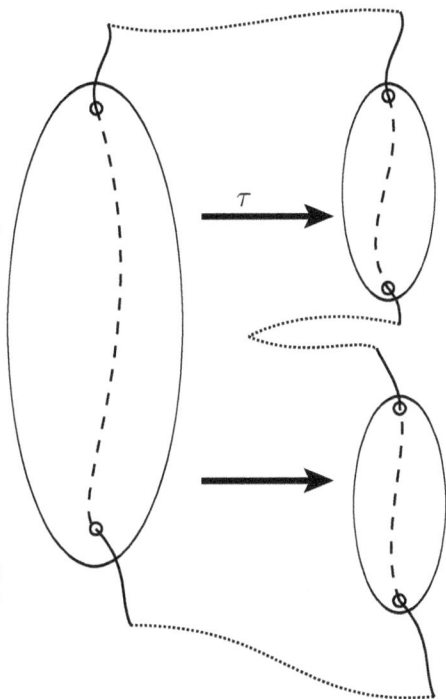

**Figure 13.6:** The splitting of a "geo-metric" string worldsheet, in which energy is conserved. Proper time travels to the right in this figure.

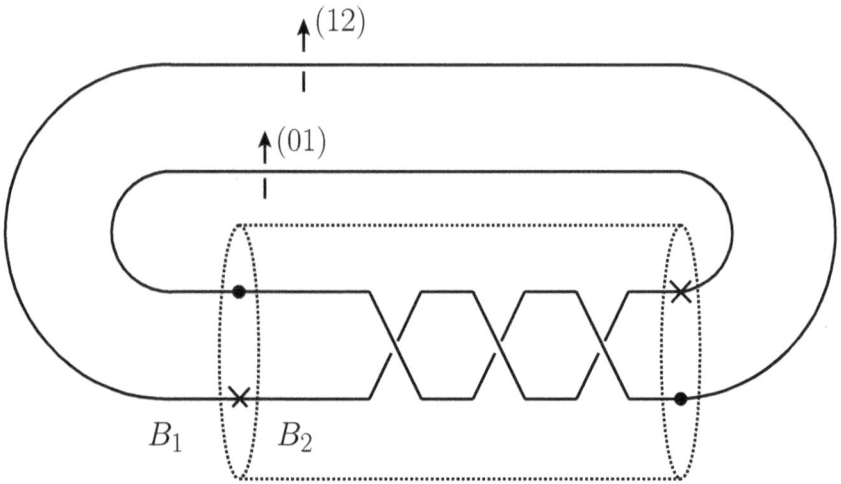

**Figure 13.7:** The trefoil in the base, with representation of the symmetric group

## 13.7 Example: The Preimage of a Trefoil in a 3-fold Branched Cover

In this final section we will present a specific example of a set of model space-times parameterized with branched covering spaces and permutation groups discussed in this paper. We will arrive at a well-defined toy model which we will then use to calculate the thermodynamic entropy of the system. We will conclude this section with a few words emphasizing that while this is an abstract example, it stands in sharp contrast to what would be possible in the standard parameterization of the gravitational degrees of freedom.

The key piece of information we need is the preimages of the 1-knots $\{\mathcal{K}_1^{(t)}\}$ in the hyperplane cross-sections of a chosen 2-knot in the base, because those will represent the singularities found on the surface $p^{-1}(\Sigma)$. For concreteness, we will be fixing the order of the cover to be 3. The inspiration for this example is the following classic example from Rolfsen [31]: *The 3-fold branched covering of $\mathbb{S}^3$ with ramification locus a (right-handed) trefoil is homeomorphic to $\mathbb{S}^3$. The ramification locus is the (2,4) torus link.*

### The Rolfsen Example

First we will recall the original construction. Represent the trefoil $\mathcal{K}$ as a braid, shown in Figure 13.7, with two generators, $a$ and $b$. The covering space will be specified with the following representation of the symmetric group $\mathcal{S}_3$,

$$\rho(a) = (01), \qquad \rho(b) = (12).$$

The base $\mathbb{S}^3$ can be split into the union of 3-balls, $B_1 \cup B_2$, and using the branched cover map $f : N^3 \to \mathbb{S}^3$ we can write the cover as $N^3 = f^{-1}(B_1) \cup$

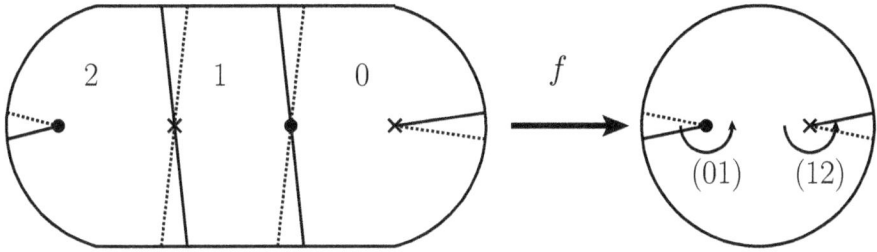

**Figure 13.8:** The branched cover over a slice through the ball $B_2$

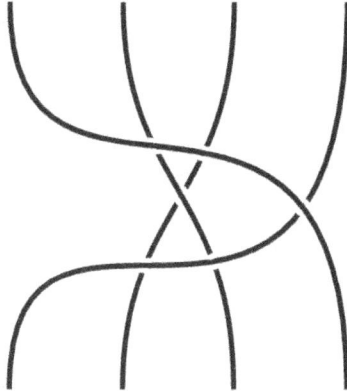

**Figure 13.9:** The braid in the covering space after a $3\pi$ twist in the base.

$f^{-1}(B_2)$. The splitting of the base is indicated in the figure - the braiding is located in $B_2$. The preimage of $\mathcal{K}|_{B_1}$ is simply three copies of two arcs. We can determine the preimage of $\mathcal{K}|_{B_2}$ by tracking what happens to the arcs as the braid is twisted in the base.

Representing $B_2 = D^2 \times I$, we know that the preimage of $B_2$ will be three copies of $D^2 \times I$ sewed together in a particular way (Figure 13.8). The arrangement of the three copies of the open disk is apparent by considering how they are sewn together by the permutations. Viewing this as $B_2 \times 0$, we want to know what happens to the preimages of each arc as we travel to $B_2 \times 1$. The braid representation of the trefoil indicates that this is a $3\pi$ twist. Based on these exchanges, we can draw a picture of the braid shown in figure 13.9. This braiding is determined by simply tracking each arc in the cover; for example, the twist is clockwise, so the cross goes under the dot (passing from heavy line to dotted line) three times, and moves from the rightmost arc to the leftmost arc in the cover. The result is figure 13.9, which is a (2,4) torus link and can be checked using the braid generators, for instance.

This will be the starting point for our construction. We first want to determine the preimages of the trefoil under *all* possible 3-fold branched covers, then construct a set of model 4-manifolds branched over a particular 2-knot which has only trefoil or unknot singularities.

Before going any further, we should point out that a key feature of the previous example is that the branched cover had a trivial fundamental group, and therefore the class of any particular knot can be well-defined there. If the branched cover $p : N^3 \to \mathbb{S}^3$ has $\pi_1(N) \neq 0$, we would not be able to unambiguously define what the knot class is. We would have to contend with questions like "how do arcs of the knot wrap around paths $\gamma \in \pi_1(N)$?". We could resolve this issue by simply restricting our attention to cases where $\pi_1(N) = 0$, which can be checked by applying a version of the Reidemeister-Schreier due to Fox[19] (we will refer to this as "Fox's algorithm"). However, the manifolds $N^3$ will be embedded in the spacetime model $M^4$, so we actually need to be concerned with the topological structure of $M^4$ (because elements of $\pi_1(N)$ could be contractible in $M$). The actual topological conditions are that the preimages of the knot $\mathcal{K}$ in $M$ must be embedded in a topologically trivial sphere $\mathbb{S}^3 \subset M$. However, this will all be subsumed by simply requiring that the knot group of the branched covers is trivial, which we will need to check for each case.

## Admissible Covers

A particular representation $\rho : G \to S_3$ is only admissible if the images of the group elements in $S_3$ satisfy the relations of the knot presentation, in this case the trefoil:

$$G = (a, b | aba = bab).$$

First, we can count the number of possible covers because they will be given by pairs

$$(\rho_1(a), \rho_2(b)) = (\rho_1, \rho_2) \in S_3 \times S_3,$$

for 36 possible cases. We will break these into categories, and for each category we will check to see if the covers described by these representations are admissible.

- Cat 1: Transpositions which share a single sheet, as in Rolfsen's original example. These are of the form $(np) \times (nq), q \neq p$, and there are 6 of them. These covers are all admissible:

$$(np)(nq)(np) = (nq)(np)(nq) \to (n)(pq) = (n)(pq).$$

- Cat 2: One of the elements is the identity, $(1) \times \rho$. There are twelve of these, and in order to be admissible they must satisfy

$$(1)\rho(1) = \rho(1)\rho \to \rho = \rho^2.$$

So only the case $\rho = (1)$ is admissible; call it Cat 2a.

- Cat 3: Pairs of identical transpositions, $(pq) \times (pq)$ or $(pqm) \times (pqm)$. There are six of these, and they are all admissible.

- Cat 4: A 3-cycle and a 2-cycle. We have to work out each of these cases, but none of them are admissible.

$$(012)(01)(012) = (01)(2) \neq (01)(012)(10) = (021),$$

$$(012)(02)(012) = (02)(1) \neq (02)(012)(02) = (021),$$

$$(012)(12)(012) = (0)(12) \neq (12)(012)(12) = (021),$$

$$(021)(01)(021) = (01)(2) \neq (01)(021)(01) = (012),$$

$$(021)(02)(021) = (02)(1) \neq (02)(021)(02) = (012),$$

$$(021)(12)(021) = (0)(12) \neq (12)(021)(12) = (012).$$

Taking into account the order of the permutations, this is 12 cases.

- Cat 5: Pairs of 3-cycles, of which there are two, neither of which are admissible. For example,

$$(012)(021)(012) = (012) \neq (021)(012)(021) = (021)$$

**Checking $\pi_1(M \setminus p^{-1}(\mathcal{K}_2)) = 0$**

Now that we know which of these covers are admissible (Cat 1, Cat 2a, and Cat 3), we will check that the branched covering space corresponding to each does indeed have a trivial fundamental group. At this stage we have to specify a particular 2-knot, and we will pick the simplest possible example, with 1 singular point represented by a trefoil. The cross sections of this knot are shown in Figure (13.10). However, following the discussion in §13.5, we must ensure that the Gauss-Bonnet theorem is satisfied. Fortunately, torus knots have the property that $(p, q) \simeq (q, p)$, so the angle deficits satisfy a kind of duality transformation, $\alpha \leftrightarrow 1/\alpha$. So our trefoil knot will have $+2/3$, and the two knots at the bottom can be $(1, 3)$ tori, with angular deficits $-1/3$ each (respecting the orientation), making this 2-knot a torus. Note that although there appears to be a significant amount of freedom here, this is the only choice for a trefoil and two unknots consistent with Gauss-Bonnet. Naturally, the trefoil could split into an arbitrary number $m$ of unknots represented as torus knots $(1, n_i)$ as long as

$$\frac{2}{3} + \sum_i \left( \pm \frac{1}{n_i} \right) = 0,$$

counting orientations. A presentation for the fundamental group of the complement of this 2-knot in the base can be found by setting $a = b$ in the trefoil presentation, so $\pi_1(\mathbb{S}^4 \setminus \mathcal{K}_2) = \mathbb{Z}$ (details on using the van Kampen theorem in this situation can be found in [19] and [6]).

The essence of Fox's algorithm ([19] and also see [14]) is a homomorphism $\phi$ between a presentation of the fundamental group of the knot complement in the base to the fundamental group of the $n$-fold cover times a free group:

$$\phi : \pi_1(\mathbb{S}^3 \setminus \mathcal{K}) \to H * F_{n-1}.$$

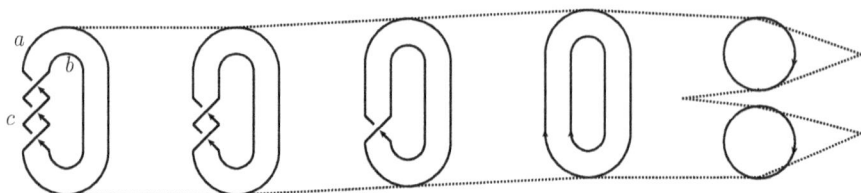

**Figure 13.10:** The particular branch set, represented with knot cross-sections.

For a word $u = x_{j_1}^{\epsilon_1} x_{j_2}^{\epsilon_2}...$, we have a set of words associated to each sheet $0, 1, ..., n-1$ of the cover,

$$u_\alpha = \phi_\alpha(u) = x_{j_1 \alpha_1}^{\epsilon_1} x_{j_2 \alpha_2}^{\epsilon_2}...,$$

where $\alpha_k = \rho_k(\alpha)$ and the permutation elements $\rho_k$ are given by the following rule:

$$\epsilon_k = \begin{cases} +1, & \rho_k = \rho(x_{j_{k-1}}^{\epsilon_{k-1}})...\rho(x_1^{\epsilon_1}) \\ -1, & \rho_k = \rho(x_{j_k}^{\epsilon_k})...\rho(x_1^{\epsilon_1}) \end{cases}$$

Here the $\rho(x)$ are the topological labels on each edge. Acting on each generator and relation in $\pi_1(\mathbb{S}^3 \setminus \mathcal{K})$, the fundamental group of the cover can be recovered from $H * F_{n-1}$ by

- Removing the free group by setting elements of a chosen Schreier tree to zero, and

- Ensuring curves that represent the preimage of the branch locus are trivial. For such an element $v$ of the branch locus represented by $\rho(v) = (\beta_1...\beta_\lambda)(...)...$ (in cycle notation) we adjoin the relations

$$\phi_{\beta_1}(v)...\phi_{\beta_\lambda}(v) = 1, ...$$

To illustrate this calculation, let us expand the original Rolfsen example and determine the knot group in the cover. In other words we will determine what the fundamental group is over a 3-fold cover of the trefoil given by the representation $\rho(a) = (01)(2)$ and $\rho(b) = (0)(12)$. Under the homomorphism $\phi$, the presentation becomes

$$(a_0, a_1, a_2, b_0, b_1, b_2 | a_0 b_1 a_2 = b_0 a_0 b_1, \ a_1 b_0 a_1 = b_1 a_2 b_1, \ a_2 b_1 a_0 = b_2 a_1 b_2).$$

A Schreier tree is given by

$$w_0 = 1, \qquad w_1 = a, \qquad w_2 = a^2,$$

which under the homomorphism gives

$$w_{00} = 1, \qquad w_{10} = a_0, \qquad w_{20} = a_0 a_1.$$

So to remove the free group we set $a_0 = 1$ and $a_1 = 1$, which makes the presentation

$$(a_2, b_0, b_1, b_2 | b_1 a_2 = b_0 b_1, \ b_0 = b_1 a_2 b_1, \ a_2 b_1 = b_2 b_2).$$

To trivialize the branch locus, we set

$$a_0 a_1 = 1, \qquad a_2 = 1, \qquad b_0 = 1, \qquad b_1 b_2 = 1.$$

The presentation is then

$$(b_1, b_2 | 1 = b_1^2, \ b_1 = b_2^2, \ b_1 b_2 = 1).$$

The only way this set of relations is consistent is if $b_1 = b_2 = 1$, and the fundamental group of the complement in the cover is trivial.

Now we need to extend this to our example, with the trefoil being a cross section of the knot in figure 13.10. In the base, the knot group will be the same as the trefoil, $(a, b | aba = bab)$, but with the additional relation $a = b$. This eliminates Cat 1 from our considerations, because that requires $\rho(a) \neq \rho(b)$.

For Cat 2a, $\rho(a) = \rho(b) = (0)(1)(2)$, and the result is a disconnected space with topology $\mathbb{S}^4 \cup \mathbb{S}^4 \cup \mathbb{S}^4$. So this space is not simply connected, but since it's completely disconnected, there will be a copy of $p^{-1}(\mathcal{K}_2)$ in each component, and the knot type is well-defined in each of them.

In Cat 3, we have 2 prototype cases to cover, $(01) \times (01)$ and $(012) \times (012)$. In the first type, one sheet is always left untouched, so it will also be disconnected, with topology $N' \cup \mathbb{S}^4$. Under the homomorphism, the image of the presentation will be

$$(a_0, a_1, a_2, b_0, b_1, b_2 | a_0 b_1 a_0 = b_0 a_1 b_0, \ a_1 b_0 a_1 = b_1 a_0 b_1,$$
$$a_2 b_2 a_2 = b_2 a_2 b_2, \ a_0 = b_0, \ a_1 = b_1, \ a_2 = b_2).$$

The third relation is the disconnected piece, but with $a_2 = b_2$ we have just another copy of the base. Using the same tree as above and eliminating the $b$s, we are left with a single generator $a_2$ and no relations. Trivializing the branch locus tells us $a_2 = 1$, so that case is simply connected (although this actually means $\mathbb{S}^4 \cup \mathbb{S}^4$).

Finally, we have the 3-cycle prototype in Cat 3. The homomorphism gives us

$$(a_0, a_1, a_2, b_0, b_1, b_2 | a_0 b_1 a_2 = b_0 a_1 b_2, \ a_1 b_2 a_0 = b_1 a_2 b_0,$$
$$a_2 b_0 a_1 = b_2 a_0 b_1, \ a_0 = b_0, \ a_1 = b_1, \ a_2 = b_2),$$

which is connected. Using the same tree and removing the $b$s, we are again left with a single generator, $a_2$, which becomes trivial when taking the branch locus into account:

$$a_0 a_1 a_2 = 1.$$

Now that we have determined that the knot complement group in each cover is trivial, we need to determine the preimage of the trefoil and each unknot.

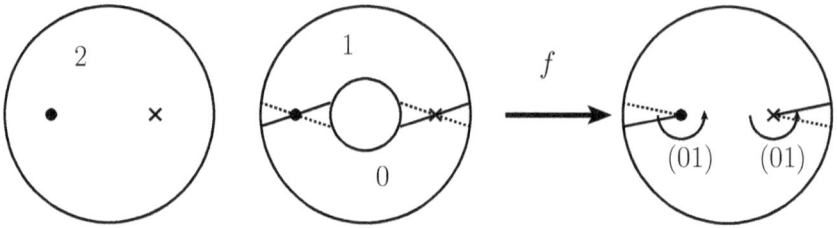

**Figure 13.11:** An example of a cover from Cat 3, for two equal transpositions.

## Preimages of $(1,3)$-Torus Knots

The preimages of the $(1,3)$ torus knots (equivalent to unknots) are straight-forward to determine. Since there is no "twist", like that found in the braid representation of $(2,3)$, we simply get a copy of a $(1,3)$ in each sheet of the cover, the location of which is determined by the permutation label. For instance, for $\rho(a) = \rho(b) = (0)(12)$, there is one copy for sheet 0 and one copy for the sheet linked by the permutation $(12)$.

## Preimages of $(2,3)$-Torus Knots

For the preimages of the trefoil, we will need to follow the technique outlined in Rolfsen's original example. Starting with Cat 2, we have to consider each sheet separately. Since there are three disjoint copies of the base, the preimage will be 3 copies of the trefoil.

For Cat 3, we'll do the transposition case first, shown in figure 13.11. Since we know this cover is embedded in a topologically trivial manifold $M^4$, the hole in the annulus will go away. Under the clockwise $3\pi$ twist in the base, both pairs of arcs simply twist around each other three times, resulting in two copies of the $(2,3)$ torus (see figure 13.7).

For pairs of identical 3-cycles, the slice through the cover is shown in Figure 13.12. There is only a single preimage of each arc, so this is a braid on a single generator. Under the $3\pi$-twist, the cross travels over the dot 3 times and this is a single copy of the $(2,3)$ torus.

| Permutation Class | Number | Torus Type of the Preimage of $(2,3)$ | Torus Type of the Preimage of $(1,3)$ |
|---|---|---|---|
| $(1) \times (1)$ | 1 | $(2,3) \cup (2,3) \cup (2,3)$ | $(1,3) \cup (1,3) \cup (1,3)$ |
| $(np) \times (np)$ | 3 | $(2,3) \cup (2,3)$ | $(1,3) \cup (1,3)$ |
| $(npq) \times (npq)$ | 3 | $(2,3)$ | $(1,3)$ |

So finally, we have a complete list of states for this particular model. In the end, we just get generalizations of the 2-knot shown in figure 13.10. For instance, for the covers represented by the permutation class $(pq) \times (pq)$, the 2-knot is a locally flat surface with two $(2,3)$-torus singularities and four $(1,3)$-torus singularities. Notice we don't actually know how these cross sections are

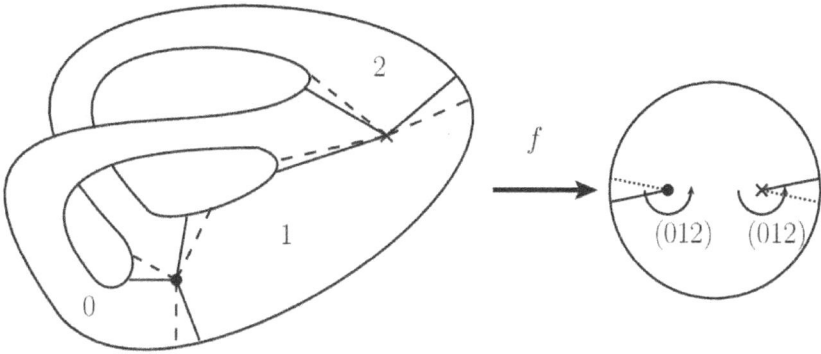

**Figure 13.12:** An example of a cover from Cat 3, for two equal 3-cycles.

precisely arranged without a closer inspection of the crossing regions in the base, but we don't need that to determine the partition function (13.7).

The partition function can be split into the contribution from the base and the contribution from the cover,

$$Z = \left( \exp\left[ \frac{3\kappa}{\hbar} \int R dV \right] \right) \left( \sum_{(i,j,g)} \exp\left[ 4\pi \frac{\kappa}{\hbar} \sum_k (1 - \alpha_k) \int \delta(\Sigma(i,j)) dV \right] \right)$$

$$= Z_0 \bar{Z},$$

where we have set $\Lambda = 0$. Of course, this means the action is negative-definite and the partition function formally divergent. This could be avoided by using an Einstein metric on the base sphere, but that modification is easy to make from our more simplified and direct approach.

Considering the radius of the base sphere $R$ as a constant, the first piece can be determined explicitly[3],

$$Z_0 = \exp\left( \frac{36\pi^2 \kappa}{\hbar} R^2 \right).$$

To work out the second piece, we will first assume that the covering maps are isometric; that is, the metric on $p^{-1}(\Sigma)$ in the covers matches the metric in the base $\Sigma$. In that case, $A_\Sigma = \int \delta(\Sigma) dV$ in the sum. Now calling the number of singularities on the $i$th preimage of each 2-knot $n_i$, we can write the sum over the angle deficits as

$$\sum_{k=1}^{n_i} (1 - \alpha_k) = n_i - \chi(\Sigma_i) = n_i,$$

---

[3]One possibly interesting generalization of this would be to consider covering maps in which the radius $R$ of the base space is allowed to vary freely. This could correspond to an expanding universe with topology change, such as [4]

since in our particular case we have $\chi(\Sigma_i) = 0$. This piece now looks like

$$\bar{Z} = \sum_i \exp\left(\frac{4\pi\kappa n_i}{\hbar} A_\Sigma\right).$$

It is pretty clear that the area $A_\Sigma$ is playing the role of the classical (inverse) temperature in this system. With that in mind, we can cast this partition function in the usual form by defining a physical constant $\gamma$ with units of energy per unit area to play the role of the Boltzmann constant, and define

$$\beta = \gamma A_\Sigma, \qquad E_i = \frac{4\pi\kappa}{\hbar\gamma} n_i,$$

so our partition function looks like

$$Z = \sum_i g_i \exp(E_i\beta),$$

with $g_i$ denoting the degeneracy of the "energy level" $E_i$, where the energy now just depends on the number of singularities in the preimage of the knot. Using standard techniques from statistical mechanics, we can calculate the free energy as $F = -\frac{1}{\beta} \ln(Z)$, and the entropy is

$$S = \frac{36\pi^2\kappa}{\hbar} R^2 + \ln\left(\sum g_i \exp(E_i\beta)\right) + \frac{\sum E_i g_i \beta \exp(E_i\beta)}{\sum \exp(E_i\beta)} = S_0 + S'.$$

The first term in this expression is dependent on the size of the base sphere, while the second two terms only depend on the 2-knots in the branched cover. We can study the scaling properties of this system by varying the ratio of the physical sizes, $\delta = R^2/A_\Sigma$, and setting the units to be $\kappa/(\hbar\gamma) = 1$. When $\delta \approx 1$, the relative contributions to the entropy are approximately equal when $\beta \approx 0.01$. As $\delta \to 0$, the temperature at which this transition occurs decreases dramatically, see figure 13.13.

We would like to contrast this calculation with an equivalent one using more conventional approaches to semiclassical gravity. Say we want to start off with a very physically reasonable model space (say, a 4-sphere), and consider the set of states which are "close" to it - topologically the same but have different geometries. Of course, since the smooth Poincaré conjecture is unsolved in dimension 4, we have no idea how many smooth 4-spheres there are which are gauge-inequivalent to the standard one (indeed, if the conjecture is true then the sphere has a unique smooth structure and such transitions would be forbidden). More specifically, the classification problem in 4-dimensions is unsolved, and we have no idea how a full characterization of 4-manifolds might be presented. Compared to our example, which lacks some physical meaning but is very well-presented as a statistical model, the advantage is clear. It is this stark contrast which is the primary motivation for this particular presentation of spacetime.

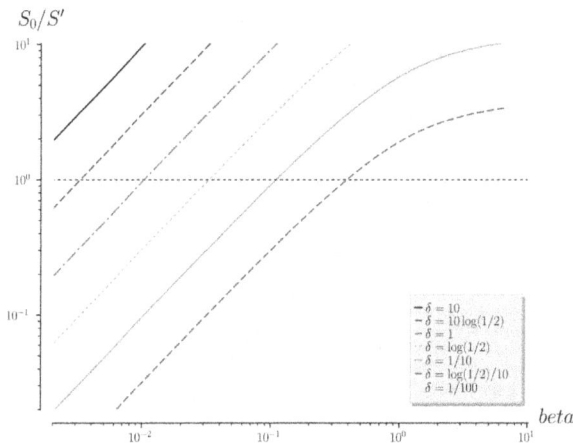

**Figure 13.13:** A plot of the temperature parameter $\beta$ against the ratio of the entropy contributions from the base sphere $S_0$ and the spacetime model $S'$.


## 13.8  Summary

A major feature of our modern understanding of the universe is the importance of the spacetime model. Fundamentally geometric objects, the necessary complexity of the spacetime models used in general relativity presents a major challenge when trying to study them either in detail or generally. In this work, we reviewed a proposal to reparameterize the spacetime model into a branched covering space over a model sphere, and applied the construction to the case in which the branching set was a 2-knot.

There are several advantages to this approach. The first is the ability to study general classes of smooth 4-manifolds as model spacetimes. Although some of the physical characteristics of the spacetimes have been lost, the lost information is replaced with an algorithmic method to construct all such spacetimes. In this way calculations requiring a sum-over-states can be well-posed, and the presence of exotic smooth structure is incorporated in a natural way. In addition, and in stark contrast to the usual approaches, the topology of the spacetime model can be explicitly tracked and studied.

When the branched covers are presented over 2-knots as they have been here, a number of abstract features of these models become more concrete. In particular, the singularities in the covering space are made explicit, and are directly related to the knot and the covering presentation in the base. This is easiest to see when the knots are chosen to be tori. It becomes possible to choose a set of kinematic states for the gravitational field and perform semiclassical calculations without having to directly solve the classification problem, which has stumped mathematicians for decades.

There are a number of directions in which this exploration could be continued. For the specific example presented here, it would be interesting to take

Rolfsen's example to it's natural conclusion, and work out a generic way to determine the preimage of any knot in a branched cover. As far as we know, this is a completely open question and a possible source of rich mathematical knowledge. If this was done, it would be possible to study a very large class of model spacetimes in a statistical manner, such as what we presented in §13.7. Even in the more restricted case that we've presented, other questions could be answered via the statistical physics/QFT interpretation. For instance, one could explore the tunneling probability between topologically inequivalent states. Or, by adding a source term $J_{ab}$ to the generating function, *e.g.*

$$Z = \exp(S_0 + J^{ab}h_{ab}),$$

one could perform loop calculations, study ghost contributions, *etc.* And since the topology is being tracked by the construction, we would finally have a model for spacetime in which the topology is determined dynamically, rather then *a priori*. Perhaps this would provide some clue as to what happens to spacetime topology when one passes to the quantum realm, but more generally it would expand the characteristic background independence of the gravitational field to include both the geometry *and* the topology.

Before we conclude this paper, we would also like to point out a higher-order theme to this work, which is a further exploration of the deep connection between physics and mathematics. The primary inspiration for this has been the discovery and exploration of exotic smooth structures in dimension 4, which utilized methods from both differential topology and geometry but also gauge theory and analysis. In this work, we are suggesting a new connection between the rich and well-studied mathematical field of knot theory and general relativity. Specifically, the appearance and importance of the non-locally flat points in our approach demonstrates that without a deep understanding of 1- and 2-knots on one hand and action integrals on the other, this exploration would have been impossible. More generally, we would like to propose the following paradigm; *if spacetime is a fundamentally geometric object, all alternative representations of the geometry must be viewed as equivalent and should be explored and evaluated for use in physical calculations.*

Thus, in the end, this study is about taking the geometric nature of the gravitational field seriously. Since the manifold structure which allows us to perform calculations and verify our cosmological models necessarily depends on all aspects of this geometry - points, sets, topology, and smooth structure - alternative presentations which also include this data represent fertile grounds for productive research. For the branched covering reparameterization, we trade off physical intuition in favor of well-posed questions, but are left with interesting toy models to examine for clues into the deeper mysteries of the four-dimensional world. It is our hope that this project will inspire other members of the community to explore this and other geometrically-motivated reparameterization of the gravitational field, to not only gain a more complete understanding of our spacetime models, but to also foster new connections between theoretical physics, observations, and mathematics.

# Acknowledgments

This work was partially supported by a 2017 Faculty Development grant from Merrimack College. The author would also like to thank the organizers of the First Hermann Minkowski Meeting on the Foundations of Spacetime Physics for the opportunity to present this work to those in attendance.

# References

[1] J. W. Alexander. Note on Riemann spaces. *Bull. Amer. Math. Soc.*, 26(8):370–372, 1920.

[2] T. Asselmeyer-Maluga and C. H. Brans. *Exotic smoothness and physics*. World Scientific Publishing Co. Pte. Ltd., Hackensack, NJ, 2007.

[3] T. Asselmeyer-Maluga and J. Krol. Exotic smooth R^4, geometry of string backgrounds and quantum D-branes. *ArXiv e-prints*, June 2010.

[4] T. Asselmeyer-Maluga and J. Król. How to obtain a cosmological constant from small exotic $R^4$. *Physics of the Dark Universe*, 19:66–77, Mar. 2018.

[5] B. Bode and M. R. Dennis. Constructing a polynomial whose nodal set is any prescribed knot or link. *ArXiv e-prints*, Dec. 2016.

[6] J. Boersema and E. Whitaker. Knots in four dimensions and the fundamental group. *Rose-Hulman Undergraduate Mathematics Journal*, 4(2), 2003.

[7] C. H. Brans. Localized exotic smoothness. *Classical Quant. Grav.*, 11(7):1785–1792, 1994.

[8] G. Burde and H. Zieschang. *Knots*, volume 5 of *de Gruyter Studies in Mathematics*. Walter de Gruyter & Co., Berlin, 1985.

[9] S. Carlip. Dimension and dimensional reduction in quantum gravity. *Classical and Quantum Gravity*, 34(19):193001, Oct. 2017.

[10] R. H. Crowell and R. H. Fox. *Introduction to knot theory*. Based upon lectures given at Haverford College under the Philips Lecture Program. Ginn and Co., Boston, Mass., 1963.

[11] B. Csikós, L. Pyber, and E. Szabó. Diffeomorphism Groups of Compact 4-manifolds are not always Jordan. *ArXiv e-prints*, Nov. 2014.

[12] D. Denicola, M. Marcolli, and A. Zainy al-Yasry. Spin foams and noncommutative geometry. *Classical Quant. Grav.*, 27(20):205025, Oct. 2010.

[13] C. L. Duston. Exotic smoothness in four dimensions and Euclidean quantum gravity. *Int. J. Geom. Methods Mod. Phys.*, 8(3):459–484, 2011.

[14] C. L. Duston. Topspin Networks in Loop Quantum Gravity. *Classical Quant. Grav.*, 29:205015, 2012.

[15] C. L. Duston. *Exotic smoothness, branched covering spaces, and quantum gravity*. PhD thesis, The Florida State University, 2013.

[16] C. L. Duston. Semiclassical partition functions for gravity with cosmic strings. *Classical and Quantum Gravity*, 30(16):165009, Aug. 2013.

[17] C. L. Duston. The Fundamental Group of a Spatial Section Represented by a Topspin Network. *ArXiv e-prints*, Aug. 2013.

[18] C. L. Duston. Using cosmic strings to relate local geometry to spatial topology. *International Journal of Modern Physics D*, 26:1750033–583, 2017.

[19] R. H. Fox. A quick trip through knot theory. In *Topology of 3-Manifolds and related topics*, pages 120–167. Prentice-Hall, 1962.

[20] D. V. Fursaev and S. N. Solodukhin. Description of the Riemannian geometry in the presence of conical defects. *Phys. Rev. D (3)*, 52(4):2133–2143, 1995.

[21] R. E. Gompf and A. I. Stipsicz. *4-manifolds and Kirby calculus*, volume 20 of *Graduate Studies in Mathematics*. American Mathematical Society, Providence, RI, 1999.

[22] H. W. Hamber. *Quantum Gravitation: The Feynman Path Integral Approach*. Springer-Verlag, Berlin, 2009.

[23] S. W. Hawking. The path-integral approach to quantum gravity. In S. W. Hawking & W. Israel, editor, *General Relativity: An Einstein centenary survey*, pages 746–789, 1979.

[24] H. M. Hilden. Three-fold branched coverings of $S^3$. *Amer. J. Math.*, 98(4):989–997, 1976.

[25] M. Iori and R. Piergallini. 4-manifolds as covers of the 4-sphere branched over non-singular surfaces. *Geom. Topol.*, 6:393–401, 2002.

[26] M. Marcolli and A. Zainy al-Yasry. Coverings, correspondences, and noncommutative geometry. *Journal of Geometry and Physics*, 58:1639–1661, Dec. 2008.

[27] J. Milnor. On the 3-dimensional Brieskorn manifolds $M(p,q,r)$. *Ann. of Math. Studies, No. 84*, pages 175–225, 1975.

[28] J. M. Montesinos. Three-manifolds as 3-fold branched covers of $S^3$. *Quart. J. Math. Oxford Ser. (2)*, 27(105):85–94, 1976.

[29] R. Piergallini. Four-manifolds as 4-fold branched covers of $S^4$. *Topology*, 34(3):497–508, 1995.

[30] V. V. Prasolov and A. B. Sossinsky. *Knots, links, braids and 3-manifolds*, volume 154 of *Translations of Mathematical Monographs*. American Mathematical Society, Providence, RI, 1997.

[31] D. Rolfsen. *Knots and links*. Publish or Perish, Inc., Berkeley, Calif., 1976. Mathematics Lecture Series, No. 7.

[32] D. Ruberman. A polynomial invariant of diffeomorphisms of 4-manifolds. In *Proceedings of the Kirbyfest (Berkeley, CA, 1998)*, volume 2 of *Geom. Topol. Monogr.*, pages 473–488. Geom. Topol. Publ., Coventry, 1999.

[33] M. Salvetti. On the number of nonequivalent differentiable structures on 4-manifolds. *Manuscripta Math.*, 63(2):157–171, 1989.

[34] A. Scorpan. *The wild world of 4-manifolds*. American Mathematical Society, Providence, RI, 2005.

[35] J. Seade. *On the topology of isolated singularities in analytic spaces*, volume 241 of *Progress in Mathematics*. Birkhäuser Verlag, Basel, 2006.

[36] T. Thiemann. *Modern canonical quantum general relativity*. Cambridge Monographs on Mathematical Physics. Cambridge University Press, Cambridge, 2007.

[37] A. Vilenkin and E. P. S. Shellard. *Cosmic strings and other topological defects*. Cambridge Monographs on Mathematical Physics. Cambridge University Press, Cambridge, 1994.

[38] A. Zorich. Flat surfaces. In *Frontiers in number theory, physics, and geometry. I*, pages 437–583. Springer, Berlin, 2006.

[39] B. Zwiebach. *A first course in string theory*. Cambridge University Press, Cambridge, 2004.

# 14 CAN MINKOWSKI SPACETIME RESOLVE QUANTUM SUPERPOSITION?

G. N. ORD

**Abstract**    Minkowski Spacetime is a mathematical encoding of a few physical principles. The clarity of its origin places it in stark contrast to quantum mechanics, a theory whose interpretation continues to evolve. The outer scale of spacetime extends far beyond the usual characteristic scales of quantum mechanics while the inner scale domain of Minkowski space is less well known. However, it is anticipated that the classical and quantum paradigms clash on small scales, with quantum mechanics taking precedence.

The question in the title is about whether we can find evidence that spacetime itself implicates a transition from the world of objects in spacetime to the world of quantum processes. In this paper we explore the circumstances under which the answer to the question is yes. We find that attaching a discrete world-signal to worldlines forces a reassessment of the implications of the relativity postulate on small scales. In the new context it has deeper non-local consequences than usually supposed, while foreshadowing the algorithmic aspect of quantum mechanics and the emergence of the superposition principle. The result indicates that Minkowski space and quantum propagation share a common origin.

## 14.1  Introduction

It is natural to question why a classical statistical mechanic might have something of interest to say to an audience focussed specifically on classical spacetime physics. Surely, *the* central attraction of both special and general relativity is the clarity and apparent solidity of the principles that spacetime physics encodes. There seems to be little need for the uncertainties of probability in a spacetime description of objects. There *is* a need for uncertainty if we admit quantum mechanics, but then why should classical statistical mechanics be

C. Duston, M. Holman (Eds), *Spacetime Physics 1907 - 2017. Selected peer-reviewed papers presented at the First Hermann Minkowski Meeting on the Foundations of Spacetime Physics, 15-18 May 2017, Albena, Bulgaria* (Minkowski Institute Press, Montreal 2019). ISBN 978-1-927763-48-3 (softcover), ISBN 978-1-927763-49-0 (ebook).

useful, since it does not come into play until the *algorithm* of quantum mechanics has been applied, and the Born postulate invoked. Classical probability in the context of quantum mechanics is a theory who's 'ghost', the wavefunction, seems more strongly tied to spacetime than probability theory itself. It does not appear obvious that classical statistical mechanics has anything to say about special relativity that would shed light on quantum mechanics.

To counter this appearance slightly, it is worth noting that historically both Schrödinger's equation and the path integral formulation of quantum mechanics evolved from deBroglie's extraction of the wavelength that bears his name[1]. This wavelength was obtained from the photoelectric effect and Einstein's formulation of special relativity. Feynman pointed this out in the path-integral approach that he developed in the 1940's. Here is a quote from his 'Lectures in Physics'. He is motivating the appearance of a phase factor associated with a particle path.

> We should remark, however, that when we say we can add two amplitudes of different wave number together to get a beat-note that will correspond to a moving particle, we have introduced something new *something that we cannot deduce from the theory of relativity*[1]. We said what the amplitude did for a particle standing still and then deduced what it would do if the particle were moving. But we cannot deduce from these arguments what would happen when there are two waves moving with different speeds. If we stop one, we cannot stop the other. So we have added tacitly the extra hypothesis that not only is [the amplitude for a state of fixed energy] a possible solution, but that there can also be solutions with all kinds of $p$'s for the same system, and that the different terms will interfere.  (R. P. Feynman [2])

The phrase *something that we cannot deduce from the theory of relativity* is all about the superposition principle and ultimately the whole algorithm of quantum mechanics. It suggests that despite using the Lorentz transformation in order to obtain path-dependent phase, classical special relativity overlooks something that quantum mechanics exposes. This paper shows that Minkowski space misses quantum mechanics by assuming that worldlines are smooth in spacetime frames that are themselves assumed massive enough to be used as reference frames. This effectively removes mass as a background property of worldlines. Mass as a property of the associated particle then has to be reinserted using dynamical arguments.

As an alternative to this picture, we shall show that if we replace the worldline concept with a 'world-signal' that functions as a 2-bit digital clock in a two dimensional Minkowski space, we can see the emergence of 'wavefunctions' and the quantum algorithm directly from special relativity coupled to probabilistic counting arguments. This potentially clarifies the origin and meaning of 'wave-particle duality', tying it directly to Minkowski spacetime.

---

[1]Emphasis by this author.

In the following section we shall go through the argument of special relativity as it applies to two-bit clocks in two dimensional spacetimes, relying heavily on statistical mechanical versions of spacetime diagrams. In section 3 we shall use a mathematical model to derive the Schrödinger equation, noting its evolution from special relativity and its relation to the diffusion equation.

The final section discusses the calculation, the relation to the Dirac equation and the change in perspective that this approach brings. The ideas that this paper discusses have evolved from consideration of the Feynman chessboard model [3-6] that come under the category of 'emergent quantum mechanics' [7]. The author's contributions to this evolving picture may be found in references [8-10], with the link to Minkowski space particularly in [12].

## 14.2    Two-Bit Clocks in Minkowski Space

It is customary in the relativity community to believe that we *live* in a relativistic world. Our senses detect 'Newtonian' mechanics only because they are too coarse to resolve time dilation and length contraction. Thus, Newtonian mechanics is viewed as a useful approximation and is not fundamental to the physical world in the same way as is relativity.

In juxtaposition to this view and despite the historical importance of relativity in the discovery of quantum mechanics, practically all texts discuss quantum mechanics as a quantization of classical Hamiltonian mechanics. This is then 'extended' to include relativistic quantum mechanics, the validity of the result being conceptually tested by taking a 'non-relativistic' limit to confirm contact with the Pauli or Schrödinger equations.

This is of course conceptually an inversion of what we expect to be true. Non-relativisitc quantum mechanics and Newtonian mechanics should both follow as approximations from relativity, suitably amended to resolve both the classical and quantum world. How to amend the elegant progression from Einstein's postulates to Minkowski spacetime is the question we tackle here, from the perspective of statistical mechanics.

Just as relativists see Newtonian mechanics as an approximation, statistical mechanicians see real numbers and continua in general as approximations. Statistical mechanics counts *first*, and takes a continuum limit later. So for example the worldline concept that is associated with point particles in Minkowski space, placing *events* arbitrarily close together will mark our point of departure from the standard view. Since in any case we know that smooth worldlines are not supported by quantum mechanics, our first task will be to replace them by a discrete alternative.

Instead of scale-free worldlines, we investigate the idea of a worldline *that has countable periodic events*. In Fig[14.1] we picture a feature encountered when requiring countable events attached to worldlines. Between any two events in a two dimensional spacetime is a spacetime *area*. We can think of the area as being 'causal' in the sense that points within these areas can be causally connected to the successive events on the worldline that imply the inter-event area. The Euclidean area between the events in this picture is

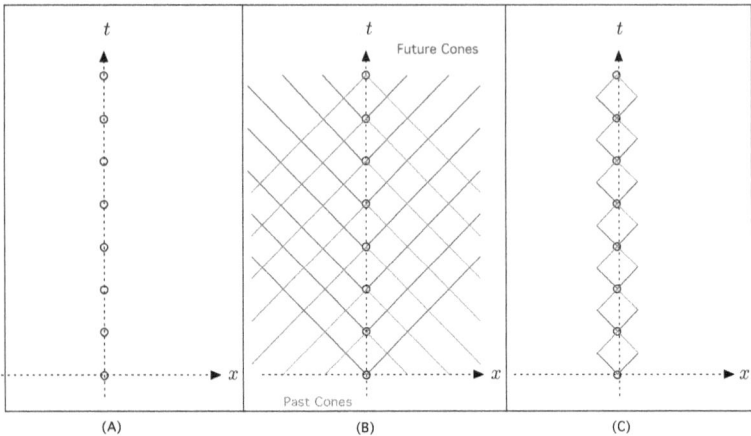

**Figure 14.1:** A spacetime diagram of a particle in its rest frame with periodic 'events'. Here $c$ is 1 and in (B) we draw the future and past light-cones for each event. In (C) we observe that the intersection of successive forward and backward light-cones form a chain of spacetime areas.

*invariant under Lorentz boosts,* so it is a useful visual tool to remind us that the proper time between two events is *a length concept derived from invariant areas.*

The periodic aspect of the chain of causal areas in Fig[14.1] gives us two distinct advantages over plain, unmarked worldlines.

1. The frequency of the events that we shall represent by a parameter $m$ gives us a label to distinguish classes of 'marked' worldlines. This frequency will ultimately be associated with mass.

2. If we assume that events serve to distinguish past and future on a spacetime diagram, we automatically introduce aspects of signal processing into Minkowski space, a feature that will induce restrictions not present with unmarked worldlines.

In a sense, taking a statistical approach, we are thinking of replacing the 'fundamental' concept of a worldline by a discrete version for which the actual worldline is an *approximation* Fig[14.2].

Here are the usual postulates used to introduce the kinematics of spacetime, rearranged for convenience.

1. Associated with any particle is a worldline. (The One.)

2. All inertial frames are equivalent. (The Many.)

3. *The speed of light is constant in all frames.*

Here the first postulate, often de-emphasized in special relativity as being obvious, is labeled as *The One* as it represents what we mean by a single point particle that persists in time. The second postulate, labeled *The Many* is the

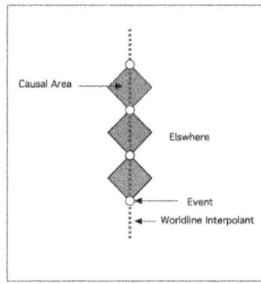

**Figure 14.2:** Worldlines as interpolants. To bring statistical mechanics into the approach to a continuum, we consider countable events. Here each event distinguishes a local past and future light-cone. The area colouring makes the successive area sequence into a one-bit clock, the two bit-states distinguishing between forward and backward light cones.

postulate with immense power. To a statistical mechanic, it associates with the first postulate an uncountably large ensemble of equivalent systems. While the first postulate is about an object, the second is about a whole equivalence class of *images* of that object.

The third postulate is the one introduced by Einstein that changes everything. In Minkowski's words:

> Henceforth space by itself, and time by itself, are doomed to fade away into mere shadows, and only a kind of union of the two will preserve an independent reality. (Minkowski [3])

Fixing $c$ as an invariant speed completely changes the relationship between *The One* and *The Many* postulates. For example in Fig[14.3] we see a worldline of a particle in its rest frame, lasting 10 units of time. In Fig[14.4] we see that the lightspeed postulate is manifest on the spacetime diagram by a stretching of the worldlines that are rotated by Lorentz boosts. Without the third postulate, space and time are independent and the multiple images of the worldline at rest would all end at $t = 10$. There would be no time dilation.

Time dilation and length contraction appear naturally in the context of our three postulates, but in the interest of probing consequences of the last two postulates from a discrete perspective consider replacing the first postulate with:

> 1'. *Associated with any particle is a binary world-signal where two bits of information distinguish the four spacetime areas between two successive pairs of events.*

For clarity, we shall call particles that do this 'two-bit clocks', clock-particles or just clocks. They are distinct from classical point particles by virtue of having 'world-signals'. The qualitative difference is that time dilation is made locally apparent by the presence of discrete markings on a worldline.

Visually, using only one bit to determine colour, the analog of figures Fig[14.3] and Fig[14.4] are Fig[14.5] and Fig[14.6] respectively. The point to

228

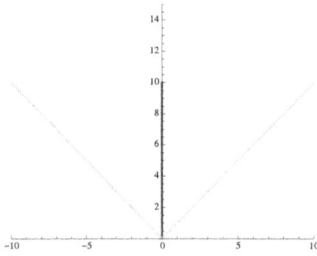

**Figure 14.3:** A worldline, creation at the origin and annihilation at $t = 10$. The forward lightcone represents the boundary between spacelike and timelike regions from the origin.

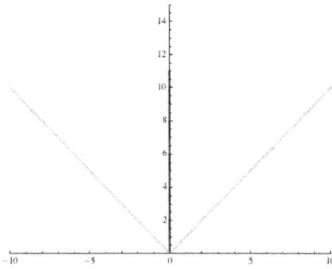

**Figure 14.4:** The spacetime diagram of the previous figure with the images of the worldline boosted to various velocities. Note the appearance of time dilation, a consequence of the third postulate.

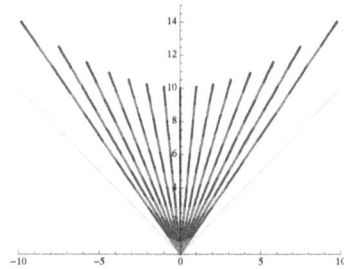

**Figure 14.5:** A world-signal, creation at the origin and annihilation at $t = 10$. The two colours represent projection of successive causal areas onto the worldline.

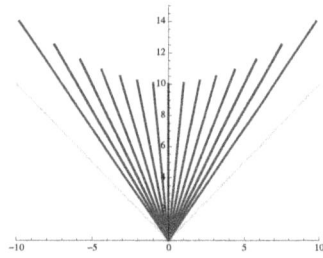

**Figure 14.6:** Lorentz boosts of the world-signal in the previous figure. Note that time dilation is 'visible' on the time-scale of unity and not just over the time between creation and annihilation.

notice is that by marking the 'one' worldline with a periodic pattern as we have done between figures Fig[14.3] and Fig[14.5], we have induced a hyperbolic pattern in Fig[14.6] *not* apparent in Fig[14.4]. That is, we can see the effect of time dilation *locally on small scales* with a discrete signal. If we pursue this a little further with images by considering a larger collection of relative velocities, we can follow the implication of the local pattern at fixed $t$.

Fig[14.7] illustrates the power of the relativity postulate as it engages with the light-speed postulate and the world-signal postulate. Notice that the entire region in the future light-cone of the origin inherits a full wave-pattern from a *single* world-signal! This is Special Relativity's version of wave-particle duality. The difference between this incarnation and the quantum context is that this version has a clear and precise provenance that we can identify and explore directly. The wave pattern is a direct manifestation of the relativity postulate constrained by a fixed $c$ with the worldline replaced by the world-signal.

To make more direct contact with the idea of waves in quantum mechanics consider fixing $t$ and looking at the pattern induced as a function of $x$. In Fig[14.8] we have coloured the line $t = 10$ according to the colours of the rays that intersect it in the previous figure. Representing the colours blue and red in the previous two figures by $+1$ and $-1$ respectively, in Fig[14.9] we show the colour map induced on the line $t = 10$ from Fig[14.8]. The resulting waveform looks a little like a square wave wrapped on a cylinder. We call this graph a 'History-Map' because it is a projection of the history of the particle generating the world-signal onto a line parallel to the $x$-axis at fixed $t$. As a waveform, *it represents a single particle's history.* In the language of statistical mechanics, the waveform is the result of the *ensemble of all possible inertial frames synchronized at the origin.* This is *very different* from the sorts of ensembles that we would use in, say, the Wiener or Path integrals. Here, there is *one* rest-frame path and an infinity of *boosted images of that same path.* There is no stochastic difference between the ensemble members, there is only a difference in the relative velocity of the inertial frames. However, it is worth noting that we can make contact with statistical mechanics with one extra step. If we *square* the amplitude of the signal in Fig[14.9] we will just get 1 for all values of $x$ between $\pm t$ so if we normalize the History-Map by multiplying it by a factor of $1/\sqrt{2t}$, the result is a function whose square is the uniform distribution over the interval $[-ct, ct]$ in the spatial domain. The information contained in this result is that all relative velocities for the clock particle between $\pm c$ are equally likely.

Fig[14.9] has a family resemblance to a figure familiar in a quantum context and to check this we look farther out in the pattern at about $t = 100$. We do this so that we can look more closely at relatively small velocities. Fig[14.10] compares the History-Map with the real part of the Feynman non-relativistic

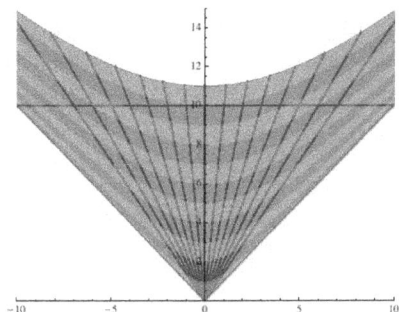

**Figure 14.7:** A world-signal with images over many velocities super-imposed. The resulting 'wave' pattern displays the familiar hyperbolae of fixed proper time. A line parallel to the $x$-axis at $t = 10$ samples the one-dimensional subspaces at different points in the particle's history.

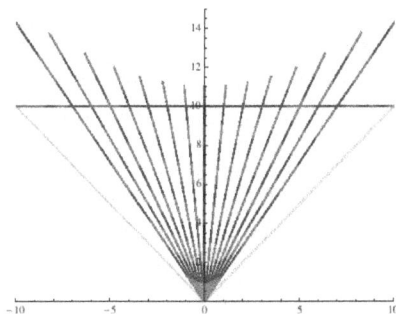

**Figure 14.8:** The previous image removing most of the boosted images just to emphasize that the horizontal line at fixed $t$ inherits colour from the one dimensional subspaces that cross it. Along this line, the closer you get to the light-cones from the origin, the earlier the signal's history is sampled.

230

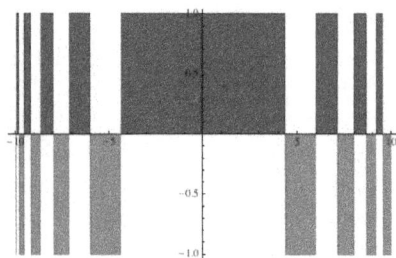

**Figure 14.9:** A two-bit clock at fixed time. From the previous two figures, blue is represented by +1, red by -1. We call such graphs 'History-Maps'

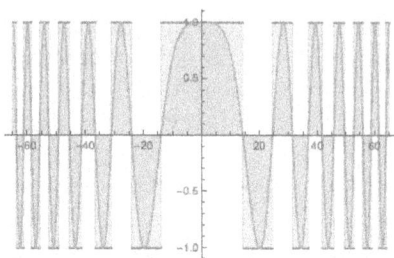

**Figure 14.10:** The History-Map of a Binary clock is the variable frequency square wave. The smooth curve is the real part of the Feynman propagator.

free particle propagator[4]. It is clear from the figure that the History-Map and the propagator are perfectly registered, this being accomplished by choice of a single frequency (the Compton frequency).

Fig[14.10] is simultaneously obvious and deeply surprising. On one hand, Feynman's non-relativistic propagator *has to* have some 'knowledge' of special relativity so that it can be 'found' in the non-relativistic limit of relativistic quantum mechanics. The 'obvious' connection here is the non-relativistic limit of the classical relativistic Lagrangian that forms the basis of the path integral. What is surprising is that the History-Map is a *classical object,* directly out of, and implied by special relativity. We know that it is not a Probability Density Function, but is related to one by taking a square. We also know that the resulting PDF is not a sum over different particle paths, *it is the manifestation of a single clock-particle's particular history.* The only stochastic basis of the PDF is the complete lack of knowledge of the velocity of the particle's co-moving frame. The History-Map carries the information of unknown relative velocity but known frequency (mass) into the future. Whether the resulting PDF has physical meaning and whether it has a linear superposition principle remains to be seen.

It is worth noting at this point that although Fig[14.10] appears to be anything but a coincidence, the very fact that the History-Map is a purely classical object immediately casts doubt on any claim that it can be a direct progenitor of the wavefunction. As far as we know, objects from the classical world are themselves unaffected by probabilistic descriptions. Coins land with either the head or tail face upward. They do not exist in superposition states between the two until observed. The classical world of objects that we think of as being described well by special relativity does not appear to have any room for wavefunctions, except through the ad hoc procedures of quantization. Surely classical probability should rule in Minkowsky space!

Just to restate our position, the argument at this point is that Fig[14.10] presents us with two clearly similar graphs from *very different* origins. The History-Map is simply a classical description of a digital clock that ticks in a two-dimensional Minkowski space, the map itself being a direct manifestation of the relativity postulate with fixed $c$ and a marked worldline. In contrast,

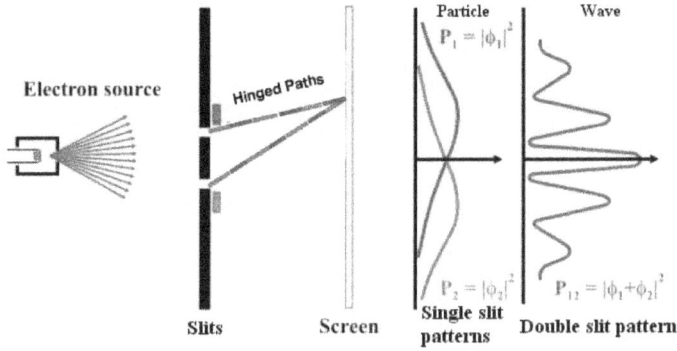

**Figure 14.11:** A schematic of the Young double slit experiment for electrons.

the Feynman propagator is a mathematical object that is useful in quantum mechanical calculations. While its physical origin is unknown, it is commonly *believed* to be independent of special relativity. Whatever quantum propagators or wavefunctions represent or do not represent in the physical world, there is at least universal agreement that their superposition principle *completely dominates* the superposition principle that would normally govern the PDFs that they imply. The demonstration of this by the Young double-slit experiment was popularized by Feynman. We review his characterization of the experiment as the failure of the probability concept as it applies to actual particles.

Fig[14.11] shows a schematic of the double slit experiment for electrons. As Feynman was careful to point out[2], the experiment is as remarkable for what fails to happen as it is for what actually does happen.

From a classical standpoint, point particles like electrons go through either the upper or lower slit. That is, passage through slit A or B are *disjoint* events for a single particle. The additivity axiom for probability then *requires* that probabilities for passage through A or B simply add. This means that sending individual particles through the apparatus will simply add the probabilities of going through either slit. We would expect that the 'Particle' result in the figure would apply. We know that under certain circumstances this *does not happen* and we get interference effects. Under these circumstances, either the probability model fails entirely, or the statement 'passage through slits A or B are *disjoint* events' must be false.

This failure of the superposition principle for probabilities is remarkable by itself but the other remarkable feature is that if we think of particles propagating as waves, using a linear superposition principle for those waves and then using the Born postulate to construct PDFs, we get a description that agrees with experiment! This makes absolutely no sense from the perspective of classical Newtonian mechanics or classical statistical mechanics and the question that we would like to ask is 'Does this make any sense in Minkowski space?'

Notice that in Fig[14.11] we have coloured two possible piecewise inertial paths between the source and the observation screen. If we think of drawing

232

the analog of the History-Map of a particle on the observation screen we have to specify what we mean by the HM of the particle given that there are in general no strictly inertial paths between the source and the screen. Instead there are two different classes of piecewise inertial paths going between the two, differing in the slit through which they travelled. We return to the digital clock picture.

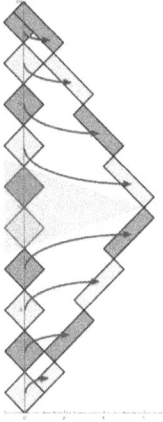

**Figure 14.12:** The twin paradox for clock particles. In the case illustrated, time dilation is such that the 'slow' clock of the rocket twin misses exactly one cycle of the clock. This means that the History-Map of the two paths agree where they cross and we call the two histories inertial equivalents.

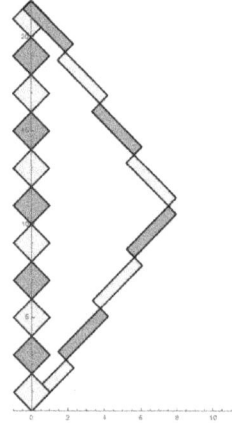

**Figure 14.13:** Here a case is illustrated where the rocket twin does not lose an integral number of periods from the stationary clock. The last half of the rocket twin history is not a boost image of the last half of the stationary clock history and we say that the two histories are inertial inequivalents.

Consider Fig[14.12] and Fig[14.13]. In the first figure we compare a stationary clock with one that has instantaneously had its velocity reversed halfway through a trip away from the origin and back. There are two interpretations of this spacetime diagram.

In the first interpretation there are two separate clocks and this is an illustration of the 'rocket-twin paradox'. The fact that they both begin in the blue state and end in the red state is immaterial, and would not be noticed in usual formulations of special relativity that do not consider world-signals.

The second interpretation is that the first half of the 'moving' path is a boost image of the initial history of the stationary clock and the last half of the moving path is a boost image of the last part of the stationary clock. The two paths disagree in the number of 'links' in the chain of causal areas but at the origin where they depart and meet they are boost images of each other.

Notice that the second interpretation does not apply to the case illustrated in Fig[14.13]. There the first leg of the trip away from the origin is a boost image of the history of the stationary particle but the second leg is not a boost image of the later history of the stationary particle. In this case the stationary clock and its rocket twin *have to be distinct clocks.*

In our construction of the History-Map for the free particle, we were careful to note that the whole map was simply a manifestation of the history of a single particle. The HM was constructed from the Lorentz Boost images of a single world-signal. No signals from other particle-clocks were involved. This suggests that if we are going to extend the idea of History-Maps to piecewise inertial frames then, where world-signals intersect, signals with opposite colour in our spacetime diagrams must be removed from consideration. So pairs like the one illustrated in Fig[14.13] that can only be interpreted as distinct clocks must be eliminated. We call such pairs inequivalent. The pair illustrated in Fig[14.12] on the other hand satisfies the condition that where the signals intersect, the local histories are images of each other and belong in an ensemble generating the History-Map of a single clock-particle.

This all suggests a simple algorithm for 'finding' the History-Map of a single particle that is not free.

- *Associate a 'parity' of $\pm 1$ with each of the four 'ticks' of the discrete clock.*

- *Construct the History-Map of a single particle by allowing all piecewise-inertial paths but adding parity rather than the usual characteristic function signifying inclusion/exclusion.*  (I)

The above algorithm will keep the pair of paths in Fig[14.12] while eliminating the pair in Fig[14.13]. It is sufficient to produce the Dirac equation in two dimensions, a result that is implicit in the Feynman Chessboard model and will be made explicit in a future work. For our purposes we can test this in the context of the double slit experiment. Considering Fig[14.11], imagine the setup illustrated with the 'hinged paths' representing the two-colour world-signals illustrated in Fig[14.5]-Fig[14.9]. We can replace the double slits by idealized slits that have a fixed finite separation but negligible width so the signals at the detectors pass through only two points representing the two slits. We can then compare the result of the History-Map using the recipe (I), with the usual calculation from Feynman's non-relativistic propagator. The result is displayed in Fig[14.14].

Fig[14.14] is very suggestive. The real difference between the History-Map and the propagator for this double slit experiment is continuity. Note that the propagator result is just a 'smoothed' version of the History-Map result. The digital aspect of the History-Map persists through to the fringes while the propagator is continuous throughout. The advantage of the History-Map is that we know *exactly* what it represents and in particular, its relation to single particles explains the relation to the resulting PDF. In the context of quantum

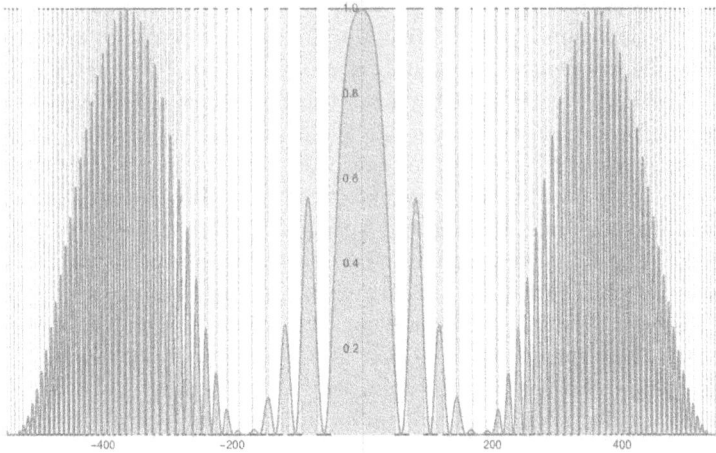

**Figure 14.14:** The contribution to the probability of detection from the real part of the Feynman propagator to the three central fringes of an idealized double slit experiment. The light blue area is the same experiment using the History-Map generated by the parity filter. The propagator and History-Map both pick out the interference fringes. The Feynman calculation is quantum mechanics. The History-Map calculation is Minkowski spacetime with worldlines replaced by digital world-signals.

mechanics, Feynman paths are part of a calculational scheme with no indication as to what the paths themselves represent. We have an ensemble of paths associated with individual particles, but the association is formal. History-Maps in contrast are direct consequences of Minkowski space. The association between *one* particle and an infinity of paths *is the relativity postulate itself.* Association of a digital clock with worldlines allows us to see this, literally and metaphorically. In the next section we add in a stochastic element to see how the Schrödinger equation emerges in a statistical continuum limit.

## 14.3   A 'Diffusive' Model

From figures [10] and [14] we can see that History-Maps provide a discrete underpinning of the wavefunction-to-PDF algorithm of quantum mechanics. The binary elimination of pairs of inequivalent paths heralded by the recipe ( I ) apparently provides a framework underlying the quantum mechanical superposition principle and interference. What is not in evidence at this point is the appearance of continuity.

The whole basis for the appearance of interference fringes in Fig[14.14] is the use of an exclusive OR counting device to eliminate pairs of paths that end in opposite parity. This is illustrated in Fig[14.15] and Table 1.

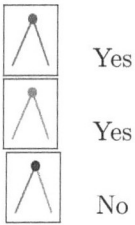

**Figure 14.15:** The use of $\sigma = \pm 1$ in counting paths to implement an exclusive or. In terms of the colour used in figures [7-9] this provides a filter for Lorentz equivalent paths.

| $\chi = (\sigma_A + \sigma_B)/2$ | | |
|---|---|---|
| $\sigma_A \backslash \sigma_B$ | 1 | -1 |
| 1 | 1 | 0 |
| -1 | 0 | -1 |

**Table 14.1:** The XOR for pairs of paths using the 'Ising spin variable' $\sigma = \pm 1$. Notice that $\chi^2 \in \{0, 1\}$ is the usual binary characteristic function used for counting in probability!

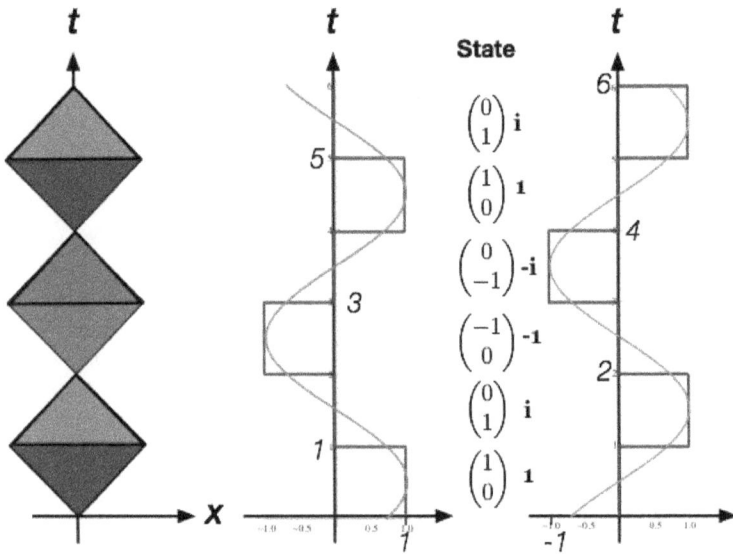

**Figure 14.16:** Encodings of the two bit digital clock. On the left is a sequence of causal areas of a two bit digital clock. The four states repeat and are counted by a two component vector where the indicator function is the Ising spin variable $\sigma = \pm 1$ considered in the previous figure. The clock ticks through the four states creating two interleaved square-wave signals. The four states are also related to the fourth roots of unity and the unit imaginary. Smoothed out versions of the components resemble two trig functions $\pi/2$ out of phase.

Returning to the digital clock, Fig[14.16] illustrates how the clock encodes a digital signal. The clock switches states when the forward light-cone of the previous event intersects the backward light-cone of the subsequent event. One way to imitate this is to consider random walks on a lattice and look for direction changes in the walk, tying state changes to direction changes. We would not expect this to work well for coarse lattice spacings, but we know that for diffusion the scale change has to employ diffusive scaling where the step size in $t$ scales as the *square* of the step size in $x$. This means that the mean free speed of the particle increases indefinitely as the lattice is refined. As a result, as the lattice is refined, lattice speeds exceed $c$ and the state change with direction change gets built into the maximum speed on the lattice, essentially building in a limit that $c$ goes to infinity. To see this we simply have to track random walks on a square lattice, giving them state changes on change of direction. To this end define densities of Clock-Particles on a lattice by $p_\mu(m\delta, s\epsilon)$, $(\mu = 1, 2, 3, 4)$. These represent the probabilities that a C-particle leaves a space time point $(m\delta, s\epsilon)$ in state $\mu$ $(m = 0, \pm 1, \ldots; \ s = 0, 1, \ldots)$. The difference equations for $p_\mu$ are

$$
\begin{aligned}
p_1(m\delta, (s+1)\epsilon) &= \tfrac{1}{2}p_1((m-1)\delta, s\epsilon) + \tfrac{1}{2}p_4((m+1)\delta, s\epsilon) \\
p_2(m\delta, (s+1)\epsilon) &= \tfrac{1}{2}p_2((m+1)\delta, s\epsilon) + \tfrac{1}{2}p_1((m-1)\delta, s\epsilon) \\
p_3(m\delta, (s+1)\epsilon) &= \tfrac{1}{2}p_3((m-1)\delta, s\epsilon) + \tfrac{1}{2}p_2((m+1)\delta, s\epsilon) \\
p_4(m\delta, (s+1)\epsilon) &= \tfrac{1}{2}p_4((m+1)\delta, s\epsilon) + \tfrac{1}{2}p_3((m-1)\delta, s\epsilon).
\end{aligned}
$$

These equations express the conservation of the number of particles over time, with half of them maintaining their direction and state at a time step, and half changing direction and state. The four states refer to the two spatial directions and the two possible 'states' for each step illustrated in Fig[14.16].

Expressing the coupled equations in matrix form with shift operators $E_x^{\pm 1}$ and $E_t$ such that

$$
\begin{aligned}
E_x^{\pm 1} \, p_i(m\delta, s\epsilon) &= p_i((m \pm 1)\delta, s\epsilon) \quad \text{and} \\
E_t \, p_i(m\delta, s\epsilon) &= p_i(m\delta, (s+1)\epsilon).
\end{aligned}
$$

The difference equations may then be written as

$$
E_t \, P(m\delta, s\epsilon) = \frac{1}{2}
\begin{bmatrix}
E_x^{-1} & 0 & 0 & E_x \\
E_x^{-1} & E_x & 0 & 0 \\
0 & E_x & E_x^{-1} & 0 \\
0 & 0 & E_x^{-1} & E_x
\end{bmatrix}
P(m\delta, s\epsilon)
$$

where $P(m\delta, s\epsilon)$ is a column vector of the $p_\mu$.

Now consider a change of variables:

$$
z_1 = \frac{p_1 + p_3}{2}, \quad z_2 = \frac{p_2 + p_4}{2} \tag{14.3}
$$

and

$$
\phi_1 = \frac{p_1 - p_3}{2}, \quad \phi_2 = \frac{p_2 - p_4}{2}. \tag{14.4}
$$

The $z_k$ just represent probabilities, partitioned by direction. The $\phi_k$ record *parity* in the system, partitioned by direction. In terms of counting paths, the $\phi_k$ record the net number of paths that are 'Lorentz equivalent' using the $\pm 1$ filtering process of the C-clock signal illustrated in Fig[14.15] and Table 1.

The change of variables block diagonalizes the shift matrix to give:

$$
E_t
\begin{bmatrix}
z_1 \\
z_2 \\
\phi_1 \\
\phi_2
\end{bmatrix}
= \frac{1}{2}
\begin{bmatrix}
E_x^{-1} & E_x & 0 & 0 \\
E_x^{-1} & E_x & 0 & 0 \\
0 & 0 & E_x^{-1} & -E_x \\
0 & 0 & E_x^{-1} & E_x
\end{bmatrix}
\begin{bmatrix}
z_1 \\
z_2 \\
\phi_1 \\
\phi_2
\end{bmatrix}. \tag{14.5}
$$

The upper block gives a discrete form of the diffusion equation.

$$
E_t
\begin{bmatrix}
z_1 \\
z_2
\end{bmatrix}
= \frac{1}{2}
\begin{bmatrix}
E_x^{-1} & E_x \\
E_x^{-1} & E_x
\end{bmatrix}
\begin{bmatrix}
z_1 \\
z_2
\end{bmatrix} \tag{14.6}
$$

the lower block is:

$$
E_t
\begin{bmatrix}
\phi_1 \\
\phi_2
\end{bmatrix}
= \frac{\alpha}{2}
\begin{bmatrix}
E_x^{-1} & -E_x \\
E_x^{-1} & E_x
\end{bmatrix}
\begin{bmatrix}
\phi_1 \\
\phi_2
\end{bmatrix} \tag{14.7}
$$

where a normalization constant $\alpha$ has been inserted. This is required since this is a filtering process that will eliminate most paths. The paths being eliminated dominate those remaining so the ratio of the two quickly decays in the continuum limit. However the two types of paths 'live' in different eigenspaces so we can choose to track only the filtered paths.

Consider now the generating function (discrete Fourier transform)

$$\phi_k(p, s\epsilon) = \sum_{m=-\infty}^{\infty} \phi_k(m\delta, s\epsilon) e^{-ipm\delta} \tag{14.8}$$

Using eqn(14.7) the shift in time is

$$\begin{pmatrix} \phi_1(p, (s+1)\epsilon) \\ \phi_2(p, (s+1)\epsilon) \end{pmatrix} = T \begin{pmatrix} \phi_1(p, s\epsilon) \\ \phi_2(p, s\epsilon) \end{pmatrix} = T^{s+1} \begin{pmatrix} \phi_1(p, 0) \\ \phi_2(p, 0) \end{pmatrix} \tag{14.9}$$

where $T = \frac{\alpha}{2} \begin{pmatrix} e^{-ip\delta} & -e^{ip\delta} \\ e^{-ip\delta} & e^{ip\delta} \end{pmatrix}$ is the transfer matrix.

To take a continuum limit large powers of $T$ are needed. The eigenvalues of $T$ are

$$\lambda_\pm = \frac{\alpha}{\sqrt{2}} e^{\pm i\pi/4} \left(1 \pm i \frac{p^2 \delta^2}{2} + O(\delta^4)\right). \tag{14.10}$$

To extract a continuum limit it is necessary to choose $\alpha = \sqrt{2}$ and to make sure the powers are taken through a sequence of integers that are 0 mod $8^2$.

$$\{\delta \to 0, \ \epsilon \to 0, \ \frac{\delta^2}{\epsilon} \to 2D, \ m\delta \to x, \ s\epsilon \to t\} \tag{14.11}$$

With the mod 8 restriction applied, $\lim_{\delta \to 0} \lambda_\pm^s = e^{\pm ip^2 Dt}$ and the resulting 'propagator' is

$$\lim_{s \to \infty} \Phi(p, s\epsilon) = \begin{pmatrix} \cos(p^2 Dt) & -\sin(p^2 Dt) \\ \sin(p^2 Dt) & \cos(p^2 Dt) \end{pmatrix} \Phi(p, 0) \tag{14.12}$$

To find a more familiar form, write

$$\begin{aligned} \psi_+(p, t) &= \left( i\phi_1(p, t) + \phi_2(p, t) \right)/2 \\ \psi_-(p, t) &= \left( -i\phi_1(p, t) + \phi_2(p, t) \right)/2, \end{aligned} \tag{14.13}$$

take $\psi_\pm(p, 0) = \frac{1}{\sqrt{2}}$ and transform back to position space to give

$$\Psi(x, t) = \begin{pmatrix} \frac{e^{ix^2/4Dt}}{\sqrt{4\pi i Dt}} & 0 \\ 0 & \frac{e^{-ix^2/4Dt}}{\sqrt{-4\pi i Dt}} \end{pmatrix} \frac{1}{\sqrt{2}} \begin{pmatrix} 1 \\ 1 \end{pmatrix}. \tag{14.14}$$

Here, it is apparent that the two components of $\Psi$ satisfy conjugate Schrödinger equations. *Note that there is no quantum mechanics here, just*

---

[2]This model contains the analog of zitterbewegung. However since the continuum limit also takes $c$ to infinity the zitterbewegung prevents convergence. The 0 mod 8 restriction applies a stroboscopic limit that directly filters out the analog of the Compton frequency, leaving in the next available characteristic frequency at the analog of the de Broglie wavelength.

*a form of 'relativistic filtering'.* This is an implementation of a statistical averaging over History-Maps to obtain a continuum limit. Note also that the arrival of complex numbers is not a formal quantization. The use of complex numbers in (14.13) is a convenience to display a familiar form. The interference effects that complex numbers mimic originate from the parity variable definition (14.4), an implementation of (I).

## 14.4   Discussion

The question in the title, "Can Minkowski spacetime see quantum superposition?" metaphorically addresses the peculiarities of two separate theories. On one hand, Minkowski space with its clear emergence from Einstein's postulates, misses quantum mechanics entirely. On the other hand, the quantum superposition principle is seemingly at odds with the whole concept of 'object' in classical physics, including special relativity.

This paper builds a bridge between these two peculiarities. The argument proposed is that Minkowski spacetime can reproduce and explain quantum superposition by replacing point particles by digital clocks that produce world-signals rather than just worldlines. The bridge between the *objects* of special relativity and the *processes* of quantum mechancs then becomes the postulate on the equivalence of inertial frames. The world-signal of a particle (the local *object*) gives rise to a unique History-Map at fixed $t$ (an associated wave) whose change in time is the *process*. The History-Map, extended to allow analysis of piecewise inertial paths in pairs, has its own superposition principle in which the analog of the Born rule makes sense. As suggested by the diffusive model, averaging History-Maps over ensembles of like particles gives rise to the analog of wavefunctions.

The direct answer to the question in the title is then that Minkowski spacetime can indeed see quantum superposition, provided only that the continuum limit implicit in the worldline of a particle is postponed, allowing the implications of the relativity postulate to affect particles on the scale of the Compton wavelength. To 'see' superposition, spacetime has to resolve features of the worldline on the Compton scale. By marking time evolution on this scale, the stretching feature of the Lorentz transformation is sufficient to create History-Maps for each particle and these form the precursors of wavefunctions, linking Minkowski space to quantum propagation.

## References

[1] Louis de Broglie. The wave nature of the electron. *Nobel Lecture*, 1929.

[2] R. P. Feynman, Leighton R., and Sands M. *The Feynman Lectures on Physics*, volume 3 volumes. Addison–Wesley, 1964.

[3] Hermann Minkowski. Space and time. In Hermann Minkowski Hendrik A. Lorentz, Albert Einstein and Hermann Weyl, editors, *The Principle of*

*Relativity: A Collection of Original Memoirs on the Special and General Theory of Relativity*, pages 75–91. Dover: New York, 1952.

[4] Richard P. Feynman and A. R. Hibbs. *Quantum Mechanics and Path Integrals*. New York: McGraw-Hill, 1965.

[5] B. Gaveau, T. Jacobson, M. Kac, and L. S. Schulman. Relativistic extension of the analogy between quantum mechanics and brownian motion. *Physical Review Letters*, 53(5):419–422, 1984.

[6] T. Jacobson and L. S. Schulman. Quantum stochastics: the passage from a relativistic to a non-relativistic path integral. *J. Phys. A*, 17:375–383, 1984.

[7] L. H. Kauffman and H. P. Noyes. Discrete physics and the Dirac equation. *Phys. Lett. A*, 218:139, 1996.

[8] Gerhard Grössing, editor. *EmerQM 11, 13, 15: Emergent Quantum Mechanics*, Journal of Physics: Conference Series Vol. 361, 504, 701.

[9] G.N. Ord. Quantum mechanics in a two-dimensional spacetime: What is a wavefunction? *Annals of Physics, Issue 6, June 2009, Pages 1211-1218*, 324(6):1211–1218, June 2009.

[10] G. N. Ord. Which came first, spacetime or clocks. *Journal of Physics, Conference Series*, **504**, 2014.

[11] G.N. Ord and R.B. Mann. How does an electron tell the time? *International Journal of Theoretical Physics*, 51(2):652–666, September 2011.

[12] G.N. Ord. Superposition as a relativistic filter. *Int J Theor Phys*, 56:2243, 2017.

[13] D. G. C. McKeon and G. N. Ord. Time reversal and a stochastic model of the Dirac equation in an electromagnetic field. *Can J. Phys.*, 82(1), 2004.

[14] G. N. Ord. A stochastic model of maxwell's equations in 1+1 dimensions. *Int J Theor Phys.*, page 263, 1995.

# 15 WHO ASKED FOR CAUSALITY?

PAUL G.N. DE VEGVAR

**Abstract**   How could classical causality originate from acausal quantum gravity? We investigate this question within the relational framework
of background independent gravity. The hypothesis that *classical* spacetime might be non-local just above Planckian lengths reveals how micro-causality spontaneously arises at longer scales.
Hopf algebra methods describing *matter induced, commutatively* deformed diffeomorphism symmetries of spacetime produce the requisite Lorentz invariant classical non-localities. Surprisingly standard
model particles can be precluded from generating those deformations, instead that substance is either a viable dark matter candi-
date or a right-handed neutrino. This approach incorporates dark
matter without invoking extra dimensions, supersymmetry, strings,
holography, grand unified theories, mirror worlds, or modified Newtonian mechanics.

## 15.1   Motivation

How do the long-range correlations of classical spacetime emerge from background independent nonpertubative quantum gravity? Those correlations are
reflected in the causal structure of a fixed classical spacetime, specifically that
local observables with supports at spacelike separation commute, known as
micro-causality (see Fig. 1). They are particularly crucial [1] given the con-
spicuous absence of experimental clues in the quantum gravitational regime.
Generally quantum spin foams do not support such relationships, and "causality" is either assumed at the outset [2][3] or just involves vertex amplitudes describing a local orientation [4]. Within canonical quantum gravity the problem
presents an even greater conceptual challenge since that approach is "timeless."
Consequently the consensus hope is that micro-causality will arise from some
yet to be developed semi-classical limit of quantum gravity, however to date
that remains lacking. Thus it is mysterious that causality plays such pivotal
role throughout physics, but is so far elusive from a background independent

C. Duston, M. Holman (Eds), *Spacetime Physics 1907 - 2017. Selected peer-reviewed
papers presented at the First Hermann Minkowski Meeting on the Foundations of
Spacetime Physics, 15-18 May 2017, Albena, Bulgaria* (Minkowski Institute Press,
Montreal 2019). ISBN 978-1-927763-48-3 (softcover), ISBN 978-1-927763-49-0 (ebook).

242

quantum gravity perspective.

Fig. 1: Micro-causality means local observables $\hat{A}$, $\hat{B}$ with supports at space-like separation obey $[\hat{A}, \hat{B}] = 0$.

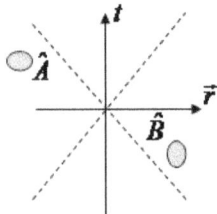

An unexpected connection between on-shell micro-causality and non-locality sheds some light on this puzzle utilizing the relational framework for canonical gravity along with Lieb-Robinson methods from solid-state physics [5]. In this semi-classical approach, spacetime is taken as classical in the sense that its gravitational degrees of freedom are not promoted to operators and their quantum fluctuations are ignored, while matter fields on spacetime are treated quantum mechanically. As discussed below, classical non-locality on a proper length scale $\xi >> L_P \simeq 1.6 \times 10^{-35}$ m (Planck length) can lead to micro-causality for lengths above $\xi$ [6]. It is then no longer necessary for quantum spacetime to be causal, even in some classical limit. Commutatively deformed diffeomorphisms (diffs) are proposed as an origin of this classical non-locality [7]. Here and throughout this contribution we will present an overview of the physical aspects, readers interested in greater mathematical detail are encouraged to consult the references. The outline of the subsequent sections is as follows: Section 2 reviews the relational framework, this is followed by its application to non-locality, Lieb-Robinson bounds, and causal structure in section 3. Section 4 introduces commutatively deformed diffs, and then section 5 describes the construction of those deformations and estimates of their non-locality length $\xi$. Next, section 6 presents the relationships with established theory and experiment. Section 7 discusses the matter that produces the deformations, and their astrophysical consequences are delineated in section 8. Section 9 outlines the known limitations of the theory together with some recent improvements. The article concludes with a summary in section 10.

## 15.2 Review of relational framework for canonical gravity

The relational formalism emphasizes that spacetime is constituted from background independent relationships between objects rather than being a physical substance [8][9][10][11]. One purpose of the relational framework is to construct Dirac (gauge invariant) observables from gauge variant (partial) observables, which is briefly reviewed here. This begins from the classical phase space $\mathcal{M}$ description of a re-parametrization invariant system whose dynamics is described by a (canonical) Hamiltonian consisting entirely of a linear combination of constraints. In the relational framework, quantization occurs on the reduced phase space. We start by describing the classical formalism. Consider then a set of first-class constraints $C_I = 0$ with $I \in \mathcal{I}$, an arbitrary index set. For the case of canonical 4-dimensional general relativity, the index $I$ includes both a continuous 3-coordinate index $y(I)$, labeling points on the 3-dimensional manifold $\Sigma$, as well as a discrete index $i(I)$. The latter ranges from 0 through $N_c - 1$, and labels the $N_c$ first-class constraints (gauge conditions) at each point (see Fig. 2).

Fig. 2: Phase space variables for the relational framework

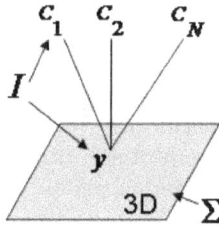

These constraints satisfy the first-class condition $\{C_I, C_J\} = f_{IJ}^K C_K$, where generally $f_{IJ}^K$ may be a structure function, possibly depending on phase space functions. We have assumed all second-class constraints have previously been handled by taking $\mathcal{M}$ to be the surface in phase space where they all vanish, and that the Poisson bracket used above is the Dirac bracket. Next select a set of gauge variant phase space functions $T^I$, $I \in \mathcal{I}$ called clock functions or clock variables that coordinatize the gauge orbit of any point in phase space within a neighborhood of the (classical) constraint surface (shell) $\bar{\mathcal{M}} \doteq \{m \in \mathcal{M} | C_I(m) = 0, \forall I \in \mathcal{I}\}$. The $T^I$ might include matter (non-gravitational) degrees of freedom. One constructs classical gauge invariant (Dirac) observables $\mathcal{O}[f](\tau_I)$ from gauge variant phase space functions $f$ on the gauge slice where $T_I$ take the values of real clock parameters $\tau_I$. More precisely:

(A) Given a gauge variant function $f(q^A, p_A)$, and (B) A set $\tau = \{\tau^I\}_I$ of (real) clock parameters/clock settings, (C) Gauge flow all variables until $T^I = \tau^I$ for all $I$ (flow clocks to settings). (D) Define the (classical) Dirac observable $\mathcal{O}[f](\tau)$ as the value of $f$ when the gauges are flowed as per (C). That is, $\mathcal{O}[f](\tau) \doteq$ (reading of $f$ when clocks $T^I$ read settings $\tau^I$). This is a relational type of gauge fixing, and the set of parameters $\tau = \{\tau^I\}_I$ may be pictured as a "multi-fingered time."

The formalism simplifies considerably if one can choose canonical coordinates so that the clock variables $T^I$ are themselves some canonical coordinates. Then one has a complete set of canonical pairs partitioned as $(q^a, p_a)$ and $(T^I, P_I)$, where the $P_I$ are the canonical momenta conjugate to the $T^I$. Hence in a local neighborhood of the constraint surface one can write the constraints as the equivalent set

$$\tilde{C}_I = P_I + h_I(q^a, p_a, T^J) \approx 0, \tag{15.1}$$

and setting $P_I = -h_I(q^a, p_a, T^J)$ formally solves the constraints. The notation $\approx$ denotes on shell equality.

Next define

$$H_I(\tau) = H_I(Q^a(\tau), P_a(\tau), \tau) \doteq \mathcal{O}[h_I](\tau) \approx h_I(Q^a(\tau), P_a(\tau), \tau). \tag{15.2}$$

If $f$ is any phase space function depending only on $q_a, p_a$, but not on $T^I, P_I$, one has

$$\frac{\partial}{\partial \tau^I} \mathcal{O}[f](\tau) \approx \{H_I(\tau), \mathcal{O}[f](\tau)\}. \tag{15.3}$$

That is, the $H_I(\tau)$ generate the $\tau$-parametrized classical gauge flow of $f$ on the constraint surface. So if one specializes to a parametrization invariant dynamical system whose canonical Hamiltonian vanishes, one may refer to the $H_I(\tau)$ as the ($\tau$ dependent) physical "Hamiltonians." It is important to realize that both (15.3) as well as its integrability condition hold only on-shell [10]. So we will henceforth limit ourselves to on-shell physics.

## 15.3 Non-locality, Lieb-Robinson bounds, and causal structure

The $H_I(\tau)$ are conventionally taken as ultra-local, meaning $H_I$ and $H_J$ possess no common canonical variables when $y(I) \neq y(J)$. Alternatively, ultra-local $H_I$ depend only on canonical fields or their spatial gradients of any finite order at $y(I)$, so that either way supp $\hat{H}_I$ is a single point on $\Sigma$. What happens for non-(ultra)local $H_I$? Introducing an external time $t$, hired as synchronizing conductor of the clock symphony orchestra: $\tau_I = \tau_I(t)$, $t$-evolution of quantized observables $\hat{\mathcal{O}}[f](t)$ becomes generated by "patchy Hamiltonians" $\hat{H}_X$ having possibly overlapping support patches $X$ on $\Sigma$. More precisely, the on-shell "patchy" $t$-evolution (gauge flow) of a quantized Dirac observable (in $\hbar = 1$ units) is

$$\frac{\mathrm{d}}{\mathrm{d}t} \hat{\mathcal{O}}[f](t) = i \left[ \sum_X \hat{H}_X(t), \hat{\mathcal{O}}[f](t) \right]. \tag{15.4}$$

This quantum evolution can be shown to remain anomaly free for monotonic flow $d\tau_I(t)/dt \geq 0$, the relational analog of global hyperbolicity; however, unlike the latter, *no strong global background causal structure is presumed*. [6]

The patchy evolution (15.4) coming from non-locality permits application of Lieb-Robinson bounds, originally used to study spin systems.[5] We briefly describe those bounds here. One spatially discretizes $\Sigma$, and in the semi-classical regime a 3-metric $d(i,j)$ for sites $i,j \in \Sigma$ may be introduced. For non-relativistic $t$-independent $\hat{H}_X$ the Lieb-Robinson condition requires: For all sites $i$ there are two positive reals $\mu, s$ such that

$$\sum_{X \ni i} ||\hat{H}_X|| \, |X| \exp[\mu \, \mathrm{diam}(X)] < s, \qquad (15.5)$$

where $\mathrm{diam}(X)$ is the maximum distance $d(i,j)$ between sites $i,j$ in patch $X$, and $|X|$ is the number of sites in $X$ (see Fig. 3).

Fig. 3: Schematic geometry of a single Hamiltonian patch $Z$

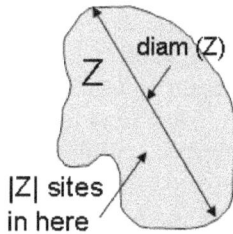

Roughly, the condition (15.5) means the interactions described by the $\hat{H}_X$ drop off at least exponentially with 3-distance $d$. Most commonly [12] the Lieb-Robinson bound is cast into the following form: Suppose the condition (15.5) holds and one is given two observables $\hat{A}_X(0), \hat{B}_Y(0)$ having supports $X, Y$ respectively at $t = 0$, then there is a constant $v_{LR} \doteq 2s/\mu$ such that for $\ell = \mathrm{dist}(X,Y)$ (the minimum distance between pairs of points in $X$ and $Y$), and $\ell \geq v_{LR} t$,

$$|| \, [\hat{A}_X(t), \hat{B}_Y(0)] \, || \leq \frac{v_{LR} \, |t|}{\ell} g(\ell) \, |X| \, ||\hat{A}_X(0)|| \, ||\hat{B}_Y(0)||, \qquad (15.6)$$

and $g(\ell)$ decays exponentially with $\ell$. One can show this means the $t$-evolved operator $\hat{A}_X(t)$ can be approximated by $\hat{A}_X^\ell(t)$ supported on the set of sites within distance $\ell = v_{LR}|t|$ of $X$ by an error whose norm is bounded by $\ell^{-1} v_{LR} |t| g(\ell) |X| \, ||A_X(0)||$. That is, supp $\hat{A}_X^\ell(t)$ is an effective $t$-dependent support for any Dirac operator $\hat{A}_X(t) = \hat{\mathcal{O}}[f](\tau_I(t))$, and norm leakage of $\hat{A}_X(t)$ outside the "light cone" is exponentially small. The "speed of light" is just the Lieb-Robinson velocity $v_{LR}$ and has been observed in the laboratory. However, if $\hat{H}_Z$ has range 0, the discrete equivalent of ultralocality, then $\mu$ is undefined since $\mathrm{diam}(X) = 0$, and there is no more Lieb-Robinson light-cone.

This emergent causal structure may be extended to $t$-dependent Hamiltonians $H_I(\tau(t))$ and rendered fully relativistic. The mathematics becomes complicated by the dependence of the 3-metric $d(x,y)$ on both external time $t$ and the gauge choice $\tau^I(t)$ (choice of "multi-fingered time"). A Lieb-Robinson bound derived by Nachtergaele, Vershynina, and Zagrebnov in 2011 [13] can be applied to address these issues. Here we summarize the results [6] of that analysis. $v_{LR}$ is invariant under smooth gauge transformations $\tau^I(t) \mapsto \tilde{\tau}^I(t)$, as well as being spatio-temporally uniform. Due to the $t$-dependence of $d$, the Lieb-Robinson light-cone becomes defined by local differential changes $\delta t$ in $t$. The operator effective supports are (3+1)-diffeomorphism (diff) invariant; that is, they remain unchanged under separate smooth redefinitions $t \mapsto \tilde{t}(t)$ and coordinate reparametrizations of the 3-manifold $\Sigma$. The local light cones can also be integrated into causal tubes for observables reminiscent of causal curves on locally Minkowskian spacetimes. The proper length $\xi \doteq \mu^{-1}$ may be interpreted as a typical patch size or non-locality scale above which micro-causality emerges spontaneously, *even if the underlying quantum description of spacetime is acausal*. Gradients of any finite order in the Hamiltonians $H_I$ produce $\xi$ proportional to the spatial discretization "lattice constant" $b$ (generate cut-off dependent effects), and so are inadmissible as physical sources of non-locality. To make contact between the spatial discretization used in the Lieb Robinson method and the continuum, it is necessary to assume the limit $b \to 0$ exists and is well-defined. It is important to bear in mind that gravitational degrees of freedom are not promoted to operators and their quantum fluctuations are ignored, while matter (non-gravitational) fields on spacetime are treated quantum mechanically. Quantum spacetime fluctuations of proper length $\delta d$, whose description lies outside this semi-classical approach, will disrupt gravitationally mean field Lieb-Robinson micro-causality. The requirement for micro-causality to survive the quantum-classical tug-of-war is $\xi >> \delta d \simeq L_P$, $L_P$ being the Planck length. This is because one must be able to introduce a physical 3-metric $d(x,y)$ on $\Sigma$ in order to use Lieb-Robinson techniques.

As shown by Dittrich [10][11], given any choice of multi-fingered time (gauges) $\tau^I(t)$, on-shell one can classically embed the phase space picture into a 4-manifold $\mathscr{S}$ with a Lorentzian 4-metric. The 3-manifold $\Sigma$ of phase space is mapped into a space-like slice (sub-manifold of $\mathscr{S}$) at $t = 0$. The tangent bundle of $\mathscr{S}$ may also be smoothly partitioned in a $(3 + 1)$-diff invariant way using its Lorentzian 4-metric to define local null directions and so generate a 4-metric based null-cone. The relational framework-Lieb-Robinson bound local light-cone should coincide with (or bound) the 4-metric null-cone, but this has not yet been explicitly established. The 4-metric null-cone, however, does not address the important issue of observable commutator leakage outside the light cone, which is the crux of the relational framework Lieb-Robinson bound.

All the commonly used classical gravitational actions, such as Einstein-Hilbert, Arnowitt-Deser-Misner and Holst, including possible scalar matter fields, possess ultra-local $h_I$ [14]. In fact, all the known interactions in the standard model of particle physics are ultra-local as well. However, a field theory is not required to be ultra-local, just that laboratory measurements so far are consistent with ultra-locality at accessible energies, and this experimentally implies the bound $\xi << (1 \text{ TeV})^{-1} \sim 10^{-19}$ m. Could our familiar micro-causality be hinting at small-scale non-local but still classical spacetime? In the following sections we discuss a possible physical origin for classical non-locality inspired by ideas from non-commutative geometry.

# 15.4   Introduction to commutatively deformed diffs

For several decades noncommutative manifolds have been an active area of mathematics research and many of these methods have been adopted by theoretical physicists to study the quantum properties of spacetime at the Planck scale [15][16][17][18]. Ideas from Hopf algebras, deformed diffeomorphisms, and quantum Lie algebras have been utilized to investigate models of quantum spacetime and field theory where the coordinates $x^\mu$ are promoted to non-commuting operators obeying $[\hat{x}^\mu, \hat{x}^\nu] = i\hat{\theta}^{\mu\nu}$. There has also been research into non-commutatively deformed classical spacetimes [19][20], which introduce a non-local star product of objects $f, g$ living on a still classical manifold, meaning $f \star g \neq g \star f$. Here we apply Hopf algbras to commutatively deformed curved classical manifolds, where $f \star g = g \star f$. The commutative deformation approach to classical manifolds has just recently begun to be explored by researchers [21][22][23] who studied flat spacetime. Specifically we examine *commutatively* deformed 4-dimensional *curved* Lorentzian manifolds where the non-local action of the (3+1)-diffeomorphism (diff) symmetries is described by Hopf algebras possessing a suitable Drinfeld twist. The deformed diffs' non-local action on the physical fields differs from the pointwise action in undeformed classical general relativity. Aside from the commutative $\star$-product, it is sufficient to consider Hopf algebras with twists satisfying an Abelian constraint on their vector field generators [20][24]. Imposing background independence requires those generators to be self-consistently related to matter fields. As a result it is found that the subtly deformed, but still classical, theory of spacetime naturally produces a non-locality length $\xi$ which can be larger than the Planck length $L_P$, and so spacetime spontaneously acquires micro-causality at longer lengths via the Lieb-Robinson route. Unexpectedly, the requisite matter fields (particle zoo) may be dark matter candidates, among other twist induced cosmological consequences.

Rather than provide a formal development of Hopf algebras and deformed diffs on classical manifolds which may be found in [20], here is a brief intuitive motivation: Deformations alter how diffs *act* on objects in spacetime. Let $v^\mu(x)$ be a 4-vector field generating a diff, its action on objects denoted by $\triangleright$. For undeformed diffs: $[v^\mu \triangleright (\text{object})](x)$ depends *only* on $v^\mu(x)$ and object at $x$ and its infinitesimal neighborhood, no other fields are involved. The "range" of the diff operation vanishes, so its action is ultra-local. Familiar examples include the Lie derivative $\mathcal{L}_v$ or the covariant derivative $\nabla_v$ acting on tensors. By contrast, deformed diffs act as $[v^\mu \triangleright_\Psi (\text{object})](x)$ which depends on $v$ and the object over some range $\xi \neq 0$ about $x$ (non-locality). Moreover, this action can be controlled by other fields $\Psi(x)$. Changing the diff action implies alterations of the Leibniz Rule (formally the co-product), multiplication, and contraction of vectors and forms. Hopf algebra methods are expressed in a coordinate-free language which maintains background independence. One can define deformed tensors, and for the special cases considered here (Abelian twists with pointwise independent vector generators) deformed metric tensors, inverse metric, Levi-Civita connections (Christoffel symbols), covariant derivatives, curvature tensors, and so on may also be consistently constructed. However those are *not* the same objects as their undeformed analogs. Finally one seeks to construct actions for *all* the physical interactions invariant under *deformed diffs*. Hopf methods applied to quantum spacetime naturally utilize non-commutative products, however for describing classical non-local spacetime when the primary goal is just to produce non-locality, the commutative twisted function product offers numerous technical simplifications. For classical spacetime, geometric and/or gravitational fields come from a commutative $C^\star$ algebra since there are no uncertainty principle or interference effects related to their measurement. Specifically, for functions (scalars) $f(x)$, $g(x)$ one sets

$$(f \star g)(x) = (g \star f)(x) = \exp\left[(\lambda/2)\,\theta^{ab}\left(X_a^\mu \mathcal{L}_\mu\right) \otimes \left(X_b^\nu \mathcal{L}_\nu\right)\right](f(x), g(x)), \quad (15.7)$$

the well-known Drinfel'd differential twist. Here $\mathcal{L}$ is the Lie derivative, generally there are $N$ vector field twist generators $X_a$ $a, b \in [1, N]$, $\theta^{ab}$ is a dimensionless real symmetric matrix (which assures $\star$-commutativity), and $\lambda$ is a dimensionful real coupling constant. Imposing $[X_a, X_b] = 0$ (Abelian twist) implies $\star$-associativity. Eqn. 15.7 is the *unique* non-trivial associative, commutative, unital ($f \star 1 = f$), suitably real differential twist in generically curved spacetime with a minimal number of derivatives. One constructs deformed actions for all interactions from the undeformed ones by replacing the standard product $\cdot$ of tensor components by $\star$-products throughout. Then Eqn. (15.7) inserts arbitrarily many factors of the differential operator

$$(\lambda/2)\,\theta^{ab}\left(\overleftarrow{X_a \mathcal{L}}\right) \otimes \left(\overrightarrow{X_b \mathcal{L}}\right) \quad (15.8)$$

into the Lagrangian densities. It may be formally demonstrated that the deformed and undeformed Hopf algebras are non-isomorphic (see the Appendix for a further discussion). The deformed and undeformed diffs possess inequivalent (Woronowicz) Lie algebras, and those different diff symmetries mean different physics.

## 15.5 Construction of commutative twists and non-locality estimates

Where do the twist's vector generators $X_a$ come from? From the undeformed field equation $G_{\mu\nu} = 8\pi G\, T_{\mu\nu}$, we are familiar with manifold curvature "coming from" matter stress-energy. Analogously, imposing background independence and Lorentz invariance requires manifold twist and the $X_a$ to be comprised from some matter/geometry fields $\Psi$. We will refer to these $\Psi$ as Grönewold-Moyal fields or simply "GM matter." Such a construction circumvents Kijowski's Universal General Relativity theorem [25] because the variation $\delta_\Psi(\star) \neq 0$ while $\delta_\Psi(\cdot) = 0$. Since we want to keep spacetime classical, the $X_a$ involve either quantum expectation values of (combinations of)

GM fields or GM fields evaluated at their action's stationary points. Counting degrees of freedom shows the number $N$ of $X_a$ obeys $N \leq 2$, and only one $X_a$ is dynamically independent. If $N = 2$ one constructs one $X_M$ from GM matter, and the other $X_T$ ($X_M$'s "twin") is determined by the Abelian condition $[X_M, X_T] = 0$ and boundary values. Hence we only need to construct one vector generator.

Let's try some explicit possibilities. (a) Try to build the argument of the exponential in eqn. (15.7) from $g_{\mu\nu}$. For generically curved spacetime this is not an Abelian twist, and so it is generally non-associative. (b) Setting $X^\mu$ to be (linear combinations of) the tetrad $e_I^\mu$ violates Lorentz invariance. (c) Making $X$ entirely from gauge fields $A_\mu$ or $F_{\mu\nu}$ violates gauge invariance or $\star$-commutativity. (d) Constructing $X$ from a set of Dirac fermions $\psi_\ell$ ($\ell$ labelling species) as a (conserved) gauge current leads to

$$X^\mu(\text{fermion}) = \binom{\text{Re}}{\text{Im}} \left\langle \sum_\ell Q_\ell \, \bar{\psi}_\ell \cdot \gamma^I e_I^\mu \cdot \psi_\ell \right\rangle. \tag{15.9}$$

This is the *only* way to obtain a gauge invariant $\star$-product without negatively normed states. The appearance of a gravitational degree of freedom such as the tetrad inside the $X_a$ and hence inside $\star$ means self-consistency (recursion) plays a role: the $\star$-product also enters the Euler-Lagrange (field) equation for the tetrad. If $X$ is taken to be a gauge current, then it can be shown [7] that $Q_\ell$ must be a GM$-U(1)$ charge, which is *not necessarily the standard* $U(1)$ charge. Alternatives for $Q_\ell$ by taking $X$ to be a non-gauge current include $Q_\ell = 1$, $(m_\ell)^n$ (powers of mass), $B_\ell$, $(B-L)_\ell$, etc. There is a similar expression for scalars, namely

$$X^\mu = \binom{\text{Re}}{\text{Im}} \sum_\ell i q_\ell \, \phi_\ell (D_\mu \phi)_\ell^\dagger, \tag{15.10}$$

for $U(1)$ charges $q_\ell$, which may then be likewise extended to other $Q_\ell$, and $D_\mu$ is the gauge covariant derivative.

What do expressions (15.9) and (15.10) imply about $\xi$'s order of magnitude? Simple dimensional analysis taking the mass dimension of the fermion fields to be $3/2$ leads to the twist coupling obeying $\lambda \simeq (M_\lambda)^{-8}$, and for $Q \neq m^n$ one finds

$$\xi/L_P \sim 151 \, (M_\lambda/3 \times 10^3 \, \text{TeV})^{-4} \, (M_\psi/\text{TeV})^{-3} \, |Q|. \tag{15.11}$$

Here $M_\psi$ denotes the lightest GM particle mass, and this estimate applies equally to both fermions and mass 1 scalar fields $\phi$. Hence there exists a robust $(M_\lambda, M_\psi)$ window which yields $1 \ll \xi/L_P \simeq 10^2 - 10^5$, a "non-locality sweet spot." This semi-classical non-locality range can produce longer scale Lieb-Robinson micro-causality. Even though this simple dimensional analysis has ignored any possible anomalous mass dimensions for the GM fields, those complications are not expected to affect the *existence* of such a semi-classical window for $\xi$. TeV scale laboratory measurements so far only restrict

$\xi/L_P \ll 6 \times 10^{15}$. The observation of a $3 \times 10^8$ TeV cosmic ray on 15 October, 1991 over Dugway Proving Ground, Utah in the Fly's Eye Cosmic Ray Detector might be used to bound $\xi/L_P \lesssim 4 \times 10^6$. However, it is unclear that those observations can be used to place constraints on causal propagation distances like $\xi$.

## 15.6 Consequences of commutative deformations

The commutative nature of the twist means its effects are distinct from its more renowned non-commutative cousins. For example, there are no UV/IR mixing phenomena for commutative twists. From a theoretical viewpoint, the standard model of particle physics will have unchanged (a) quantum numbers, (b) gauge as well as discrete (C, P, T) symmetries, and (c) anomaly cancellations. The principal fiber bundle foundation for gauge theory will, however, alter the product of group or Lie valued functions over spacetime from the standard product to the commutative $\star$-product. In the gravitational sector, the Weak Equivalence Principle together with its extension to gravitationally bound systems will both be unaffected. Geometry remains described by (deformed) $g_{\mu\nu}$ or $e_I^\mu$ fields, so deformed general relativity is still a metric theory of gravity, but the deformed diff equivalence classes are affected by the conserved GM matter currents.

The deformed (torsion-free) Einstein-Hilbert gravitational action with cosmological constant $\Lambda$ reads

$$S_{EH}^\star = \frac{1}{2\kappa} \int_{\mathcal{M}} \mathrm{d}^4x \, |g^\star|^{1/2} \star (R^\star - 2\Lambda), \quad \text{with} \tag{15.12}$$

$$\kappa = 8\pi G/c^4. \tag{15.13}$$

There are no Gibbons-Hawking-York terms since we are taking $\partial\mathcal{M} = \emptyset$ for simplicity. The classical deformed gravitational field equations then become

$$R_{\mu\nu}^\star - \frac{1}{2} R^\star \star g_{\mu\nu}^\star + \Lambda g_{\mu\nu}^\star + \Delta_{\mu\nu}(\lambda) = \kappa T_{\mu\nu}^\star \quad \text{with} \tag{15.14}$$

$$T_{\mu\nu}^\star = -2 \frac{\delta \mathcal{L}_M^\star}{\delta g_\star^{\mu\nu}} + g_{\mu\nu}^\star \star \mathcal{L}_M^\star. \tag{15.15}$$

The last term on the left hand side of equation (15.14) arises from the dependence of $\star$ on the inverse metric tensor, and it has leading order $\mathcal{O}(\lambda^1)$. In equation (15.15) $\mathcal{L}_M^\star$ is the deformed Lagrangian density for *all* matter (non-metric tensor) fields, including the twist producing matter fields. $T_{\mu\nu}$ still obeys a deformed version of energy momentum conservation.

Regarding the matter and Yang-Mills actions, those may also be rendered invariant under deformed diffs by replacing the pointwise multiplication by $\star$

throughout. Hence the classical electromagnetic (Maxwell) action is

$$S_{EM}^{\star} = -(1/4) \int_{\mathcal{M}} \mathrm{d}^4 x\, |g^{\star}|^{1/2} \star F_{\mu\nu}^{\star} \star F_{\star}^{\mu\nu}, \tag{15.16}$$

$$F_{\mu\nu}^{\star} = \nabla_{\nu}^{\star} A_{\mu}^{\star} - \nabla_{\mu}^{\star} A_{\nu}^{\star} = A_{\mu,\nu}^{\star} - A_{\nu,\mu}^{\star}, \tag{15.17}$$

where $\nabla_{\mu}^{\star}$ denotes the deformed spacetime covariant derivative.

What are the expected consequences of a commutative twist for laboratory experiments and astronomical observations? The actions for all the known interactions will be deformed, and taking the "non-locality sweet spot" range of $\xi/L_P \sim 10^2 - 10^5$ leads to fractional changes in the following sorts of measurements: Atomic spectra ($10^{-46} - 10^{-40}$); nuclear transition energies ($10^{-36} - 10^{-30}$); particle cross sections at 1 TeV ($10^{-28} - 10^{-22}$); surface gravitational potential ($10^{-58}$ on Earth, $10^{-51}$ at the 10 km horizon radius of a black hole); and finally cosmological scale factor $a$ at Big Bang nucleosynthesis ($|\delta a/a| \sim 10^{-58\pm4}$). Taken together these estimates show that "sweet spot non-locality" is so far experimentally admissible, but laboratory tests are exceedingly difficult clearly due to its near Planckian scale.

## 15.7   What is GM matter?

We can partially address this question by assembling some clues. First, GM matter must be stable in some form to maintain micro-causality. The previous estimate for $\xi/L_P$ eqn. (15.11) is also informative. One can then readily deduce the following: (1) GM matter is not all the standard particles, nor all the leptons. (2) $Q_\ell$ is not the standard $U(1)$ charge. (3) The mass models $Q_\ell = (m_\ell)^n$, are not viable for $n \geq 1$. (4) Requiring weak $SU(2)_L$ invariance (i.e. deformed gravity does not violate $SU(2)_L$) implies GM matter cannot be the proton, neutron, electron, or any of the left-handed neutrinos. (5) Setting $Q_\ell = B_\ell$ leads to rest mass 1-10 TeV Standard Model baryons behaving acausally. If we may make the relatively mild assumption that those baryons will propagate causally in not so distant future measurements, one concludes: GM matter is either (a) a sterile right-handed neutrino or (b) some set of nonstandard particles, at least one of which is stable.

We now discuss in slightly more detail how these points follow from the numerical estimate eqn. (15.11) for $\xi$. Suppose the $X_a$ were composed entirely from Standard Model fields by taking $Q_\ell$ to be either the particle's (dimensionless) standard $U(1)$ electric charge or unity. The twist generated by the electron field has to have $\xi/L_P > 1$ to possess a classical non-locality length. This implies $M_\lambda < 200$ GeV. But then the twist produced by the mass 174 GeV top quark would yield a non-locality scale $\xi$ longer than $4.3 \times 10^{-19}$ m or (460 GeV)$^{-1}$ at its Compton radius $\Lambda_C$, making it nonlocal and problematic for it to be causally behaved at experimentally accessible energies since one no longer has $\xi \ll \Lambda_C$. Next consider a Standard Model mass model with $Q_\ell = m_\ell$. The analog of estimate (15.11) for such a model is

$$\left(\frac{\xi}{L_P}\right)_{\text{mass}} \sim (151.)\,(M_\lambda/6 \times 10^2\,\text{TeV})^{-5}\,(M_\phi/\text{TeV})^4. \qquad (15.18)$$

Applying this to the electron to first constrain $M_\lambda$ by $\xi/L_P > 1$ then yields $\xi$ for the Higgs of order $1.3 \times 10^{-13}$ m, also long enough to make the Higgs start to act nonlocally in experiments. Standard Model mass models with $Q_\ell = (m_\ell)^n$, with fixed power $n > 1$, only make this problem worse. Likewise, considering a fermionic model coupling only to electronic lepton number, and using a mass of 0.32 eV for the electron neutrino to bound $M_\lambda$, leads to an experimentally unacceptable $\xi$ in excess of $1.3 \times 10^{-15}$ m for the electron. However, a Standard Model fermionic model coupling to baryon number $B$ survives a similar numerical trial when applied to the proton and top quark. Consequently, either all the Standard Model baryons produce a twist or none, and the mesons generate none.

It is also possible to use symmetries to further restrict the possible $Q_\ell$ entering the $X_a$. In particular, let us examine the weak interaction and its $SU(2)_L$ invariance. One finds that $Q_\ell$ must be none of: the electron number(s), the neutrino number(s), the proton number, or the neutron number; otherwise the deformed action would be $SU(2)_L$ variant, and so would the deformed Einstein equation. This happens since the (left-handed parts of) the u and d quarks, as well as the electron and its neutrino, gauge transform as $SU(2)_L$ doublets. Similarly $Q_\ell = (B - L)_\ell$ is ruled out by the numerical analysis for the electronic lepton number just discussed. However $Q_\ell$ could still be the sterile (right handed) neutrino number or the baryon number $B_\ell$ for species $\ell$ (within the Standard Model).

An alternative route might be to rule out experimentally the possibility that twist producing matter is comprised of Standard Model baryons. Using the estimate (15.11), one finds if all (and equivalently by the considerations just discussed, any) Standard Model baryons generate a twist through $Q_\ell = B_\ell$, then baryons having rest masses $M$ exceeding a non-locality mass $M_{\text{NL}}$ will find it problematic to act according to a local field theory that respects microcausality. This is because for $M \gtrsim M_{\text{NL}}$, the particle's own twist produced nonlocality scale $\xi(M)$ becomes a significant fraction of its Compton wavelength $\Lambda_C(M)$. Specifically, one obtains

$$M_{\text{NL}} \simeq (56.5\,\text{TeV})\,(\xi(M)/\Lambda_C(M))^{1/4}\,(\xi(\text{proton})/L_P)^{-1/4}. \qquad (15.19)$$

For instance, $\xi(\text{proton})/L_P \simeq 10^2$ and $\xi(M)/\Lambda_C(M) \simeq 0.1$ imply $M_{\text{NL}} \simeq 10.0$ TeV, and $\xi(\text{proton})/L_P \simeq 10^5$ together with $\xi(M)/\Lambda_C(M) \simeq 0.1$ yield $M_{\text{NL}} \simeq 1.8$ TeV. These energies still lie mostly beyond present day accelerator laboratory capabilities; but as baryons of higher rest mass are studied and found to continue to behave as law abiding citizens of Standard Model local quantum field theory, then the bounds excluding Standard Model baryonic matter as twist producing particles become tighter. Of course such a continuation of baryonic micro-causality into the TeV range by itself would not constitute positive experimental evidence for commutatively deformed general relativity. The baryons' rest masses $M$ generally (but not monotonically) increase with their total angular momentum $J$, so one expects eventually to find baryons with $M(J) \gtrsim M_{NL}$, which will violate micro-causality if the baryons are twist producing. If we anticipate baryonic micro-causality to hold through rest masses of a few tens of TeV, that would only leave a right-handed neutrino or a non-standard GM matter sector as twist generating possibilities.

## 15.8 Astrophysical implications of commutative deformations

The conclusion of the previous section begs one to ask: could GM particles be dark matter? GM matter will clump gravitationally since it does contribute to the stress-energy $T_{\mu\nu}$. However GM matter is not *standard* $U(1)$ charged, so it will be electromagnetically "dark" as far as astronomical observations are concerned. Comparison with the observed dark matter abundance [26] shows GM fermions must interact weakly and have mass $M_\psi \simeq 1 - 2$ TeV, remarkably consistent with the previous "sweet spot" estimate eqn. (15.11). GM matter most naturally describes dark matter as a stable non-standard $U(1)$-gauge self-interacting plasma accompanied by "dark-$U(1)$ photons," an observationally viable model that has been studied phenomenologically by several researchers. [27][28][29] [30][31][32][33]

Commutatively deformed general relativity could have several cosmological consequences aside from GM matter possibly being dark matter. Foremost among these is the existence of an acausal classical epoch in early Universe. When the Hubble radius obeyed $L_P \ll H^{-1}(a) \lesssim \xi$ the Universe was in an acausal era while spacetime was still described as a classical metric theory. During this epoch, all matter and radiation within the current Hubble radius (the entire observable Universe) resided inside $\xi$ and so interacted *without* micro-causality. The "sweet spot" $\xi/L_P \sim 10^2 - 10^5$ implies this acausal epoch occurred when the scale factor $a$ was in the range $a \lesssim 10^{-30\pm1}$. These are typical end of inflation values, moreover this estimate is independent of twist details or even the mechanism producing the non-locality.

How old is GM matter? Some form of GM matter has been maintaining micro-causality ever since the end of the acausal epoch. That would make GM matter and its ancestors a truly ancient family of matter. From the point of view of commutatively deformed general relativity, dark matter may be a relic of non-locality still lurking about in GM disguise!

Deformations will also affect classical electromagnetism and perturb the Maxwell equations, see eqn. (15.16). If one models photon propagation over

$\gtrsim 10^9$ Ly cosmological distances through a uniform isotropic GM particle gas by using the measured dark matter number density, and includes the "sweet spot" sized commutative twist through first order, one finds immeasurably small dispersion or absorption. This is consistent with the 2013 Fermi-LAT $\gamma$-ray observations of 94 GeV cosmological photons from $\gamma$-ray Bursters [34], which produced a null result. A finite $\xi$ also breaks the conformal invariance of the undeformed Maxwell equations, which could be significant for generating primordial magnetic fields in the early Universe.[35]

# 15.9 Limitations of commutatively deformed general relativity

The differential expression for the Drinfel'd twist in eqn. (15.7) raises several issues. These relate to the Euler-Lagrange equations derived from undeformed actions with standard products replaced by the $\star$-product and to the renormalizability of those field theories. The role of the differential twist is to introduce *arbitrarily* many powers of $\Delta L \doteq -(\lambda/2)\theta^{ab}(X_a^\mu \mathscr{L}_\mu) \otimes (X_b^\nu \mathscr{L}_\nu)$ into the undeformed Lagrangian. Here the $X_a$ are classicized fields, and $\lambda$ has canonical mass dimension $-8$ for scalar or spinor GM matter. This will produce an Euler-Lagrange equation which is a partial differential equation (PDE) of infinite order, and therefore having no predictive power. This is similar to the PDE governing time evolution of the Wigner function in the phase space formulation of quantum mechanics. Although one may be tempted to terminate the series expansion of the differential twist and thereby obtain a finite order PDE, strictly speaking that would merely describe a local field theory approximating the desired fully non-local one.

One way to address this is to abandon the differential form of the twist in favor of an integral kernel expression for the commutative $\star$-product. Such an approach to deformations has been developed for the translationally invariant case of Minkowski spacetime by Galluccio, Lizzi and Vitale [22]. Unfortunately since generally curved spacetimes are not translationally invariant, that method cannot be directly applied to gravitational physics. However, an integral kernel expression for the commutative $\star$-product has recently been constructed that is suitable for generically curved spacetimes.[36] It reduces to that of Galluccio et al. for flat spacetime, possesses a vector generator $X$ like that introduced earlier, and can also be shown to reduce to the differential expression of the twist (eqn. 15.7) in a suitable limit. The resulting non-linear integro-differential Euler-Lagrange equation is of finite order in derivatives and yields the same order of magnitude physical consequences as the differential expression for the twist.

Next consider renormalizability of the differential twist. Even though the single insertion $\Delta L$ with coupling constants has overall mass dimension zero by construction, eventually the Lie derivatives $\mathscr{L}_\mu$ act on the other fields in the Lagrangian to generate terms in the deformed Lagrangian with sufficient number of derivatives to become non-renormalizable. That is, those terms will have an overall coupling constant of negative mass dimension. Hence the deformed Lagrangian describes a non-renormalizable effective quantum field theory, even if one started with a renormalizable undeformed action. However, renormalizability is *not* a fundamental requirement for a physical theory. As lucidly discussed by S. Weinberg in "Is Renormalizability Necessary?"[37], as long as one includes in the Lagrangian all of the infinite number of interactions allowed by the symmetries, then there will be counterterms available to cancel every UV divergence. The twist will do this because of the systematic way it inserts arbitrarily many factors of $\Delta L$ into the undeformed action. Then on dimensional grounds, the terms having couplings with negative mass dimension $g_i \simeq M_i^{\Delta_i}$, with $\Delta_i < 0$ and $M_i$ some mass characterizing the $i$-th interaction, will have their effects suppressed for momenta $k \ll M_i$ by a factor $(k/M_i)^{-\Delta_i} \ll 1$. Einstein-Hilbert gravity is well known to be non-renormalizable, with an characteristic energy scale of order the Planck energy $E_P$. So what is the effective energy scale of the deformations being considered here? The deformation's single insertions are of the form $\xi^2(\mathscr{L}F_1) \cdot (\mathscr{L}F_2)$, where $F_1, F_2$ are factors of the Lagrangian outside this twist's single action. Therefore the energy scale of the deformations are on the scale $\xi^{-1} \sim (10^{-5} - 10^{-2})E_P$, that is close to, but not at, the Planck scale $E_P \sim 8 \times 10^{16}$ TeV. Notice this is *not* the other scales entering the twist: $M_\lambda \sim 10^3$ TeV or $M_\phi \sim$ TeV. The non-renormalizable effects (quantum corrections) for momenta $k << \xi^{-1}$ are suppressed by a factor $(k\xi)^2 \simeq (10^{-28} - 10^{-22})(k/\text{TeV})^2$. However, as $k$ approaches $\xi^{-1}$ from below, and starts to exceed it, those quantum corrections become non-perturbative. This arises from the new *quantum* character of the non-locality starting at those scales: the $X_a$ there can no longer be taken as coming from classical stationary points or semi-classical expectation values as this model does. Quantum corrections to the twist start to become non-negligible. The spacetime metric itself becomes fully quantum at the still higher energy $E_P$. This transformation also occurs for the integral kernel expression for the $\star$-product. The emergence of new physics at a scale corresponding to that of non-renormalizability is roughly similar to what occurs at the electroweak scale of about 300 GeV, and its description is well beyond the scope of this article.

These ideas may be made more specific. Let $\ell$ be some physical length characterizing non-twist interactions, such as a curvature scale, the Hubble radius, or some inverse particle 4-momentum. There are two distinct regimes of physical behavior. Regime (A): $L_P \ll \xi \ll \min \ell$. This is the domain of the analytically tractable post-acausal epoch, where all particle or field momenta are much smaller than $\xi^{-1}$. The twist's self-consistency (recursion) can be handled perturbatively, and its quantum corrections are strongly suppressed at these low energies. Regime (B): $L_P \ll \ell \lesssim \xi$. Here one enters the physically interesting but analytically challenging acausal epoch. Deformations become non-perturbative, and the twist is strongly recursive and difficult to calculate. This strange land could also be named The Big Squeeze: there is no more scale separation between $\ell$ and $\xi$, and the effects of quantum corrections on all interactions start to become non-negligible as the twist begins morphing into a quantum object.

## 15.10  Summary

Classical non-locality on the length scale $\xi \sim (10^2 - 10^5) L_P$ produces emergent Lieb-Robinson micro-causality on longer scales. This resolves background independent quantum gravity's long-range classical correlation puzzle. Commutatively deformed general relativity with a matter induced twist, where both the mathematical objects and the diff symmetries differ from their undeformed counterparts, can generate that classical non-locality sweet spot. There are minimal changes to standard particle theory and immeasurable perturbations to $\lesssim$ TeV measurements. This exploration into the origins of micro-causality uncovered two surprises: The theory implies a non-standard GM matter sector which can viably describe dark matter. It also predicts an early acausal cosmological epoch. However, analysis of what happened during that peculiar era calls for novel ideas and techniques. We began our journey with one mystery, only to confront another.

## 15.11  Acknowledgements

The author wishes to express his gratitude to the organizers of the First Hermann Minkowski Meeting on the Foundations of Spacetime Physics, May 15-18, 2017, in Albena, Bulgaria, for giving him the opportunity to present these findings to fellow researchers exploring related areas.

## 15.12  Appendix: Can commutative $\star$-diffs be undone by field redefinitions?

**Concern (Gel'fand-Naimark theorem):** For any commutative $C\star$ algebra $\mathcal{A}$ of functions (fields) with unit, there is a (Gel'fand) function isomorphism which is a $\star$-isometry (norm and $\star$-preserving) from $\mathcal{A}$ to the commutative algebra of continuous $\mathbb{C}$-valued functions with **pointwise multiplication**.

**Answer Step (1):**

Construct a Hopf algebra isomorphic to deformed one that has a $\star$-Lie bracket of vectors (deformed diffs), it obeys

$$[v, w]_\star - [v, w] = \sum_{n=1}^{\infty} (1/n!)(-\lambda/2)^n [\theta^{a_1 b_1} \cdots \theta^{a_n b_n}] \times$$
$$\times [(X_{a_1} \cdots X_{a_n} \cdot v), (X_{b_1} \cdots X_{b_n} \cdot w)], \tag{15.20}$$

with generally non-zero RHS. Hence

$$[v, w]_\star \neq f(v, w)[v, w],\qquad(15.21)$$

and the $\lambda \neq 0$ $\star$-diffs have a Lie algebra non-isomorphic to the undeformed $(\lambda = 0)$ Lie algebra. The $X_a$ are **conserved** currents of particle fields, so cannot generally be taken to vanish almost everywhere by a field redefinition.

**Answer Step (2):**

One can then demonstrate [7]: There is no mapping of vectors (diffs) $v \to m(v)$ such that

$$[v, w]_\star = [m(v), m(w)].\qquad(15.22)$$

There is no Gel'fand **vector (diff)** isomorphism, even if there is a Gel'fand **function (field)** isomorphism. Moreover, this is reflected in the fact that the deformed and undeformed Hopf algebras can be shown to be non-isomorphic.

More intuitively this states: The **deformed action** of vector $v$ on an object is NOT the **undeformed action** by some morphed $m(v)$, such as $m(v)(x) = v(x^\mu + \Delta^\mu(x))$ for some smooth "morphing" $\Delta^\mu(x)$.

**Conclude: Commutative $\star$-diffs cannot be undone by field redefinitions.**

Deformed and undeformed diffs lead to distinct diff equivalence classes of metric tensors, and hence to distinct geometries and physics.

# References

[1] L. Smolin, "The Classical Limit and the Form of the Hamiltonian Constraint in Non-Perturbative Quantum General Relativity," (1996), arXiv: 9609034.

[2] F. Markopoulou and L. Smolin, *Nucl. Phys. B* **508**, 409 (1997), arXiv: 9702025; F. Markopoulou and L. Smolin, *Phys. Rev. D* **58**, 084032 (1998), arXiv: 9712067.

[3] E. R. Livine and D.R. Terno, *Phys. Rev. D* **75**, 084001 (2007), arXiv: 0611135.

[4] E. R. Livine and D. Oriti, *Nucl. Phys. B* **663**, 231 (2003), arXiv: 0210064; D. Oriti, "The Feynman Propagator for Quantum Geometry: Spin Foams, Proper Time, Orientation, Causality, and Timeless Ordering," in Procs. of DICE 2004 Workshop, "From Decoherence and Emergent Classicality to Emergent Quantum Mechanics," arXiv:0412035.

[5] E. H. Lieb and D. W. Robinson, *Comm. Math. Phys.* **28**, 251 (1972).

[6] P.G.N. de Vegvar, "Nonultralocality and causality in the relational framework of canonical quantum gravity," *Phys. Rev. D* **93** (2016) 104038, arXiv: 1508.02482.

[7] P.G.N. de Vegvar, "Commutative deformations of general relativity: non-locality, causality, and dark matter," *Euro. Phys. J.* C, 77(1) (2017) 1-26, arXiv: 1605.06011.

[8] K. V. Kuchař, *J. Math. Phys.*, **13**, 768 (1972).

[9] C. Rovelli, *Class. Quant. Grav.* **8** 1895 (1991); C. Rovelli, *Quantum Gravity*, (Cambridge University Press, 2004).

[10] B. Dittrich, *Gen. Rel. Grav.* **39**, 1891 (2007), arXiv: 0411013.

[11] B. Dittrich, *Class. Quant. Grav.* **23**, 6155 (2006), arXiv: 0507106.

[12] M. B. Hastings, "Locality in Quantum Systems," in Lecture Notes from Les Houches Summer School 2010, arXiv: 1008.5137.

[13] B. Nachtergaele, A. Vershynina, and V. Zagrebnov, *AMS Contemporary Mathematics* **552**, 161 (2011), arXiv: 1103.1122.

[14] M. X. Han, *Class. Quant. Grav.* **27**, 245015 (2010), arXiv: 0911.3436.

[15] D.V. Ahluwalia, "Quantum Measurement, Gravitation, Locality," *Phys. Lett.* B **339** (1994) 301 and 303, arXiv: 9308007.

[16] S. Doplicher, K. Fredenhagen, J.E. Roberts, "The Quantum Structure of Spacetime at the Planck Scale and Quantum Fields," *Commun. Math. Phys.* **172** (1995) 187.

[17] M.R. Douglas and N.A. Nekrasov, "Noncommutative Field Theory," *Rev. Mod. Phys.* **73** (2001) 977, arXiv: 0106048.

[18] A.H. Chamseddine, "Deforming Einstein's Gravity," *Phys. Lett.* B **504** (2001) 33, arXiv: 009153.

[19] P. Aschieri, M. Dmitrijevic, F. Meyer, J. Wess,"Noncommutative Geometry and Gravity," *Class. Quant. Gravity* **23** (2006) 1883, arXiv: 0510059.

[20] A. Schenkel, "Noncommutative Gravity and Quantum Field Theory on Noncommutative Curved Spacetimes," Ph.D. Thesis, University of Würzburg (2012) unpublished, arXiv: 1210.1115.

[21] F. Lizzi and P. Vitale, "Gauge and Poincaré Invariant Regularizations and Hopf Symmetries," *Mod. Phys. Lett.* A **27** (2012) 125097, arXiv:1202.1190.

[22] S. Galluccio, F. Lizzi, and P. Vitale, "Translational Invariance, Commutation Relations, and Ultraviolet/Infrared Mixing," *JHEP* **0909** (2009) 054, arXiv: 0907.3540.

[23] F. Ardalan, H. Arfaei, M. Ghasemkhani, and N. Sadooghi, "Gauge Invariant Cutoff QED," *Phys. Scripta* **03** (2013) 035101.

[24] P. Aschieri and L. Castellani, "Noncommutative Gravity Solutions," *J. Geom. Phys.* **60** (2010) 375, arXiv: 0906.2774.

[25] J. Kijowsky, "Universality of the Einstein theory of gravitation," (2016) arXiv: 1604.05052.

[26] G. Bertone, D. Hooper, and J. Silk, "Particle Dark Matter: Evidence, Candidates, and Constraints," *Phys. Repts.* **405** (2005) 279.

[27] L. Ackerman, M.R. Buckley S. Carroll, and M. Kamionkowski,"Dark Matter and Dark Radiation," *Phys. Rev. D* **79** (2009) 023519, arXiv: 0810.5126.

[28] D.V. Ahluwalia, C.Y. Lee, D. Schritt, and T.F. Wubin, "Dark Matter and Dark Gauge Fields," in *Dark Matter in Astroparticle and Particle Physics*, DARK 2007, Procs. 6-th Intl. Heidelberg Conf. (2007), Klapdor-Kleingrothaus and G.F. Lewis Eds., p. 198, arXiv: 0712.4190.

[29] B.A. Dobrescu, "Massless Bosons other than the Photon," *Phys. Rev. Lett.* **94** (2005) 151802, arXiv: 0411004.

[30] S.S. Gubser and P.J.E. Peebles, "Cosmology with a Dynamically Screened Scalar Interaction in the Dark Sector," *Phys. Rev. D* **70** (2004) 123510, arXiv: 0407097.

[31] J.L. Feng and J. Kumar, "The WIMPless Miracle: Dark Matter Particles without Weak-Scale Masses or Weak Interactions," *Phys. Rev. Lett.* **101** (2008) 231001, arXiv: 0803.4196.

[32] J.L. Feng, H. Tu, and H.B. Yu, "Thermal Relics in Hidden Sectors," *JCAP* **08** (2008) 10043, arXiv:0808.2318.

[33] M. Pospelov, A. Ritz, and M.D. Voloshin, "Bosonic Super-WIMPs as keV-scale Dark Matter," *Phys. Lett. B* **662** (2008) 53, arXiv: 0807.3279.

[34] V. Vasileiou for Fermi LAT, "Constraining Lorentz Invariance Violation with *Fermi*," *Nature* **462** (2009) 7271, arXiv: 1002.0349.

[35] K. Subramanian,"The origin, evolution and signatures of primordial magnetic fields," *Repts. Prog. Phys.* **79** article id 076901 (2016), arXiv: 1504.02311.

[36] P.G.N. de Vegvar, to be published.

[37] S. Weinberg, *The Quantum Theory of Fields,* Vol. 1, pp. 516-525 (Cambridge University Press, Cambridge, 1995).

# 16 Quantum Groups, Non-Commutative Lorentzian Spacetimes and Curved Momentum Spaces

I. Gutierrez-Sagredo,[1] A. Ballesteros, G. Gubitosi, F.J. Herranz

**Abstract**   The essential features of a quantum group deformation of classical symmetries of General Relativity in the case with non-vanishing cosmological constant $\Lambda$ are presented. We fully describe (anti-)de Sitter non-commutative spacetimes and curved momentum spaces in $(1+1)$ and $(2+1)$ dimensions arising from the $\kappa$-deformed quantum group symmetries. These non-commutative spacetimes are introduced semiclassically by means of a canonical Poisson structure, the Sklyanin bracket, depending on the classical $r$-matrix defining the $\kappa$-deformation, while curved momentum spaces are defined as orbits generated by the $\kappa$-dual of the Hopf algebra of quantum symmetries. Throughout this construction we use kinematical coordinates, in terms of which the physical interpretation becomes more transparent, and the cosmological constant $\Lambda$ is included as an explicit parameter whose $\Lambda \to 0$ limit provides the Minkowskian case. The generalization of these results to the physically relevant $(3+1)$-dimensional deformation is also commented.

## 16.1   Introduction

Recent research in phenomenological aspects of quantum gravity suggests that Plank scale deformations of classical symmetries of General Relativity are to be expected when introducing quantum theory in the picture. An example is provided by Deformed Special Relativity theories (formerly introduced in [1, 2, 3] and further developed in [4, 5, 6, 7, 8, 9]) in which Planck energy,

---

[1]Based on the contribution presented at the "First Hermann Minkowski Meeting on the Foundations of Spacetime Physics" held in Albena, Bulgaria, May 15-18, 2017.

C. Duston, M. Holman (Eds), *Spacetime Physics 1907 - 2017. Selected peer-reviewed papers presented at the First Hermann Minkowski Meeting on the Foundations of Spacetime Physics, 15-18 May 2017, Albena, Bulgaria* (Minkowski Institute Press, Montreal 2019). ISBN 978-1-927763-48-3 (softcover), ISBN 978-1-927763-49-0 (ebook).

or equivalently Planck length, is introduced as a second relativistic invariant (besides the speed of light) which modifies the classical symmetries of the system. These kind of theories have attracted a considerable amount of attention during last years due to the observable phenomenology they could provide [10]. In this setting, quantum groups [11, 12, 13] (in which the deformation parameter is related with the Planck energy) are natural candidates to replace classical Lie groups of spacetime symmetries. These quantum deformations not only imply the emergence of non-commutative spacetimes (see for example [14, 15, 16, 17, 18, 19, 20, 21, 22]) but also induce a non-trivial structure of momentum space [23, 24, 25, 26] (a long forgotten idea whose intuition is clearly stated in the so-called Born Reciprocity [27]). In relation with this non-trivial structure of momentum space, some concrete phenomenology has been described: firstly, curvature of momentum space induces a dual red-shift [28], i.e. an energy dependent distance covered in a given time by a free massless particle. Secondly, the so-called dual-gravity lensing [29], i.e. an energy dependent direction from which a particle emitted by a given source reaches a faraway detector.

The aim of this paper is, firstly, to review some of the key features of the $\kappa$-deformation for de Sitter and anti-de Sitter spaces, in such a way that the well-known results for the flat case are smoothly recovered by taking the limit $\Lambda \to 0$. Secondly, we will summarize recent investigations in which momentum spaces arising from de Sitter, anti-de Sitter and Poincaré symmetries have been explicitly constructed [30, 31]. By presenting these two different features of the same quantum deformation in a unified way we attempt to clarify their relations and common origins.

In the first part of the paper we will present the coboundary Lie bialgebra structure which characterizes the $\kappa$-deformation, which is defined by a certain classical $r$-matrix. The cocommutator defined by this $r$-matrix endows the original Lie algebra with a Lie bialgebra structure, which is the tangent counterpart of the $\kappa$-deformed algebra and thus identifies the deformation. Such $\kappa$-Poisson-Hopf algebra for the (1+1) and (2+1)-dimensional cases (which is the semi-classical version of the full quantum algebra) will be explicitly presented.

At the Lie group level, the $\kappa$-Poisson-Lie structure on the (anti-)de Sitter and Poincaré groups will be then given by the so-called Sklyanin bracket. By using exponential coordinates of the second kind on these Lie groups, we will give explicit expressions for the fundamental Poisson brackets of the subalgebra generated by the spacetime coordinates. We stress that the so-obtained non-commutative (Poisson homogeneous) spacetime is invariant under the Poisson-Lie group associated with the Poisson-Hopf algebra of deformed symmetries in the sense that the group action on the non-commutative Poisson spacetime defines a Poisson map.

In the second part of the paper we will pay attention to the dual quantum group and its relation with the construction of momentum space. By computing the dual Poisson-Lie group to the Poisson-Hopf algebra of deformed spacetime symmetries, we will give a full description of the associated $\kappa$-deformed momentum space. Moreover, we will prove that a certain orbit of this dual group generates (half of) de Sitter space in $2(n + 1) - 1 = 2n + 1$ dimensions. The explicit expressions of this construction will be presented for the $(1 + 1)$ and $(2 + 1)$-dimensional cases ($n = 1$ and $n = 2$, respectively).

The central role played in this paper by the (2+1)-dimensional case is due to its relevance in Quantum Gravity, where it is often considered as a suitable toy model which incorporates some conceptual key points that a full quantum gravity theory is supposed to address (see [32] for an excellent introduction to the topic). The fundamental reason for this is that (2+1)-dimensional general relativity is a topological theory, which in more physical terms means that there are no gravitational waves. This implies that all possible solutions of Einstein's field equations are locally isometric to the three model spacetimes (Minkowski, de Sitter or anti-de Sitter) depending only on the value of the cosmological constant. In this way, the study of these three possibilities gives a full understanding of (2+1) general relativity, modulo global topology. In fact, as proved in [33, 34], general relativity in (2+1) dimensions admits a reformulation as a Chern-Simons gauge theory, with the isometry group playing the role of the gauge group. This has motivated a lot of recent work regarding the compatibility of this Chern-Simons action with quantum group symmetries and their associated Poisson-Lie counterparts. For a detailed account of this subject, see [35, 21, 36, 37] and references therein.

The structure of the paper is the following. In section 16.2 we will summarize the (anti-)de Sitter and Poincaré classical symmetries of Lorentzian spacetimes, as well as the mathematical setting of quantum groups and the essential features of the $\kappa$-deformation. Then we will describe in detail the $(1 + 1)$-dimensional case, including its non-commutative spacetime (section 16.3) and its momentum space (section 16.4). In the same way, in sections 16.5 and 16.6 their $(2 + 1)$-dimensional analogues will be presented. In the closing section some comments on the generalization of this construction to the physically relevant $(3 + 1)$-dimensional case will be discussed.

Some comments on the notation are in order. We will write $\mathfrak{g}_\Lambda^{n+1}$ for the Lie algebra of isometries of the $(n+1)$-dimensional maximally symmetric spacetime $\mathcal{M}_\Lambda^{n+1}$, which depending of the value of the cosmological constant $\Lambda$ will be the so-called de Sitter $\Lambda > 0$, Minkowski $\Lambda = 0$ or anti-de Sitter $\Lambda < 0$ spacetime. Their associated Lie group will be denoted by $G_\Lambda^{n+1}$, and by $(\mathfrak{g}_\Lambda^{n+1}, \delta_\kappa)$ we will denote the Lie bialgebra associated to the $\kappa$-deformation of $\mathfrak{g}_\Lambda^{n+1}$.

# 16.2  Fundamental concepts

In this section we give an introduction to the basic concepts and mathematical structures that will be used in the rest of the paper. Firstly, we describe the spacetimes $\mathcal{M}_\Lambda^{n+1}$ as coset spaces of its isometry groups by the Lorentz subgroup. Then, we give a motivating introduction to the mathematical concept of quantum group, which will be heavily used in the rest of the paper. Finally, we introduce the $\kappa$-deformation, whose applications in the construction of non-commutative spacetimes and curved momentum spaces in $(1 + 1)$ and $(2 + 1)$-dimensions are the main topic of this paper.

## 16.2.1 Lorentzian kinematical groups in (1+1) and (2+1) dimensions

The main purpose of this paper is the study of certain quantum deformations of Minkowski, de Sitter and anti-de Sitter spacetimes (the maximally symmetric Lorentzian spacetimes with constant curvature), namely the so-called $\kappa$-deformation, and their further consequences in momentum space. Moreover, our construction is based on a deformation of the group of isometries of these spacetimes, which will result in the construction of (the Poisson version of) the so-called quantum Poincaré, de Sitter and anti-de Sitter spacetimes. For the sake of completeness we shall review here some results and fix the notation used in the rest of the paper.

Let $\Lambda$ be the cosmological constant and let $\mathcal{M}_\Lambda^{n+1}$ be the maximally symmetric Lorentzian spacetime with curvature proportional to $\Lambda$ with $n$ space dimensions (and one temporal dimension). Then denote its isometry group (its identity component to be rigorous) by $G_\Lambda^{n+1}$. The Lie algebra of this group will be denoted by $\mathfrak{g}_\Lambda^{n+1}$. From this point of view maximally symmetric spacetimes are defined as cosets $G_\Lambda^{m+1}/H^{n+1}$, where $H^{n+1}$ is the Lorentz group (the stabilizer of the origin).

In $(1+1)$ dimensions we have

$$
\mathfrak{g}_\Lambda^{1+1} = \begin{cases} \mathfrak{so}(1,2) & \text{if} \quad \Lambda < 0 \\ \mathfrak{iso}(1,1) & \text{if} \quad \Lambda = 0 \\ \mathfrak{so}(2,1) & \text{if} \quad \Lambda > 0 \end{cases}, \qquad
G_\Lambda^{1+1} = \begin{cases} SO(1,2) & \text{if} \quad \Lambda < 0 \\ ISO(1,1) & \text{if} \quad \Lambda = 0 \\ SO(2,1) & \text{if} \quad \Lambda > 0 \end{cases},
$$

$$
\mathcal{M}_\Lambda^{1+1} = \begin{cases} SO(1,2)/SO(1,1) & \text{if} \quad \Lambda < 0 \\ ISO(1,1)/SO(1,1) & \text{if} \quad \Lambda = 0 \\ SO(2,1)/SO(1,1) & \text{if} \quad \Lambda > 0 \end{cases}.
$$

The $\mathfrak{g}_\Lambda^{1+1}$ algebra written in the kinematical basis $\{P_0, P_1, K\}$ is defined by the brackets

$$[K, P_0] = P_1, \qquad [K, P_1] = P_0, \qquad [P_0, P_1] = -\Lambda K, \tag{16.1}$$

where $K$ is the generator of boost transformations, $P_0$ and $P_1$ are the time and space translation generators. Note that in the $(1+1)$-dimensional case the de Sitter ($\Lambda > 0$) and anti-de Sitter ($\Lambda < 0$) algebras and groups are isomorphic; this is no longer true in higher dimensions.

In $(2+1)$ dimensions we have

$$
\mathfrak{g}_\Lambda^{2+1} = \begin{cases} \mathfrak{so}(2,2) & \text{if} \quad \Lambda < 0 \\ \mathfrak{iso}(2,1) & \text{if} \quad \Lambda = 0 \\ \mathfrak{so}(3,1) & \text{if} \quad \Lambda > 0 \end{cases}, \qquad
G_\Lambda^{2+1} = \begin{cases} SO(2,2) & \text{if} \quad \Lambda < 0 \\ ISO(2,1) & \text{if} \quad \Lambda = 0 \\ SO(3,1) & \text{if} \quad \Lambda > 0 \end{cases},
$$

$$
\mathcal{M}_\Lambda^{2+1} = \begin{cases} SO(2,2)/SO(2,1) & \text{if} \quad \Lambda < 0 \\ ISO(2,1)/SO(2,1) & \text{if} \quad \Lambda = 0 \\ SO(3,1)/SO(2,1) & \text{if} \quad \Lambda > 0 \end{cases}.
$$

The $\mathfrak{g}_\Lambda^{2+1}$ algebra in the kinematical basis $\{P_0, P_1, P_2, K_1, K_2, J\}$ is defined by the brackets

$$
\begin{array}{lll}
[J, P_i] = \epsilon_{ij} P_j, & [J, K_i] = \epsilon_{ij} K_j, & [J, P_0] = 0, \\
[P_i, K_j] = -\delta_{ij} P_0, & [P_0, K_i] = -P_i, & [K_1, K_2] = -J, \qquad (16.2) \\
[P_0, P_i] = -\Lambda K_i, & [P_1, P_2] = \Lambda J,
\end{array}
$$

where $i, j \in \{1, 2\}$, and $\epsilon_{ij}$ is the skew-symmetric tensor with $\epsilon_{12} = 1$. Here $J$ is the generator of spatial rotations, $K_1, K_2$ are generators of boosts and $P_0, P_1, P_2$ are generators of temporal and spatial translations, respectively. Note that under the projection onto the subspace spanned by $\{P_0, P_1, K_1 \equiv K\}$ the Lie algebra $\mathfrak{g}_\Lambda^{1+1}$ (16.1) is recovered.

## 16.2.2   Quantum groups and Lie bialgebras

Quantum groups are quantizations of Poisson-Lie groups, *i.e.*, quantizations of the Poisson-Hopf algebras of multiplicative Poisson structures on Lie groups [11, 12, 13]. It is well known that Poisson-Lie structures on a (connected and simply connected) Lie group $G$ are in one-to-one correspondence with Lie bialgebra structures $(\mathfrak{g}, \delta)$ on $\mathfrak{g} = \text{Lie}(G)$ [38], where the skewsymmetric cocommutator map $\delta : \mathfrak{g} \to \mathfrak{g} \wedge \mathfrak{g}$ fulfils two conditions:

- i) $\delta$ is a 1-cocycle, *i.e.*,

$$
\delta([X, Y]) = [\delta(X),\, Y \otimes 1 + 1 \otimes Y] + [X \otimes 1 + 1 \otimes X,\, \delta(Y)], \qquad \forall\, X, Y \in \mathfrak{g}.
$$

- ii) The dual map $\delta^* : \mathfrak{g}^* \wedge \mathfrak{g}^* \to \mathfrak{g}^*$ is a Lie bracket on $\mathfrak{g}^*$.

Therefore, each quantum group $G_z$ (with quantum deformation parameter $z = \ln q$) can be associated with a Poisson-Lie group $G$, and the latter with a unique Lie bialgebra structure $(\mathfrak{g}, \delta)$.

On the other hand, the dual version of quantum groups are quantum algebras $\mathcal{U}_z(\mathfrak{g})$, which are Hopf algebra deformations of universal enveloping algebras $\mathcal{U}(\mathfrak{g})$, and are constructed as formal power series in the deformation parameter $z$ and coefficients in $\mathcal{U}(\mathfrak{g})$. The Hopf algebra structure in $\mathcal{U}_z(\mathfrak{g})$ is provided by a coassociative coproduct map $\Delta_z : \mathcal{U}_z(\mathfrak{g}) \longrightarrow \mathcal{U}_z(\mathfrak{g}) \otimes \mathcal{U}_z(\mathfrak{g})$, which is an algebra homomorphism, together with its associated counit $\epsilon$ and antipode $\gamma$ mappings. If we write the coproduct as a formal power series of maps, namely

$$
\Delta_z = \Delta_0 + z\, \delta + o[z^2],
$$

then the skew-symmetric part of the first-order deformation (in $z$) of the coproduct map is just the Lie bialgebra cocommutator map $\delta$, where we have denoted the primitive (undeformed) coproduct for $\mathcal{U}(\mathfrak{g})$ as $\Delta_0(X) = X \otimes 1 + 1 \otimes X$.

In this way, each quantum deformation becomes related to a unique Lie bialgebra structure $(\mathfrak{g}, \delta)$. More explicitly, if we consider a basis for $\mathfrak{g}$ where

$$
[X_i, X_j] = c_{ij}^k X_k,
$$

any cocommutator $\delta$ will be of the form

$$\delta(X_i) = f_i^{jk} X_j \wedge X_k,$$

where $f_i^{jk}$ is the structure tensor of the dual Lie algebra $\mathfrak{g}^*$, that will be given by

$$[\hat{\xi}^j, \hat{\xi}^k] = f_i^{jk} \hat{\xi}^i, \qquad (16.3)$$

where $\langle \hat{\xi}^j, X_k \rangle = \delta_k^j$. Notice that the cocycle condition for the cocommutator $\delta$ implies the following compatibility equations among the structure constants $c_{ij}^k$ and $f_k^{ij}$:

$$f_k^{ab} c_{ij}^k = f_i^{ak} c_{kj}^b + f_i^{kb} c_{kj}^a + f_j^{ak} c_{ik}^b + f_j^{kb} c_{ik}^a.$$

The connection of these structures with non-commutative spacetimes arises when $G$ is a group of isometries of a given spacetime (for instance $G_\Lambda^{n+1}$ for the spacetime $\mathcal{M}_\Lambda^{n+1}$). Then $X_i$ will be the Lie algebra generators and $\hat{\xi}^j$ will be the local coordinates on the group. If we have a non-trivial deformation of $\mathcal{U}(\mathfrak{g})$, then the cocommutator $\delta$ is non-vanishing and the commutator (16.3) among the spacetime coordinates associated to the translation generators of the group will be non-zero. This is just the way in which non-commutative spacetimes arise from quantum groups. Moreover, higher-order contributions to the non-commutative spacetime (16.3) can be obtained from higher orders of the full quantum coproduct $\Delta_z$.

It is worth mentioning that in many cases the 1-cocycle $\delta$ is found to be a coboundary

$$\delta(X) = [X \otimes 1 + 1 \otimes X, r], \qquad \forall X \in \mathfrak{g}, \qquad (16.4)$$

where the classical $r$-matrix $r = r^{ij} X_i \wedge X_j$, is a solution of the modified classical Yang–Baxter equation

$$[X \otimes 1 \otimes 1 + 1 \otimes X \otimes 1 + 1 \otimes 1 \otimes X, [[r, r]]] = 0, \qquad \forall X \in \mathfrak{g},$$

and the Schouten bracket is defined as

$$[[r, r]] := [r_{12}, r_{13}] + [r_{12}, r_{23}] + [r_{13}, r_{23}],$$

where $r_{12} = r^{ij} X_i \otimes X_j \otimes 1$, $r_{13} = r^{ij} X_i \otimes 1 \otimes X_j$, $r_{23} = r^{ij} 1 \otimes X_i \otimes X_j$. Recall that $[[r, r]] = 0$ is just the classical Yang–Baxter equation. In these coboundary cases the $r$-matrix identifies both the Lie bialgebra and the quantum deformation completely.

### 16.2.3 The $\kappa$-deformation of Lorentzian kinematical groups

In the rest of this paper we will concentrate on spacetime and momentum space deformations induced by a certain well-known deformation of Lorentzian symmetries, the so-called $\kappa$-deformation. The $\kappa$-deformation was originally introduced in [39, 40, 41] for the case of vanishing cosmological constant and the associated non-commutative spacetime was also constructed [14] (see also [18]). This non-commutative spacetime has as its main feature the non-commutativity between space $\hat{x}^i$ and time $\hat{x}^0$ coordinates, whereas space coordinates commute:

$$[\hat{x}^i, \hat{x}^0] = \frac{1}{\kappa}\,\hat{x}^i, \qquad [\hat{x}^i, \hat{x}^j] = 0, \qquad \Lambda = 0.$$

Thus this non-commutativity is controlled by a parameter called $\kappa$ (in terms of the usual deformation parameter $q = \ln(1/\kappa)$), which is somehow related with the Planck length or Planck energy, in such a way that the limit $\kappa \to \infty$ recovers the classical Minkowski spacetime and the Poincaré group.

Some years later, this construction was generalized to the case of non-vanishing cosmological constant in $(1 + 1)$ and $(2 + 1)$ dimensions, giving rise to the so-called $\kappa$-de Sitter and $\kappa$-anti-de Sitter groups [42, 43, 44] and their associated non-commutative spacetimes [21, 22]. Only recently [45] the Poisson version of the $\kappa$-(anti-)de Sitter algebra in $(3 + 1)$ dimensions has been constructed. It has been made clear in these studies that the interplay between the cosmological constant $\Lambda$ and the deformation parameter $z \equiv 1/\kappa$ is quite intricate and the explicit construction of deformed symmetries in $(3+1)$ dimensions is highly non-trivial. [2] Moreover, the associated $(3+1)$ non-commutative spacetime is still work in progress [46].

Mathematically, the $\kappa$-deformation can be defined as a Hopf algebra structure on $\mathcal{U}_z(\mathfrak{g}_\Lambda^{n+1})$ in the direction of the coboundary Lie bialgebra $(\mathfrak{g}_\Lambda^{n+1}, \delta_\kappa)$, defined by an $r$-matrix that we will call $r^{n+1}$. The explicit form of this $r$-matrix is dimension dependent, but the structure is always similar: it contains products of boost and translation generators as well as products of rotations weighted by the square root of the cosmological constant. In the case of vanishing cosmological constant it can be written in our kinematical basis as

$$r_0^{n+1} = z \sum_{i=1}^{n} K_i \wedge P_i\,. \tag{16.5}$$

In the case of non-vanishing cosmological constant $\Lambda$ this $r$-matrix is not modified when $n = 1$ or $n = 2$, but in the $(3 + 1)$-dimensional case a new term must be added [31, 45, 47, 48], resulting in

$$r_\Lambda^{3+1} = z \sum_{i=1}^{3} K_i \wedge P_i + z\sqrt{-\Lambda}\, J_1 \wedge J_2\,. \tag{16.6}$$

This last term implies a deformation of the three-dimensional rotation subalgebra, which cannot be avoided by any generalization of the $\kappa$-Poincaré algebra (see [46] for a complete discussion).

---

[2] In our notation $\kappa = 1/z$, so that the commutative limit is $z \to 0$. The notation $\kappa$-deformation is here maintained due to historical reasons, although we could as well speak about $z$-deformation.

## 16.3 The $\kappa$-$G_\Lambda$ quantum group in (1+1) dimensions

In this section we will present the Poisson-Hopf algebra structure on the universal enveloping algebra $\mathcal{U}_z(\mathfrak{g}_\Lambda^{1+1})$ induced by the $\kappa$-deformation. Firstly we give the full description of the Lie bialgebra $(\mathfrak{g}_\Lambda^{1+1}, \delta_\kappa)$ and then we present the explicit expressions for the Poisson-Hopf structure on $\mathcal{U}_z(\mathfrak{g}_\Lambda^{1+1})$. After that, we consider the Lie group $G_\Lambda^{1+1}$ and a concrete Poisson structure, the so-called Sklyanin bracket, which endows $\kappa$-$G_\Lambda^{1+1}$ with a Poisson-Lie structure. Explicit expressions for the (Poisson version of the) non-commutative spacetime are given. We conclude by performing the limit $\Lambda \to 0$ and recovering well-known expressions for $\kappa$-Minkowski spacetime.

### 16.3.1 Coboundary Lie bialgebra structure $(\mathfrak{g}_\Lambda^{1+1}, \delta_\kappa)$

The Lie algebra brackets for $\mathfrak{g}_\Lambda^{1+1}$ are given in the kinematical basis $\{P_0, P_1, K \equiv K_1\}$ in (16.1) and the $r$-matrix defining the Lie bialgebra $(\mathfrak{g}_\Lambda^{1+1}, \delta_\kappa)$ reads [47, 49]

$$r_\Lambda^{1+1} = zK \wedge P_1.$$

The associated cocommutator can be directly obtained from (16.4) and is given by

$$\delta_\kappa(P_0) = 0, \qquad \delta_\kappa(P_1) = z\, P_1 \wedge P_0, \qquad \delta_\kappa(K) = z\, K \wedge P_0. \qquad (16.7)$$

The quadratic Casimir for $\mathfrak{g}_\Lambda^{1+1}$ turns out to be

$$\mathcal{C} = P_0^2 - P_1^2 + \Lambda\, K^2. \qquad (16.8)$$

### 16.3.2 Poisson-Hopf algebra structure on $\mathcal{U}_z(\mathfrak{g}_\Lambda^{1+1})$

The Poisson version of $\mathfrak{g}_\Lambda^{1+1}$ (16.1) is defined by the Poisson brackets

$$\{K, P_0\} = P_1, \qquad \{K, P_1\} = P_0, \qquad \{P_0, P_1\} = -\Lambda\, K,$$

and the undeformed Poisson-Hopf structure is given by the primitive coproduct

$$\Delta_0(X) = X \otimes 1 + 1 \otimes X, \qquad \forall X \in \mathfrak{g}_\Lambda^{1+1}.$$

To make contact with previous results [50, 30, 31] we will work from now on in the so-called bicrossproduct-type basis [15], which is related to the previous basis by the following nonlinear redefinition of the generators

$$P_0 \to P_0, \qquad P_1 \to e^{\frac{z}{2}P_0}\, P_1, \qquad K \to e^{\frac{z}{2}P_0}\, K,$$

so that the deformed Poisson-Hopf algebra becomes

$$\{K, P_0\} = P_1, \qquad \{P_0, P_1\} = -\Lambda\, K,$$
$$\{K, P_1\} = \frac{1 - \exp(-2zP_0)}{2z} - \frac{z}{2}\, (P_1^2 - \Lambda\, K^2), \qquad (16.9)$$

with deformed coproduct

$$\Delta_\kappa(P_0) = P_0 \otimes 1 + 1 \otimes P_0,$$
$$\Delta_\kappa(P_1) = P_1 \otimes 1 + e^{-zP_0} \otimes P_1, \qquad (16.10)$$
$$\Delta_\kappa(K) = K \otimes 1 + e^{-zP_0} \otimes K.$$

It is worth noting here that a deformation at the coalgebra level (coproduct) implies a specific deformation at the Poisson algebra level (Poisson brackets), since the former has to be a Poisson algebra homomorphism with respect to the latter. This deformation also affects the quadratic Casimir (16.8), resulting in the modified Casimir

$$\mathcal{C}_z = \left( \frac{\sinh(zP_0/2)}{z/2} \right)^2 - e^{zP_0}(P_1^2 - \Lambda K^2), \qquad (16.11)$$

which could be interpreted as a deformed dispersion relation [30]. A clearer intuition for this interpretation will be given when constructing associated momentum spaces in the next section.

### 16.3.3 Non-commutative spacetime induced by a Poisson-Lie structure on $G_\Lambda^{1+1}$

Until now we have looked at the consequences of the $\kappa$-deformation for spacetime symmetries, but these deformations also affect spacetime itself (later on we will see that also momentum space is affected, so we can say that these kind of symmetry deformations lead to a whole deformed phase space). The approach followed here to study (the Poisson versions of) non-commutative spacetimes heavily relies on the following two facts:

- The $\kappa$-deformation is defined by means of a coboundary Lie bialgebra;

- Any coboundary Lie bialgebra $(\mathfrak{g} = \text{Lie}(G), \delta)$ canonically defines a Poisson-Lie structure on $G$.

This canonical Poisson-Lie structure on $G$ is the so-called Sklyanin bracket [12], given by

$$\{f, g\} = r^{ij} \left( X_i^L f X_j^L g - X_i^R f X_j^R g \right), \qquad (16.12)$$

in terms of the components $r^{ij}$ of the classical $r$-matrix defining the $\kappa$-deformation, and the left- and right-invariant vector fields $X_i^L, X_i^R$ on the Lie group $G$.

For the $(1+1)$-dimensional case under consideration, the Appendix contains the construction of $G_\Lambda^{1+1}$ using easy-to-interpret coordinates and explicit expressions for $X_i^L, X_i^R$, which are displayed in Table 16.1. Using these results, we can write an explicit expression for the (Poisson version of the) non-commutative spacetime just as the fundamental Poisson brackets associated to time and space coordinates, which read

$$\{x^1, x^0\} = z \frac{\tanh(\sqrt{-\Lambda}x^1)}{\sqrt{-\Lambda}} = z \frac{\tan(\sqrt{\Lambda}x^1)}{\sqrt{\Lambda}}. \qquad (16.13)$$

As we have emphasized above, this expression is the Poisson version of the full non-commutative spacetime with non-vanishing cosmological constant, which is not easy to be constructed by making use of Hopf algebra duality from the quantum algebra due to reordering contributions. However, in this particular case no ordering problems appear in (16.13) and the full non-commutative spacetime is given by

$$[\hat{x}^1, \hat{x}^0] = z \frac{\tanh\left(\sqrt{-\Lambda}\hat{x}^1\right)}{\sqrt{-\Lambda}} = z \frac{\tan\left(\sqrt{\Lambda}\hat{x}^1\right)}{\sqrt{\Lambda}},$$

where now $\hat{x}^0, \hat{x}^1$ are to be understood as operators acting on some suitable space of fundamental states.

### 16.3.4  Poincaré-Minkowski limit: contraction $\Lambda \to 0$

To finish with our study of spacetime consequences of the $\kappa$-deformation in $(1+1)$ dimensions, we will explicitly show how in the limit $\Lambda \to 0$ we recover the results for the well-known $\kappa$-Poincaré case. Note that the coproduct is not modified in this case (the cosmological constant only appears in the Lie algebra and the coalgebra structure is independent of $\Lambda$). The Poisson-Hopf structure in the limit $\Lambda \to 0$ is given by

$$\{K, P_0\} = P_1, \qquad \{K, P_1\} = \frac{1 - \exp(-2zP_0)}{2z} - \frac{z}{2} P_1^2, \qquad \{P_0, P_1\} = 0,$$
$$(16.14)$$

and the associated coproduct map by

$$\begin{aligned}
\Delta_\kappa(P_0) &= P_0 \otimes 1 + 1 \otimes P_0, \\
\Delta_\kappa(P_1) &= P_1 \otimes 1 + e^{-zP_0} \otimes P_1, \\
\Delta_\kappa(K) &= K \otimes 1 + e^{-zP_0} \otimes K.
\end{aligned} \qquad (16.15)$$

The deformed quadratic Casimir reads

$$C_z = \left( \frac{\sinh\left(zP_0/2\right)}{z/2} \right)^2 - e^{zP_0} P_1^2. \qquad (16.16)$$

Finally, the Poisson non-commutative spacetime is given by

$$\{x^1, x^0\} = z\, x^1$$

that can be trivially quantized giving the full non-commutative spacetime

$$[\hat{x}^1, \hat{x}^0] = z\, \hat{x}^1,$$

which is the well-known $\kappa$-Minkowski spacetime introduced in [14].

## 16.4 Curved momentum spaces in (1+1) dimensions

In this section we review some recent results concerning the effects of the presence of a cosmological constant $\Lambda$ on the momentum space of the $\kappa$-deformation. The fact that a quantum deformation could imply a non-trivial structure of momentum space was first proposed in [51]. This is related with Born's Reciprocity idea [27]. Here we use the language of Hopf algebras and quantum groups, using the quantum duality principle as the key ingredient to construct a deformed momentum space. In fact the first explicit construction of a curved momentum space using this language was that of the associated momentum space to $\kappa$-Poincaré deformation, and this construction was reviewed in [50]. The presence of spacetime curvature induced by $\Lambda$ in addition to the quantum deformation has been a long standing problem obstructing the construction of the momentum space associated to $\kappa$-(anti-)de Sitter, and this issue has been recently solved in [30]. We will see how the presence of $\Lambda$ enforces us to enlarge the momentum space by including boost generators on an equal footing as translation generators. The need of this enlarged space will become evident once the underlying Lie bialgebra is considered, since in the latter spatial translation generators and boost generators play algebraically equivalent roles, as shown in [31].

### 16.4.1 Dual Poisson-Lie group

Consider the Lie bialgebra $(\mathfrak{g}_\Lambda^{1+1}, \delta_\kappa)$ whose cocommutator is given by (16.7). If we denote by $\{X^0, X^1, L\}$ the generators dual to, respectively, $\{P_0, P_1, K\}$, the dual Lie algebra $(\mathfrak{g}_\Lambda^{1+1})^*$ is given by the Lie brackets

$$[X^0, X^1] = -z\, X^1, \qquad [X^0, L] = -z\, L, \qquad [X^1, L] = 0.$$

Note that the quantum parameter $z$ that controls the quantum deformation appears in the above Lie bracket of the dual algebra, while it only appears in the coalgebra structure of the original bialgebra (16.7). A faithful representation $\rho : (\mathfrak{g}_\Lambda^{1+1})^* \to \mathrm{End}(\mathbb{R}^4)$ of this dual Lie algebra is given by

$$\rho(X^0) = z \begin{pmatrix} 0 & 0 & 0 & 1 \\ 0 & 0 & 0 & 0 \\ 0 & 0 & 0 & 0 \\ 1 & 0 & 0 & 0 \end{pmatrix},$$

$$\rho(X^1) = z \begin{pmatrix} 0 & 1 & 0 & 0 \\ 1 & 0 & 0 & 1 \\ 0 & 0 & 0 & 0 \\ 0 & -1 & 0 & 0 \end{pmatrix},$$

$$\rho(L) = z\sqrt{\Lambda} \begin{pmatrix} 0 & 0 & 1 & 0 \\ 0 & 0 & 0 & 0 \\ 1 & 0 & 0 & 1 \\ 0 & 0 & -1 & 0 \end{pmatrix}.$$

This representation is faithful if and only if $\Lambda \neq 0$. In fact, as $\sqrt{\Lambda}$ appears, to avoid complex quantities we will assume from now on that $\Lambda > 0$, that is, the de Sitter case. It should be noted that for anti-de Sitter with $\Lambda < 0$ a completely analogous construction is possible and results are completely equivalent (see final comments in [30]).

If we introduce exponential coordinates of the second kind $\{p_0, p_1, \chi\}$ on the dual Lie group, a group element $h$ sufficiently close to the identity is given by

$$h = \exp\left(p_1 \rho(X^1)\right) \exp\left(\chi \rho(L)\right) \exp\left(p_0 \rho(X^0)\right).$$

A straightforward computation leads to the following explicit parametrization

$$h = \begin{pmatrix} \cosh(zp_0) + \frac{1}{2}z^2 e^z {}^{p_0} (p_1^2 + \Lambda \chi^2) & zp_1 & z\sqrt{\Lambda}\,\chi & \sinh(zp_0) + \frac{1}{2}z^2 e^z {}^{p_0} (p_1^2 + \Lambda \chi^2) \\ z\, e^z {}^{p_0} \, p_1 & 1 & 0 & z\, e^z {}^{p_0} \, p_1 \\ z\, e^z {}^{p_0} \sqrt{\Lambda}\,\chi & 0 & 1 & z\, e^z {}^{p_0} \sqrt{\Lambda}\,\chi \\ \sinh(zp_0) - \frac{1}{2}z^2 e^z {}^{p_0} (p_1^2 + \Lambda \chi^2) & -zp_1 & -z\sqrt{\Lambda}\,\chi & \cosh(zp_0) - \frac{1}{2}z^2 e^z {}^{p_0} (p_1^2 + \Lambda \chi^2) \end{pmatrix}.$$

We denote this dual Lie group by $(G_\Lambda^{1+1})^*$ and we have that $(\mathfrak{g}_\Lambda^{1+1})^* = \text{Lie}((G_\Lambda^{1+1})^*)$.

The multiplication law for the group $(G_\Lambda^{1+1})^*$ can be written as a co-product (see [52]) in the form

$$\begin{aligned} \Delta(p_0) &= p_0 \otimes 1 + 1 \otimes p_0, \\ \Delta(p_1) &= p_1 \otimes 1 + e^{-zp_0} \otimes p_1, \\ \Delta(\chi) &= \chi \otimes 1 + e^{-zp_0} \otimes \chi. \end{aligned} \tag{16.17}$$

Following the quantum duality principle, this coproduct is just (16.10) for $\mathcal{U}_z(\mathfrak{g}_\Lambda^{1+1})$ if dual group coordinates and the generators of the Poisson-Hopf algebra $\mathcal{U}_z(\mathfrak{g}_\Lambda^{1+1})$ are identified as follows:

$$p_0 \equiv P_0, \quad p_1 \equiv P_1, \quad \chi \equiv K. \tag{16.18}$$

Moreover, by following the technique presented in [52] it can be shown that the unique Poisson-Lie structure on $(G_\Lambda^{1+1})^*$ that is compatible with the coproduct (16.17) and has the Lie algebra $\mathfrak{g}_\Lambda^{1+1}$ (16.1) as its linearization is given by the Poisson brackets

$$\{\chi, p_0\} = p_1, \qquad \{p_0, p_1\} = -\Lambda \chi,$$
$$\{\chi, p_1\} = \frac{1 - \exp(-2zp_0)}{2z} - \frac{z}{2}(p_1^2 - \Lambda \chi^2),$$

which is exactly the Poisson-Hopf algebra $\mathcal{U}_z(\mathfrak{g}_\Lambda^{1+1})$ (16.9) under the identification (16.18). Evidently, the Casimir function for this Poisson bracket is

$$C_z = \left(\frac{\sinh(zp_0/2)}{z/2}\right)^2 - e^{zp_0}(p_1^2 - \Lambda \chi^2). \tag{16.19}$$

In this way, the composition law for the momenta with $\kappa$-de Sitter symmetry (16.10) has been reobtained as the group law (16.17) for the coordinates of the dual Poisson-Lie group $(G_\Lambda^{1+1})^*$, and the $\kappa$-de Sitter Casimir function (16.11) can be interpreted as an on-shell relation (16.19) for these coordinates.

As said before the main novelty with respect to the $\kappa$-Poincaré case described in [50] is that here $(G_\Lambda^{1+1})^*$ is three-dimensional, and the momentum space associated to the $\kappa$-deformation is parametrized by $\{p_0, p_1, \chi\}$ and not only by the momenta associated to spacetime translations. Moreover, both in the coproduct (16.17) and the Casimir function (16.19) the role of the parameters $\chi$ and $p_1$ is identical, which supports the role of the former as an additional "hyperbolic" momentum for quantum symmetries with non-vanishing cosmological constant $\Lambda$. As stated before this construction has been explicitly done for $\Lambda > 0$ but an analogue for $\Lambda < 0$ can be constructed by following the very same procedure.

## 16.4.2 Momentum space as an orbit and deformed dispersion relation

In order to give a geometric interpretation of this enlarged momentum space, we follow the procedure formerly proposed in [50] for vanishing cosmological constant $\Lambda = 0$ and recently generalized in [30] for $\Lambda \neq 0$. Consider the action $(G_\Lambda^{1+1})^* \rhd \mathbb{R}^{1,3}$ given by left multiplication, the point $(0, 0, 0, 1) \in \mathbb{R}^{1,3}$ and the orbit of $(G_\Lambda^{1+1})^*$ passing through it. We can easily describe this orbit in terms of Cartesian coordinates $S_i$ on $\mathbb{R}^{1,3}$ as the set

$$\left\{ (S_0, S_1, S_2, S_3) \in \mathbb{R}^{1,3} : -S_0^2 + S_1^2 + S_2^2 + S_3^2 = 1, \ S_0 + S_3 > 0 \right\}, \quad (16.20)$$

which is nothing but a subset of de Sitter spacetime in $(2+1)$ dimensions written as the usual embedding in $\mathbb{R}^{1,3}$. Moreover, we can use previous coordinates $\{p_0, p_1, \chi\}$ on the Lie group $(G_\Lambda^{1+1})^*$ to parametrize this space. Explicitly we can write

$$S_0 = \sinh(z p_0) + \frac{1}{2} z^2 e^{z p_0} (p_1^2 + \Lambda \chi^2),$$
$$S_1 = z e^{z p_0} p_1,$$
$$S_2 = z e^{z p_0} \sqrt{\Lambda} \chi,$$
$$S_3 = \cosh(z p_0) - \frac{1}{2} z^2 e^{z p_0} (p_1^2 + \Lambda \chi^2).$$

The fact that this orbit is not the whole de Sitter spacetime in $(2+1)$ dimensions in encoded in the condition

$$S_0 + S_3 = e^{z p_0} > 0,$$

so we can say that this orbit is half of $(2+1)$-dimensional de Sitter spacetime. Note that the orbit passing through the point $(0, 0, 0, \alpha)$, with $\alpha \neq 0$, would satisfy $-S_0^2 + S_1^2 + S_2^2 + S_3^2 = \alpha^2$, so the resulting de Sitter spacetime radius is free.

### 16.4.3 Poincaré-Minkowski $\Lambda \to 0$ limit in momentum space

We finish the study of the momentum space associated to the $\kappa$-deformation in $(1+1)$ dimensions by giving explicitly the Poisson-Hopf structure on the dual group $(G_\Lambda^{1+1})^*$ in the limit $\Lambda \to 0$ and the geometric interpretation of the orbit, obtaining in this way the known results for the $\kappa$-Minkowski momentum space [50].

The fundamental Poisson brackets are given by

$$\{\chi, p_0\} = p_1, \qquad \{\chi, p_1\} = \frac{1 - \exp(-2zp_0)}{2z} - \frac{z}{2}p_1^2, \qquad \{p_0, p_1\} = 0,$$

and the coproduct by

$$\Delta(p_0) = p_0 \otimes 1 + 1 \otimes p_0,$$
$$\Delta(p_1) = p_1 \otimes 1 + e^{-zp_0} \otimes p_1,$$
$$\Delta(\chi) = \chi \otimes 1 + e^{-zp_0} \otimes \chi.$$

The Casimir function for this Poisson structure reads

$$C_z = \left(\frac{\sinh(zp_0/2)}{z/2}\right)^2 - e^{zp_0}p_1^2.$$

Again, it should be noted that these expressions are just (16.14), (16.15) and (16.16) under the identification (16.18).

The parametrization of the orbit is now just

$$S_0 = \sinh(zp_0) + \frac{1}{2}z^2\,e^{z\,p_0}\,p_1^2,$$
$$S_1 = z\,e^{z\,p_0}\,p_1,$$
$$S_2 = 0,$$
$$S_3 = \cosh(zp_0) - \frac{1}{2}z^2\,e^{z\,p_0}\,p_1^2,$$

so we can describe it as

$$\{(S_0, S_1, 0, S_3) \in \mathbb{R}^{1,3} : -S_0^2 + S_1^2 + S_3^2 = 1,\ S_0 + S_3 > 0\}$$
$$= \{(S_0, S_1, S_3) \in \mathbb{R}^{1,2} : -S_0^2 + S_1^2 + S_3^2 = 1,\ S_0 + S_3 > 0\},$$

where the first description is the intersection of (16.20) with the hyperplane $S_2 = 0$ and the second one is the description as half of a de Sitter spacetime in $(1+1)$ dimensions. This interpretation will be important in higher dimensional generalizations of this construction.

## 16.5 The $\kappa$-$G_\Lambda$ quantum group in $(2+1)$ dimensions

In this section we present the analogous of the previous construction to the case of $(2+1)$ dimensions. The main ideas of the construction are essentially unchanged, so the emphasis is concentrated in presenting the relevant expressions in such a way that the $(1+1)$-dimensional analogues are obtained by canonical Lie algebra projection $\pi_\mathfrak{g}^{2\to1} : \mathfrak{g}_\Lambda^{2+1} \to \mathfrak{g}_\Lambda^{1+1}$. As it has been mentioned before, the $(2+1)$-dimensional case is specially interesting because of the outstanding role that quantum groups and their semiclassical counterparts, Poisson-Lie groups, play as the underlying symmetries for $(2+1)$ gravity. Moreover, the topological nature of this theory implies that a unified consideration of the solutions for positive, zero and negative cosmological constant provides a unified description of all the possible solutions, at least locally.

## 16.5.1   Coboundary Lie bialgebra structure $(\mathfrak{g}_\Lambda^{2+1}, \delta_\kappa)$

Commutation relations in a kinematical basis for $\mathfrak{g}_\Lambda^{2+1}$ are given by (16.2). The two quadratic Casimir functions for (16.2) are

$$\mathcal{C} = P_0^2 - \mathbf{P}^2 - \Lambda(J^2 - \mathbf{K}^2), \qquad \mathcal{W} = -JP_0 + K_1 P_2 - K_2 P_1, \qquad (16.21)$$

where $\mathbf{P}^2 = P_1^2 + P_2^2$ and $\mathbf{K}^2 = K_1^2 + K_2^2$. Recall that $\mathcal{C}$ comes from the Killing–Cartan form and is related to the energy of a point particle, while $\mathcal{W}$ is the Pauli–Lubanski vector.

The coboundary Lie bialgebra $(\mathfrak{g}_\Lambda^{2+1}, \delta_\kappa)$ is defined by the following $r$-matrix:

$$r_\Lambda^{2+1} = z(K_1 \wedge P_1 + K_2 \wedge P_2). \qquad (16.22)$$

Hence the associated cocommutator map is given by

$$\begin{aligned}
\delta_\kappa(P_0) &= \delta_\kappa(J) = 0, \\
\delta_\kappa(P_1) &= z(P_1 \wedge P_0 + \Lambda K_2 \wedge J), \\
\delta_\kappa(P_2) &= z(P_2 \wedge P_0 - \Lambda K_1 \wedge J), \\
\delta_\kappa(K_1) &= z(K_1 \wedge P_0 + P_2 \wedge J), \\
\delta_\kappa(K_2) &= z(K_2 \wedge P_0 - P_1 \wedge J).
\end{aligned} \qquad (16.23)$$

## 16.5.2   Poisson-Hopf algebra structure on $\mathcal{U}_z(\mathfrak{g}_\Lambda^{2+1})$

The Poisson version of $\mathfrak{g}_\Lambda^{2+1}$ (16.2) reads

$$\begin{array}{lll}
\{J, P_i\} = \epsilon_{ij} P_j, & \{J, K_i\} = \epsilon_{ij} K_j, & \{J, P_0\} = 0, \\
\{P_i, K_j\} = -\delta_{ij} P_0, & \{P_0, K_i\} = -P_i, & \{K_1, K_2\} = -J, \\
\{P_0, P_i\} = -\Lambda K_i, & \{P_1, P_2\} = \Lambda J,
\end{array}$$

and the undeformed Poisson-Hopf structure on $\mathcal{U}(\mathfrak{g}_\Lambda^{2+1})$ is given by the primitive coproduct

$$\Delta_0(X) = X \otimes 1 + 1 \otimes X, \qquad \forall X \in \mathfrak{g}_\Lambda^{2+1}.$$

We will use the bicrossproduct type basis [15]. In this basis the Poisson-Hopf algebra structure $\mathcal{U}_z(\mathfrak{g}_\Lambda^{2+1})$ is the Hopf algebra deformation with parameter $z = 1/\kappa$ given by [43, 22, 44]

$$\{J, P_1\} = P_2, \qquad \{J, P_2\} = -P_1, \qquad \{J, P_0\} = 0,$$

$$\{J, K_1\} = K_2, \qquad \{J, K_2\} = -K_1, \qquad \{K_1, K_2\} = -\frac{\sin(2z\sqrt{\Lambda}J)}{2z\sqrt{\Lambda}},$$

$$\{P_0, P_1\} = -\Lambda\, K_1, \quad \{P_0, P_2\} = -\Lambda\, K_2, \quad \{P_1, P_2\} = \Lambda\,\frac{\sin(2z\sqrt{\Lambda}J)}{2z\sqrt{\Lambda}},$$

$$\{K_1, P_0\} = P_1, \qquad\qquad \{K_2, P_0\} = P_2, \qquad\qquad (16.24)$$

$$\{P_2, K_1\} = z\,(P_1 P_2 - \Lambda K_1 K_2), \qquad \{P_1, K_2\} = z\,(P_1 P_2 - \Lambda K_1 K_2),$$

$$\{K_1, P_1\} = \frac{1}{2z}\left(\cos(2z\sqrt{\Lambda}J) - e^{-2zP_0}\right) + \frac{z}{2}\left(P_2^2 - P_1^2\right) - \frac{z\Lambda}{2}\left(K_2^2 - K_1^2\right),$$

$$\{K_2, P_2\} = \frac{1}{2z}\left(\cos(2z\sqrt{\Lambda}J) - e^{-2zP_0}\right) + \frac{z}{2}\left(P_1^2 - P_2^2\right) - \frac{z\Lambda}{2}\left(K_1^2 - K_2^2\right),$$

and deformed coproduct map

$$\Delta_\kappa(P_0) = P_0 \otimes 1 + 1 \otimes P_0, \qquad \Delta_\kappa(J) = J \otimes 1 + 1 \otimes J,$$

$$\Delta_\kappa(P_1) = P_1 \otimes \cos(z\sqrt{\Lambda}J) + e^{-zP_0} \otimes P_1 + \Lambda\, K_2 \otimes \frac{\sin(z\sqrt{\Lambda}J)}{\sqrt{\Lambda}},$$

$$\Delta_\kappa(P_2) = P_2 \otimes \cos(z\sqrt{\Lambda}J) + e^{-zP_0} \otimes P_2 - \Lambda\, K_1 \otimes \frac{\sin(z\sqrt{\Lambda}J)}{\sqrt{\Lambda}}, (16.25)$$

$$\Delta_\kappa(K_1) = K_1 \otimes \cos(z\sqrt{\Lambda}J) + e^{-zP_0} \otimes K_1 + P_2 \otimes \frac{\sin(z\sqrt{\Lambda}J)}{\sqrt{\Lambda}},$$

$$\Delta_\kappa(K_2) = K_2 \otimes \cos(z\sqrt{\Lambda}J) + e^{-zP_0} \otimes K_2 - P_1 \otimes \frac{\sin(z\sqrt{\Lambda}J)}{\sqrt{\Lambda}},$$

which explicitly depends on the cosmological constant $\Lambda$ (compare with (16.10)). The deformed counterpart of the Casimir function $\mathcal{C}$ (16.21) for this Poisson-Hopf algebra reads

$$\mathcal{C}_z = \frac{2}{z^2}\left[\cosh(zP_0)\cos(z\sqrt{\Lambda}J) - 1\right] - e^{zP_0}\left(\mathbf{P}^2 - \Lambda\mathbf{K}^2\right)\cos(z\sqrt{\Lambda}\,J)$$
$$- 2\Lambda\, e^{zP_0}\,\frac{\sin(z\sqrt{\Lambda}J)}{\sqrt{\Lambda}}\,(K_1 P_2 - K_2 P_1).$$

When constructing the associated momentum space in the next section, we will see that a given projection of $\mathcal{C}_z$ will play the role of the deformed dispersion relation.

## 16.5.3  Non-commutative spacetime induced by a Poisson-Lie structure on $G_\Lambda^{2+1}$

As in the $(1+1)$-dimensional case, the fact that the $\kappa$-deformation is defined by a classical $r$-matrix allows us to construct a Poisson structure on $G_\Lambda^{2+1}$ such that group multiplication is a Poisson homomorphism (see the Appendix for details on $G_\Lambda^{2+1}$ and the expressions of left- and right-invariant vector fields which are presented in Tables 16.2 and 16.3).

This canonical Poisson structure is given by the so-called Sklyanin bracket (16.12) on $G_\Lambda^{2+1}$, now for the classical $r$-matrix (16.22). The Poisson subalgebra for spacetime coordinates can be interpreted as a non-commutative spacetime. The fundamental Poisson brackets are given by

$$\{x^1, x^0\} = z\,\frac{\tanh\left(\sqrt{-\Lambda}x^1\right)}{\sqrt{-\Lambda}\cosh^2\left(\sqrt{-\Lambda}x^2\right)} = z\,\frac{\tan\left(\sqrt{\Lambda}x^1\right)}{\sqrt{\Lambda}\cos^2\left(\sqrt{\Lambda}x^2\right)},$$

$$\{x^2, x^0\} = z\,\frac{\tanh\left(\sqrt{-\Lambda}x^2\right)}{\sqrt{-\Lambda}} = z\,\frac{\tan\left(\sqrt{\Lambda}x^2\right)}{\sqrt{\Lambda}},$$

$$\{x^1, x^2\} = 0.$$

Quite remarkably, this Poisson non-commutative spacetime can be trivially quantized as no ordering problems arise due to commutativity of space coordinates. The full non-commutative spacetime is thus given by

$$[\hat{x}^1, \hat{x}^0] = z\,\frac{\tanh\left(\sqrt{-\Lambda}\hat{x}^1\right)}{\sqrt{-\Lambda}\cosh^2\left(\sqrt{-\Lambda}\hat{x}^2\right)} = z\,\frac{\tan\left(\sqrt{\Lambda}\hat{x}^1\right)}{\sqrt{\Lambda}\cos^2\left(\sqrt{\Lambda}\hat{x}^2\right)},$$

$$[\hat{x}^2, \hat{x}^0] = z\,\frac{\tanh\left(\sqrt{-\Lambda}\hat{x}^2\right)}{\sqrt{-\Lambda}} = z\,\frac{\tan\left(\sqrt{\Lambda}\hat{x}^2\right)}{\sqrt{\Lambda}},$$

$$[\hat{x}^1, \hat{x}^2] = 0,$$

where $\{\hat{x}^0, \hat{x}^1, \hat{x}^2\}$ are the generators of the non-commutative algebra.

## 16.5.4 Poincaré-Minkowski limit: contraction $\Lambda \to 0$

As we did in the $(1+1)$-dimensional case, we perform the limit $\Lambda \to 0$ to recover the well-known expressions for the Poisson analogue of the $\kappa$-Poincaré algebra, which reads

$$\begin{aligned}
&\{J, P_1\} = P_2, &&\{J, P_2\} = -P_1, &&\{J, P_0\} = 0,\\
&\{J, K_1\} = K_2, &&\{J, K_2\} = -K_1, &&\{K_1, K_2\} = -J,\\
&\{P_0, P_1\} = 0, &&\{P_0, P_2\} = 0, &&\{P_1, P_2\} = 0,\\
&\{K_1, P_0\} = P_1, &&\{K_2, P_0\} = P_2, &&\quad(16.26)\\
&\{P_2, K_1\} = zP_1 P_2, &&\{P_1, K_2\} = zP_1 P_2,\\
&\{K_1, P_1\} = \frac{1}{2z}\left(1 - e^{-2zP_0}\right) + \frac{z}{2}\left(P_2^2 - P_1^2\right),\\
&\{K_2, P_2\} = \frac{1}{2z}\left(1 - e^{-2zP_0}\right) + \frac{z}{2}\left(P_1^2 - P_2^2\right),
\end{aligned}$$

and the deformed coproduct map is

$$\begin{aligned}
\Delta_\kappa(P_0) &= P_0 \otimes 1 + 1 \otimes P_0, \\
\Delta_\kappa(J) &= J \otimes 1 + 1 \otimes J, \\
\Delta_\kappa(P_1) &= P_1 \otimes 1 + e^{-zP_0} \otimes P_1, \\
\Delta_\kappa(P_2) &= P_2 \otimes 1 + e^{-zP_0} \otimes P_2, \\
\Delta_\kappa(K_1) &= K_1 \otimes 1 + e^{-zP_0} \otimes K_1 + zP_2 \otimes J, \\
\Delta_\kappa(K_2) &= K_2 \otimes 1 + e^{-zP_0} \otimes K_2 - zP_1 \otimes J.
\end{aligned}$$ 
(16.27)

The deformed $\kappa$-Poincaré Casimir

$$C_z = \frac{2}{z^2}\left[\cosh(zP_0) - 1\right] - e^{zP_0}\left(P_1^2 + P_2^2\right) \tag{16.28}$$

is greatly simplified in this limit and in particular it does not depend on boost and rotation generators.

Finally, the associated non-commutative Poisson Minkowski spacetime is defined by the fundamental brackets

$$\begin{aligned}
\{x^1, x^0\} &= z\, x^1, \\
\{x^2, x^0\} &= z\, x^2, \\
\{x^1, x^2\} &= 0,
\end{aligned}$$

and the full non-commutative spacetime is given by

$$\begin{aligned}
[\hat{x}^1, \hat{x}^0] &= z\, \hat{x}^1, \\
[\hat{x}^2, \hat{x}^0] &= z\, \hat{x}^2, \\
[\hat{x}^1, \hat{x}^2] &= 0,
\end{aligned}$$

which is the $(2+1)$-dimensional version of the well-known $\kappa$-Minkowski spacetime [14].

# 16.6 Curved momentum spaces in (2+1) dimensions

In this section we describe the associated momentum space to the $\kappa$-deformation in $(2+1)$ dimensions. We carry out that by following the same procedure as in the $(1+1)$-dimensional case: we firstly construct the dual Lie algebra $(\mathfrak{g}_\Lambda^{2+1})^*$ and afterwards we make use of the quantum duality principle to obtain the dual quantum group. The momentum space will arise as an appropriate orbit of the latter acting on a suitable ambient space.

## 16.6.1 Dual Poisson-Lie group

The skew symmmetrized first-order in $z$ of the coproduct (16.25) is given by the cocommutator map (16.23). Denoting by $\{X^0, X^1, X^2, L^1, L^2, R\}$ the generators dual to, respectively, $\{P_0, P_1, P_2, K_1, K_2, J\}$, the Lie brackets defining the Lie algebra $(\mathfrak{g}_\Lambda^{2+1})^*$ of the dual Poisson-Lie group $(G_\Lambda^{2+1})^*$ are

$$
\begin{array}{lll}
[X^0, X^1] = -z\, X^1, & [X^0, X^2] = -z\, X^2, & [X^1, X^2] = 0, \\
[X^0, L^1] = -z\, L^1, & [X^0, L^2] = -z\, L^2, & [L^1, L^2] = 0, \\
[R, X^2] = -z\, L^1, & [R, L^1] = z \Lambda\, X^2, & [L^1, X^2] = 0, \\
[R, X^1] = z\, L^2, & [R, L^2] = -z \Lambda\, X^1, & [L^2, X^1] = 0, \\
[R, X^0] = 0, & [L^1, X^1] = 0, & [L^2, X^2] = 0.
\end{array}
$$

A faithful representation $\rho : (\mathfrak{g}_\Lambda^{2+1})^* \to \mathrm{End}(\mathbb{R}^6)$ of this Lie algebra for $\Lambda \neq 0$ (as in the $(1+1)$-dimensional case here we present expressions for $\Lambda > 0$, but an analogous construction is valid in the case of negative cosmological constant and the final expressions are valid for both cases) turns out to be

$$
\rho(X^0) = z \begin{pmatrix} 0 & 0 & 0 & 0 & 0 & 1 \\ 0 & 0 & 0 & 0 & 0 & 0 \\ 0 & 0 & 0 & 0 & 0 & 0 \\ 0 & 0 & 0 & 0 & 0 & 0 \\ 0 & 0 & 0 & 0 & 0 & 0 \\ 1 & 0 & 0 & 0 & 0 & 0 \end{pmatrix}, \quad
\rho(X^1) = z \begin{pmatrix} 0 & 1 & 0 & 0 & 0 & 0 \\ 1 & 0 & 0 & 0 & 0 & 1 \\ 0 & 0 & 0 & 0 & 0 & 0 \\ 0 & 0 & 0 & 0 & 0 & 0 \\ 0 & 0 & 0 & 0 & 0 & 0 \\ 0 & -1 & 0 & 0 & 0 & 0 \end{pmatrix},
$$

$$
\rho(X^2) = z \begin{pmatrix} 0 & 0 & 1 & 0 & 0 & 0 \\ 0 & 0 & 0 & 0 & 0 & 0 \\ 1 & 0 & 0 & 0 & 0 & 1 \\ 0 & 0 & 0 & 0 & 0 & 0 \\ 0 & 0 & 0 & 0 & 0 & 0 \\ 0 & 0 & -1 & 0 & 0 & 0 \end{pmatrix}, \quad
\rho(L^1) = z\sqrt{\Lambda} \begin{pmatrix} 0 & 0 & 0 & 1 & 0 & 0 \\ 0 & 0 & 0 & 0 & 0 & 0 \\ 0 & 0 & 0 & 0 & 0 & 0 \\ 1 & 0 & 0 & 0 & 0 & 1 \\ 0 & 0 & 0 & 0 & 0 & 0 \\ 0 & 0 & 0 & -1 & 0 & 0 \end{pmatrix},
$$

$$
\rho(L^2) = z\sqrt{\Lambda} \begin{pmatrix} 0 & 0 & 0 & 0 & 1 & 0 \\ 0 & 0 & 0 & 0 & 0 & 0 \\ 0 & 0 & 0 & 0 & 0 & 0 \\ 0 & 0 & 0 & 0 & 0 & 0 \\ 1 & 0 & 0 & 0 & 0 & 1 \\ 0 & 0 & 0 & 0 & -1 & 0 \end{pmatrix}, \quad
\rho(R) = z\sqrt{\Lambda} \begin{pmatrix} 0 & 0 & 0 & 0 & 0 & 0 \\ 0 & 0 & 0 & 0 & -1 & 0 \\ 0 & 0 & 0 & 1 & 0 & 0 \\ 0 & 0 & -1 & 0 & 0 & 0 \\ 0 & 1 & 0 & 0 & 0 & 0 \\ 0 & 0 & 0 & 0 & 0 & 0 \end{pmatrix}.
$$

If we denote as $\{p_0, p_1, p_2, \chi_1, \chi_2, \theta\}$ the local group coordinates which are dual to $\{X^0, X^1, X^2, L^1, L^2, R\}$, correspondingly, then a Lie group element $h$ on $(G_\Lambda^{2+1})^*$ can be written, provided it is sufficiently close to the identity, as

$$
\begin{aligned}
h = {} & \exp\left(\theta\rho(R)\right) \exp\left(p_1\rho(X^1)\right) \exp\left(p_2\rho(X^2)\right) \\
& \times \exp\left(\chi_1\rho(L^1)\right) \exp\left(\chi_2\rho(L^2)\right) \exp\left(p_0\rho(X^0)\right).
\end{aligned}
$$

Its explicit expression can be straightforwardly computed, although we omit it here for the sake of brevity. The group law for $(G_\Lambda^{2+1})^*$ can be directly derived and written as the following coproduct map for the six group coordinates:

$$
\Delta(p_0) = p_0 \otimes 1 + 1 \otimes p_0, \qquad \Delta(\theta) = \theta \otimes 1 + 1 \otimes \theta,
$$

$$
\Delta(p_1) = p_1 \otimes \cos(z\sqrt{\Lambda}\,\theta) + e^{-zp_0} \otimes p_1 + \Lambda \chi_2 \otimes \frac{\sin(z\sqrt{\Lambda}\,\theta)}{\sqrt{\Lambda}},
$$

$$
\Delta(p_2) = p_2 \otimes \cos(z\sqrt{\Lambda}\,\theta) + e^{-zp_0} \otimes p_2 - \Lambda \chi_1 \otimes \frac{\sin(z\sqrt{\Lambda}\,\theta)}{\sqrt{\Lambda}}, \tag{16.29}
$$

$$
\Delta(\chi_1) = \chi_1 \otimes \cos(z\sqrt{\Lambda}\,\theta) + e^{-zp_0} \otimes \chi_1 + p_2 \otimes \frac{\sin(z\sqrt{\Lambda}\,\theta)}{\sqrt{\Lambda}},
$$

$$
\Delta(\chi_2) = \chi_2 \otimes \cos(z\sqrt{\Lambda}\,\theta) + e^{-zp_0} \otimes \chi_2 - p_1 \otimes \frac{\sin(z\sqrt{\Lambda}\,\theta)}{\sqrt{\Lambda}}.
$$

Again, under the identification

$$p_0 \equiv P_0, \quad p_1 \equiv P_1, \quad p_2 \equiv P_2, \quad \chi_1 \equiv K_1, \quad \chi_2 \equiv K_2, \quad \theta \equiv J, \quad (16.30)$$

this is exactly the coproduct for the Poisson-Hopf algebra $\mathcal{U}_z(\mathfrak{g}_\Lambda^{2+1})$ given in (16.25), and the unique Poisson-Lie structure on $(G_\Lambda^{2+1})^*$ that is compatible with (16.29) and has the undeformed Lie algebra $\mathfrak{g}_\Lambda^{2+1}$ (16.2) as its linearization is the deformed Poisson algebra given by (16.24).

### 16.6.2 Momentum space as an orbit and deformed dispersion relation

As in the $(1+1)$-dimensional case, we consider the action $(G_\Lambda^{2+1})^* \triangleright \mathbb{R}^{1,5}$ given by left multiplication, the point $(0,0,0,0,0,1) \in \mathbb{R}^{1,5}$ and the orbit of $(G_\Lambda^{2+1})^*$ passing through it. This orbit is given in terms of Cartesian coordinates $S_i$ on $\mathbb{R}^{1,5}$ as the set

$$\left\{ (S_0, S_1, S_2, S_3, S_4, S_5) \in \mathbb{R}^{1,5} : -S_0^2 + S_1^2 + S_2^2 + S_3^2 + S_4^2 + S_5^2 = 1, \ S_0 + S_5 > 0 \right\}, \quad (16.31)$$

which is nothing but a subset of de Sitter spacetime in $(4+1)$ dimensions written as the usual embedding in $\mathbb{R}^{1,5}$. Moreover, we can use previous coordinates $\{p_0, p_1, p_2, \chi_1, \chi_2, \theta\}$ on the Lie group $(G_\Lambda^{2+1})^*$ to parametrize this space. Explicitly

$$S_0 = \sinh(zp_0) + \frac{1}{2} z^2 e^{z p_0} \left[ p_1^2 + p_2^2 + \Lambda \left( \chi_1^2 + \chi_2^2 \right) \right],$$

$$S_1 = z e^{z p_0} \left( \cos(z\sqrt{\Lambda}\,\theta)\, p_1 - \sqrt{\Lambda}\, \sin(z\sqrt{\Lambda}\,\theta)\, \chi_2 \right),$$

$$S_2 = z e^{z p_0} \left( \cos(z\sqrt{\Lambda}\,\theta)\, p_2 + \sqrt{\Lambda}\, \sin(z\sqrt{\Lambda}\,\theta)\, \chi_1 \right),$$

$$S_3 = z e^{z p_0} \left( \sqrt{\Lambda}\, \cos(z\sqrt{\Lambda}\,\theta)\, \chi_1 - \sin(z\sqrt{\Lambda}\,\theta)\, p_2 \right),$$

$$S_4 = z e^{z p_0} \left( \sqrt{\Lambda}\, \cos(z\sqrt{\Lambda}\,\theta)\, \chi_2 + \sin(z\sqrt{\Lambda}\,\theta)\, p_1 \right),$$

$$S_5 = \cosh(zp_0) - \frac{1}{2} z^2 e^{z p_0} \left[ p_1^2 + p_2^2 + \Lambda \left( \chi_1^2 + \chi_2^2 \right) \right].$$

Additionally, the condition

$$S_0 + S_5 = e^{z p_0} > 0$$

makes clear that this orbit is half of a de Sitter spacetime in $(4+1)$ dimensions.

### 16.6.3 Poincaré-Minkowski $\Lambda \to 0$ limit in momentum space

Similarly to the $(1+1)$-dimensional case, we finish by writing explicitly the Poisson-Hopf structure on the dual group $(G_\Lambda^{2+1})^*$ in the limit $\Lambda \to 0$, which gives rise to the fundamental Poisson brackets given by

$$
\begin{aligned}
&\{\theta, p_1\} = p_2, && \{\theta, p_2\} = -p_1, && \{\theta, p_0\} = 0, \\
&\{\theta, \chi_1\} = \chi_2, && \{\theta, \chi_2\} = -\chi_1, && \{\chi_1, \chi_2\} = -\theta, \\
&\{p_0, p_1\} = 0, && \{p_0, p_2\} = 0, && \{p_1, p_2\} = 0, \\
&\{\chi_1, p_0\} = p_1, && \{\chi_2, p_0\} = p_2, && \\
&\{p_2, \chi_1\} = z\, p_1 p_2, && \{p_1, \chi_2\} = z\, p_1 p_2, && \\
&\{\chi_1, p_1\} = \frac{1}{2z}\left(1 - e^{-2z p_0}\right) + \frac{z}{2}\left(p_2^2 - p_1^2\right), && \\
&\{\chi_2, p_2\} = \frac{1}{2z}\left(1 - e^{-2z p_0}\right) + \frac{z}{2}\left(p_1^2 - p_2^2\right), &&
\end{aligned}
$$

along with the following coproduct

$$
\begin{aligned}
\Delta(p_0) &= p_0 \otimes 1 + 1 \otimes p_0, \\
\Delta(\theta) &= \theta \otimes 1 + 1 \otimes \theta, \\
\Delta(p_1) &= p_1 \otimes 1 + e^{-z p_0} \otimes p_1, \\
\Delta(p_2) &= p_2 \otimes 1 + e^{-z p_0} \otimes p_2, \\
\Delta(\chi_1) &= \chi_1 \otimes 1 + e^{-z p_0} \otimes \chi_1 + z\, p_2 \otimes \theta, \\
\Delta(\chi_2) &= \chi_2 \otimes 1 + e^{-z p_0} \otimes \chi_2 - z\, p_1 \otimes \theta.
\end{aligned}
$$

The Casimir function for this Poisson structure reads

$$
C_z = \frac{2}{z^2}\left[\cosh(z p_0) - 1\right] - e^{z p_0}\left(p_1^2 + p_2^2\right).
$$

It should be noted that, again, these expressions are just (16.26), (16.27) and (16.28) under the identification (16.30).

Now the parametrization of the orbit is just

$$
\begin{aligned}
S_0 &= \sinh(z p_0) + \frac{1}{2} z^2 e^{z\, p_0}\left(p_1^2 + p_2^2\right), \\
S_1 &= z\, e^{z\, p_0}\, p_1, \\
S_2 &= z\, e^{z\, p_0}\, p_2, \\
S_3 &= 0, \\
S_4 &= 0, \\
S_5 &= \cosh(z p_0) - \frac{1}{2} z^2 e^{z\, p_0}\left(p_1^2 + p_2^2\right),
\end{aligned}
$$

so we can describe it as

$$
\begin{aligned}
&\left\{(S_0, S_1, S_2, 0, 0, S_5) \in \mathbb{R}^{1,5} : -S_0^2 + S_1^2 + S_2^2 + S_5^2 = 1,\ S_0 + S_5 > 0\right\} \\
&= \left\{(S_0, S_1, S_2, S_5) \in \mathbb{R}^{1,3} : -S_0^2 + S_1^2 + S_2^2 + S_5^2 = 1,\ S_0 + S_5 > 0\right\},
\end{aligned}
$$

where the first description corresponds to the intersection of (16.31) with the hyperplane $S_3 = S_4 = 0$ and the second one is the description as half of a de Sitter spacetime in $(2+1)$ dimensions, as it was originally obtained in [50].

## 16.7 Concluding remarks

We have presented a review of recent developments relating non-commutative spacetimes and curved momentum spaces arising from the so-called $\kappa$-(anti-)de Sitter quantum deformation, which is the analogous of the well known $\kappa$-Poincaré deformation with non-vanishing cosmological constant. Our construction treats in a unified way all the $\kappa$-deformations, regardless of the value of the cosmological constant, and thus allows for an easier interpretation of the global consequences of the quantum deformation. In particular, all the expressions are continuous with respect to the cosmological constant parameter $\Lambda$, producing in this way a uniform description of the Poincaré, de Sitter and anti-de Sitter cases.

While we focussed on $(1+1)$ and $(2+1)$ dimensional models, both the quantum spacetimes and the curved momentum spaces constructed in this paper can be generalized to the physically relevant $(3+1)$-dimensional case, as shown in [31]. Again, the cases with positive and negative cosmological constant give rise to different momentum spaces: when $\Lambda > 0$ (de Sitter spacetime) the momentum space is half of a $(6+1)$-dimensional de Sitter spacetime, while when $\Lambda < 0$ (anti-de Sitter spacetime) the momentum space can be identified with half of an $SO(4,4)$-quadric.

As it is discussed in [31] and rigorously studied in [46], the essential structural difference that arises in the $(3+1)$-dimensional case is related with the appearance of the term $z\sqrt{-\Lambda}\, J_1 \wedge J_2$ in the classical $r$-matrix (16.6), which is not present either in lower dimensional cases or in the case of vanishing cosmological constant (16.5). Again, a remarkable property of both the $(6+1)$-dimensional de Sitter spacetime and the $SO(4,4)$-quadric is that a $(3+1)$-dimensional de Sitter spacetime is obtained as an intersection with some hyperplane, giving in this way a clear geometrical interpretation of the vanishing cosmological constant limit.

We conclude by providing an outlook on a few open issues that the results presented here will allow to understand better. As mentioned in the preceding paragraph, our improved understanding of the new structures that emerge in going from lower-dimensional cases to the $(3+1)$ model allows to construct explicitly the $(3+1)$ non-commutative spacetime corresponding to the $\kappa$-deformation in presence of a non-vanishing cosmological constant $\Lambda$. This is already work in progress and will be presented in [46]. Also, the very same techniques developed here can be applied to the construction of the curved momentum spaces associated to other quantum deformations. In the same spirit, the study of non-relativistic limits of the construction here described is work in progress and will be given elsewhere. Finally, the work we presented here could be generalised to describe the momentum space associated to deformed symmetries of spacetime models that are not maximally symmetric. For example, it would be of particular interest to analyse the momentum space associated to a quantum-deformed Friedmann-Lemaître-Robertson-Walker (FLRW) geometry, which is relevant from a cosmological point of view. The case of a homogeneous and isotropic spacetime in the framework of the $\kappa$-deformation was considered from a phenomenological point of view in [53], where general dispersion relations for a particle moving on a FLRW geometry were derived. Finding a way to perform the symmetry breaking under which the static (A)dS results here presented could be connected to a FLRW-type model is indeed worthy of being investigated. A possible line of attack would rely on a procedure that was employed in [54, 55] to describe relativistically-compatible deformations of particles worldlines propagating over such a quantum FLRW geometry. The fundamental ingredient that allowed to identify such worldlines relied on describing the FLRW spacetime as a series of 'slices' of de Sitter spacetime with different cosmological constants.

## Acknowledgments

A.B., I.G-S. and F.J.H. have been partially supported by the grant MTM2016-79639-P (AEI/FEDER, UE), by Junta de Castilla y León (Spain) under grant VA057U16 and by the Action MP1405 QSPACE from the European Cooperation in Science and Technology (COST). I.G-S. acknowledges a PhD grant from the Junta de Castilla y León (Spain) and European Social Fund. G.G. acknowledges a Grant for Visiting Researchers at the Campus of International Excellence "Triangular-E3" (MECD, Spain).

## Appendix: Invariant vector fields

In this appendix we will summarize useful expressions regarding $G_\Lambda^{n+1}$. Consider a Lie group $G$ with Lie algebra $\mathfrak{g} = \text{Lie } G$ and fix a basis $\{X_i\}$ of $\mathfrak{g}$. Then

$$[X_i, X_j] = c_{ij}^k \, X_k, \quad i, j, k \in \{1, \ldots, N\},$$

where $c_{ij}^k$ are the structure constants with respect to this basis and $N = \dim \mathfrak{g}$. From a faithful representation $\rho : \mathfrak{g} \to \text{End}(V)$ of the Lie algebra $\mathfrak{g}$ on some vector space $V$, we can always introduce local coordinates on the Lie group using the so called exponential coordinates of the second kind $\{\alpha^i\}$, in terms of which an element of the group is written as

$$g = \prod_{i=1}^{N} \exp\left(\alpha^i \rho(X_i)\right).$$

We denote by $X_i^L, X_i^R$ left- and right-invariant vector fields, respectively, defined by their action on functions $f \in \mathcal{C}^\infty(G)$:

$$X_i^L f(g) = \frac{\mathrm{d}}{\mathrm{d}t}\bigg|_{t=0} f\left(g \, e^{tX_i}\right), \qquad X_i^R f(g) = \frac{\mathrm{d}}{\mathrm{d}t}\bigg|_{t=0} f\left(e^{tX_i} g\right).$$

The Lie brackets of these vector fields satisfy

$$[X_i^L, X_j^L] = c_{ij}^k \, X_k^L, \qquad\qquad [X_i^R, X_j^R] = -c_{ij}^k \, X_k^R.$$

**Table 16.1:** Left- and right-invariant vector fields on the isometry groups of the $(1+1)$-dimensional de Sitter ($\Lambda > 0$), anti-de Sitter ($\Lambda < 0$) and Minkowski ($\Lambda = 0$) spaces in terms of $\eta = \sqrt{-\Lambda}$.

---

$$X_{P_0}^L = \frac{1}{\cosh(\eta x^1)} \left( \cosh\xi\, \partial_{x^0} + \cosh(\eta x^1)\sinh\xi\, \partial_{x^1} - \eta\sinh(\eta x^1)\cosh\xi\, \partial_\xi \right)$$

$$X_{P_1}^L = \frac{1}{\cosh(\eta x^1)} \left( \sinh\xi\, \partial_{x^0} + \cosh(\eta x^1)\cosh\xi\, \partial_{x^1} - \eta\sinh(\eta x^1)\sinh\xi\, \partial_\xi \right)$$

$$X_K^L = \partial_\xi$$

---

$$X_{P_0}^R = \partial_{x^0}$$

$$X_{P_1}^R = \frac{1}{\cosh(\eta x^1)} \left( -\sin(\eta x^0)\sinh(\eta x^1)\, \partial_{x^0} + \cos(\eta x^0)\cosh(\eta x^1)\, \partial_{x^1} - \eta\sin(\eta x^0)\, \partial_\xi \right)$$

$$X_K^R = \frac{1}{\cosh(\eta x^1)} \left( \frac{\cos(\eta x^0)\sinh(\eta x^1)}{\eta}\, \partial_{x^0} + \frac{\sin(\eta x^0)\cosh(\eta x^1)}{\eta}\, \partial_{x^1} + \cos(\eta x^0)\, \partial_\xi \right)$$

---

We start with the $(1+1)$-dimensional case and consider the faithful representation for the Lie algebra $\rho : \mathfrak{g}_\Lambda^{1+1} \to \mathrm{End}(\mathbb{R}^3)$ given by

$$\rho(P_0) = \begin{pmatrix} 0 & \Lambda & 0 \\ 1 & 0 & 0 \\ 0 & 0 & 0 \end{pmatrix}, \quad \rho(P_1) = \begin{pmatrix} 0 & 0 & -\Lambda \\ 0 & 0 & 0 \\ 1 & 0 & 0 \end{pmatrix}, \quad \rho(K) = \begin{pmatrix} 0 & 0 & 0 \\ 0 & 0 & 1 \\ 0 & 1 & 0 \end{pmatrix},$$

and we introduce local coordinates on $G_\Lambda^{1+1}$ using exponential coordinates of the second kind given by

$$g = \exp\left(x^0\rho(P_0)\right) \exp\left(x^1\rho(P_1)\right) \exp\left(\xi\,\rho(K)\right).$$

Note that this coordinates $\{x^0, x^1, \xi \equiv \xi^1\}$ have a direct physical interpretation as time, space and boost (rapidity) coordinates, respectively. Left- and right-invariant vector fields in this coordinates are written in Table 16.1 in terms of a parameter $\eta = \sqrt{-\Lambda}$.

Following the very same previous procedure, we consider the $(2+1)$-dimensional case and begin with a faithful representation for the Lie algebra $\rho : \mathfrak{g}_\Lambda^{2+1} \to \mathrm{End}(\mathbb{R}^4)$, which takes the explicit form

$$\rho(P_0) = \begin{pmatrix} 0 & \Lambda & 0 & 0 \\ 1 & 0 & 0 & 0 \\ 0 & 0 & 0 & 0 \\ 0 & 0 & 0 & 0 \end{pmatrix}, \quad \rho(J) = \begin{pmatrix} 0 & 0 & 0 & 0 \\ 0 & 0 & 0 & 0 \\ 0 & 0 & 0 & -1 \\ 0 & 0 & 1 & 0 \end{pmatrix},$$

$$\rho(P_1) = \begin{pmatrix} 0 & 0 & -\Lambda & 0 \\ 0 & 0 & 0 & 0 \\ 1 & 0 & 0 & 0 \\ 0 & 0 & 0 & 0 \end{pmatrix}, \quad \rho(K_1) = \begin{pmatrix} 0 & 0 & 0 & 0 \\ 0 & 0 & 1 & 0 \\ 0 & 1 & 0 & 0 \\ 0 & 0 & 0 & 0 \end{pmatrix},$$

$$\rho(P_2) = \begin{pmatrix} 0 & 0 & 0 & -\Lambda \\ 0 & 0 & 0 & 0 \\ 0 & 0 & 0 & 0 \\ 1 & 0 & 0 & 0 \end{pmatrix}, \quad \rho(K_2) = \begin{pmatrix} 0 & 0 & 0 & 0 \\ 0 & 0 & 0 & 1 \\ 0 & 0 & 0 & 0 \\ 0 & 1 & 0 & 0 \end{pmatrix}.$$

Next we introduce local coordinates on $G_\Lambda^{2+1}$ using exponential coordinates of the second kind in the form

$$g = \exp(x^0\rho(P_0))\exp(x^1\rho(P_1))\exp(x^2\rho(P_2))\exp(\xi^1\rho(K_1))\exp(\xi^2\rho(K_2))\exp(\theta J).$$

This coordinates $\{x^0, x^1, x^2, \xi^1, \xi^2, \theta\}$ have a direct physical interpretation as time, space, boosts and rotation coordinates, respectively. In Tables 16.2 and 16.3 left- and right-invariant vector fields are given in terms of these coordinates.

**Table 16.2:** Left-invariant vector fields on the isometry groups of the $(2+1)$-dimensional de Sitter $(\Lambda > 0)$, anti-de Sitter $(\Lambda < 0)$ and Minkowski $(\Lambda = 0)$ spaces in terms of $\eta = \sqrt{-\Lambda}$.

$$X_{P_0}^L = \frac{\cosh \xi^1 \cosh \xi^2}{\cosh(\eta x^1) \cosh(\eta x^2)} \left( \partial_{x^0} - \eta \sinh(\eta x^1) \partial_{\xi^1} \right) + \frac{\sinh \xi^1 \cosh \xi^2}{\cosh(\eta x^2)} \partial_{x^1} + \sinh \xi^2 \, \partial_{x^2}$$
$$-\eta \tanh(\eta x^2) \cosh \xi^2 \, \partial_{\xi^2}$$

$$X_{P_1}^L = \left( \frac{\sinh \xi^1 \cos \theta + \cosh \xi^1 \sinh \xi^2 \sin \theta}{\cosh(\eta x^1) \cosh(\eta x^2)} \right) \left( \partial_{x^0} - \eta \sinh(\eta x^1) \partial_{\xi^1} \right)$$
$$+ \left( \frac{\cosh \xi^1 \cos \theta + \sinh \xi^1 \sinh \xi^2 \sin \theta}{\cosh(\eta x^2)} \right) \partial_{x^1}$$
$$+ \cosh \xi^2 \sin \theta \, \partial_{x^2} - \eta \tanh(\eta x^2) \left( \tanh \xi^2 \cos \theta \, \partial_{\xi^1} + \sinh \xi^2 \sin \theta \, \partial_{\xi^2} - \frac{\cos \theta}{\cosh \xi^2} \partial_\theta \right)$$

$$X_{P_2}^L = \left( \frac{\cosh \xi^1 \sinh \xi^2 \cos \theta - \sinh \xi^1 \sin \theta}{\cosh(\eta x^1) \cosh(\eta x^2)} \right) \left( \partial_{x^0} - \eta \sinh(\eta x^1) \partial_{\xi^1} \right)$$
$$+ \left( \frac{\sinh \xi^1 \sinh \xi^2 \cos \theta - \cosh \xi^1 \sin \theta}{\cosh(\eta x^2)} \right) \partial_{x^1}$$
$$+ \cosh \xi^2 \cos \theta \, \partial_{x^2} + \eta \tanh(\eta x^2) \left( \tanh \xi^2 \sin \theta \, \partial_{\xi^1} - \sinh \xi^2 \cos \theta \, \partial_{\xi^2} - \frac{\sin \theta}{\cosh \xi^2} \partial_\theta \right)$$

$$X_{K_1}^L = \frac{\cos \theta}{\cosh \xi^2} \partial_{\xi^1} + \sin \theta \, \partial_{\xi^2} + \tanh \xi^2 \cos \theta \, \partial_\theta$$

$$X_{K_2}^L = -\frac{\sin \theta}{\cosh \xi^2} \partial_{\xi^1} + \cos \theta \, \partial_{\xi^2} - \tanh \xi^2 \sin \theta \, \partial_\theta$$

$$X_J^L = \partial_\theta$$

**Table 16.3:** Right-invariant vector fields on the isometry groups of the (2+1)-dimensional de Sitter ($\Lambda > 0$), anti-de Sitter ($\Lambda < 0$) and Minkowski ($\Lambda = 0$) spaces in terms of $\eta = \sqrt{-\Lambda}$.

$$X_{P_0}^R = \partial_{x^0}$$

$$X_{P_1}^R = -\sin(\eta x^0)\tanh(\eta x^1)\,\partial_{x^0} + \cos(\eta x^0)\,\partial_{x^1} - \eta\,\frac{\sin(\eta x^0)}{\cosh(\eta x^1)}\,\partial_{\xi^1}$$

$$
\begin{aligned}
X_{P_2}^R = {}& -\frac{\sin(\eta x^0)\tanh(\eta x^2)}{\cosh(\eta x^1)}\left(\partial_{x^0} - \eta\sinh(\eta x^1)\,\partial_{\xi^1}\right) \\
& -\cos(\eta x^0)\sinh(\eta x^1)\tanh(\eta x^2)\,\partial_{x^1} + \cos(\eta x^0)\cosh(\eta x^1)\,\partial_{x^2} \\
& +\eta\left(\frac{\cos(\eta x^0)\sinh(\eta x^1)\sinh\xi^1 - \sin(\eta x^0)\cosh\xi^1}{\cosh(\eta x^2)}\right)\partial_{\xi^2} \\
& +\eta\left(\frac{\cos(\eta x^0)\sinh(\eta x^1)\cosh\xi^1 - \sin(\eta x^0)\sinh\xi^1}{\cosh(\eta x^2)\cosh\xi^2}\right)\left(\partial_\theta - \sinh\xi^2\partial_{\xi^1}\right)
\end{aligned}
$$

$$X_{K_1}^R = \frac{\cos(\eta x^0)\tanh(\eta x^1)}{\eta}\,\partial_{x^0} + \frac{\sin(\eta x^0)}{\eta}\,\partial_{x^1} + \frac{\cos(\eta x^0)}{\cosh(\eta x^1)}\,\partial_{\xi^1}$$

$$
\begin{aligned}
X_{K_2}^R = {}& \frac{\cos(\eta x^0)\tanh(\eta x^2)}{\eta\cosh(\eta x^1)}\left(\partial_{x^0} - \eta\sinh(\eta x^1)\,\partial_{\xi^1}\right) \\
& -\frac{\sin(\eta x^0)\sinh(\eta x^1)\tanh(\eta x^2)}{\eta}\,\partial_{x^1} + \frac{\sin(\eta x^0)\cosh(\eta x^1)}{\eta}\,\partial_{x^2} \\
& +\left(\frac{\sin(\eta x^0)\sinh(\eta x^1)\sinh\xi^1 + \cos(\eta x^0)\cosh\xi^1}{\cosh(\eta x^2)}\right)\partial_{\xi^2} \\
& +\left(\frac{\sin(\eta x^0)\sinh(\eta x^1)\cosh\xi^1 + \cos(\eta x^0)\sinh\xi^1}{\cosh(\eta x^2)\cosh\xi^2}\right)\left(\partial_\theta - \sinh\xi^2\partial_{\xi^1}\right)
\end{aligned}
$$

$$
\begin{aligned}
X_J^R = {}& -\frac{\cosh(\eta x^1)\tanh(\eta x^2)}{\eta}\,\partial_{x^1} + \frac{\sinh(\eta x^1)}{\eta}\,\partial_{x^2} \\
& -\frac{\cosh(\eta x^1)}{\cosh(\eta x^2)}\left(\cosh\xi^1\tanh\xi^2\,\partial_{\xi^1} - \sinh\xi^1\partial_{\xi^2} - \frac{\cosh\xi^1}{\cosh\xi^2}\,\partial_\theta\right)
\end{aligned}
$$

# References

[1] G. Amelino-Camelia, Phys. Lett. B **510** (2001) 255

[2] G. Amelino-Camelia, Int. J. Mod. Phys. D **11** (2002) 35

[3] G. Amelino-Camelia, Int. J. Mod. Phys. D **11** (2002) 1643

[4] J. Magueijo, L. Smolin, Phys. Rev. Lett. **88** (2002) 190403

[5] J. Kowalski-Glikman, S. Nowak, Phys. Lett. B **539** (2002) 126

[6] J. Kowalski-Glikman, Phys. Lett. B **547** (2002) 291

[7] J. Kowalski-Glikman, S. Nowak, Class. Quantum Grav. **20** (2003) 4799

[8] J. Lukierski, A. Nowicki, Int. J. Mod. Phys. A **18** (2003) 7

[9] A. Ballesteros, N.R. Bruno, F.J. Herranz, J. Phys. A: Math. Gen. **36** (2003) 10493

[10] G. Amelino-Camelia, Living Rev. Rel. **16** (2013) 5

[11] V.G. Drinfel'd, *Quantum Groups*, in: Gleason A V (Ed.), *Proc. Int. Cong. Math. Berkeley 1986*, (Providence: AMS) (1987) p. 798

[12] V. Chari, A. Pressley, A Guide to Quantum Groups, Cambridge University Press, Cambridge, 1994

[13] S. Majid, Foundations of Quantum Group Theory, Cambridge University Press, Cambridge, 1995

[14] P. Maslanka, J. Phys. A: Math. Gen. **26** (1993) L1251

[15] S. Majid, H. Ruegg, Phys. Lett. B **334** (1994) 348

[16] S. Zakrzewski, J. Phys. A: Math. Gen. **27** (1994) 2075

[17] J. Lukierski, H. Ruegg, Phys. Lett. B **329** (1994) 189

[18] J. Lukierski, A. Nowicki, W.J. Zakrzewski, Ann. Phys. **243** (1995) 90

[19] A. Ballesteros, F.J. Herranz, C.M. Pereña, Phys. Lett. B **391** (1997) 71

[20] A. Ballesteros, F.J. Herranz, N.R. Bruno, Phys. Lett. B **574** (2003) 276

[21] A. Ballesteros, I. Gutierrez-Sagredo, C. Meusburger, F.J. Herranz, P. Naranjo, J. Phys.: Conf. Ser. **880** (2017) 012023

[22] A. Ballesteros, F.J. Herranz, N.R. Bruno, Adv. High Energy Phys. **2017** (2017) 7876942

[23] S. Majid, Lect. Notes Phys. **541** (2000) 227

[24] G. Amelino-Camelia, S. Majid, Int. J. Mod. Phys. A **15** (2000) 4301

[25] J. Kowalski-Glikman, S. Nowak, Int. J. Mod. Phys. D **12** (2003) 299

[26] G. Gubitosi, F. Mercati, Class. Quant. Grav. **30** (2013) 145002

[27] M. Born, Proc. R. Soc. Lond. A **165** (1938) 291

[28] G. Amelino-Camelia, L. Barcaroli, G. Gubitosi, N. Loret, Class. Quant. Grav. **30** (2013) 235002

[29] G. Amelino-Camelia, L. Barcaroli, N. Loret, Int. J. Theor. Phys. **51** (2012) 3359

[30] A. Ballesteros, G. Gubitosi, I. Gutierrez-Sagredo, F.J. Herranz, Phys. Lett. B **773** (2017) 47

[31] A. Ballesteros, G. Gubitosi, I. Gutierrez-Sagredo, F.J. Herranz, Phys. Rev. D, **97** (2018) 106024

[32] S. Carlip, Quantum gravity in 2+ 1 dimensions, Cambridge University Press, Cambridge, 2003.

[33] A. Achúcarro and P. K. Townsend, Phys. Lett. B **180** (1986) 89

[34] E. Witten, Nucl. Phys. B, **311** (1988) 46

[35] C. Meusburger and B. Schroers, Class. Quant. Grav. **20** (2003) 2193

[36] C. Meusburger and B. Schroers, Nucl. Phys. B **806** (2009) 462

[37] A. Ballesteros, I. Gutierrez-Sagredo and F. J. Herranz, Class. Quant. Grav. **36** (2019) 025003

[38] V.G. Drinfel'd, Soviet Math. Dokl. **27** (1983) 68

[39] J. Lukierski, A. Nowicki, H. Ruegg, V.N. Tolstoy, Phys. Lett. B **264** (1991) 331

[40] S. Giller, P. Kosinski, J. Kunz, M. Majewski, P. Maslanka, Phys. Lett. B **286** (1992) 57

[41] J. Lukierski, H. Ruegg, A. Nowicky, Phys. Lett. B **293** (1992) 344

[42] A. Ballesteros, F.J. Herranz, M.A. del Olmo, M. Santander, J. Phys. A: Math. Gen. **26** (1993) 5801

[43] A. Ballesteros, F.J. Herranz, M.A. del Olmo, M. Santander, J. Phys. A: Math. Gen. **27** (1994) 1283

[44] G. Amelino-Camelia, L. Smolin, A. Starodubtsev, Class. Quantum Grav. **21** (2004) 3095

[45] A. Ballesteros, F.J. Herranz, F. Musso, P. Naranjo, Phys. Lett. B **766** (2017) 205

[46] A. Ballesteros, I. Gutierrez-Sagredo, F.J. Herranz, Phys. Lett. B, in press, (2019). https://doi.org/10.1016/j.physletb.2019.07.038

[47] A. Ballesteros, N. Gromov, F.J. Herranz, M.A. del Olmo, M. Santander, J. Math. Phys. **36** (1995) 5916

[48] A. Ballesteros, N.R. Bruno, F.J. Herranz, Czech. J. Phys. **11** (2004) 1321

[49] A. Ballesteros, F.J. Herranz, M.A. del Olmo, M. Santander, J. Math. Phys. **36** (1995) 631

[50] J. Kowalski-Glikman, Int. J. Mod. Phys. A **28** (2013) 1330014

[51] H.S. Snyder, Phys. Rev. **71** (1947) 38

[52] A. Ballesteros, F. Musso, J. Phys. A: Math. Theor. **46** (2013) 195203

[53] L. Barcaroli, L. K. Brunkhorst, G. Gubitosi, N. Loret, C. Pfeifer, Phys Rev D. **95** (2017) 024036

[54] A. Marciano, G. Amelino-Camelia, N. R. Bruno, G. Gubitosi, G. Mandanici and A. Melchiorri, JCAP **1006** (2010) 030

[55] G. Rosati, G. Amelino-Camelia, A. Marciano and M. Matassa, Phys. Rev. D **92** (2015) no.12, 124042

# 17  SOME DEVELOPMENTS IN THE HOLOGRAPHY THEORIES OF BLACK HOLES AND CONFORMAL FIELD

A. M. GHEZELBASH

**Abstract**    It is recently conjectured that generic non-extremal Kerr black hole could be holographically dual to a hidden conformal field theory (CFT) in two dimensions. The correspondence states that the near-horizon states of an extremal four (or higher) dimensional black hole could be identified with a certain chiral conformal field theory under the assumption that the central charges from the non-gravitational fields vanish. To understand the chiral conformal field theory, we consider the class of extremal Kerr-Sen black hole that contains three non-gravitational fields as a class of solutions in the low energy limit of heterotic string theory. We derive the expression of the conserved charges for the extremal Kerr-Sen solutions that contain dilaton, abelian gauge filed and antisymmetric tensor filed and show that the central charges from the non-gravitational fields vanish for theories including antisymmetric tensor fields. We combine the calculated central charges with the expected form of the temperature using the Cardy formula to obtain the entropy of the extremal black hole microscopically; in agreement with the macroscopic Bekenstein-Hawking entropy of the extremal black hole. Moreover, it is known that there are two CFT duals (pictures) to describe the charged rotating black holes which correspond to angular momentum J and electric charge Q of the black hole. Furthermore these two pictures can be incorporated by the CFT duals (general picture) that are generated by $SL(2, Z)$ modular group.

C. Duston, M. Holman (Eds), *Spacetime Physics 1907 - 2017. Selected peer-reviewed papers presented at the First Hermann Minkowski Meeting on the Foundations of Spacetime Physics, 15-18 May 2017, Albena, Bulgaria* (Minkowski Institute Press, Montreal 2019). ISBN 978-1-927763-48-3 (softcover), ISBN 978-1-927763-49-0 (ebook).

# 17.1   Introduction[1]

Black holes have been an interesting theoretical laboratories to better understand the nature of quantum gravity. The microscopic entropy of extremal Kerr black hole can be calculated by finding the dual chiral conformal field theory associated with the diffeomorphisms of near-horizon geometry of the Kerr black hole [1, 2, 3, 4, 5, 6, 7]. The diffeomorphisms satisfy an appropriate boundary condition at the infinity. We must note that the correspondence doesn't rely on supersymmetry and string theory, in contrast to the well known AdS/CFT correspondence [8, 9, 10, 11, 12, 13].

In [14], the authors find the entropy of dual CFT for four and higher dimensional Kerr black holes in AdS spacetimes and gauged supergravity. The dual CFT for the five-dimensional BMPV black holes has been found in [15]. Moreover the correspondence has been studied in string theory D1-D5-P and BMPV black holes in [16] and also in the five dimensional Kerr-Gödel black hole [17]. The rotating bubbles, Kerr-Newman black holes in (A)dS spacetimes and rotating NS5 branes and Kerr-Bolt spacetimes have been considered in [18, 19, 20] and [21, 22]. Moreover, in references [23, 24], the authors find the explicit form of two-point function for the conformal operators with different spins, on the near horizon of a near extremal Kerr black hole by variation of the proper boundary actions.

By a close look at these papers, we find that the central charge is computed only from the gravitational tensor field while contributions from other fields like scalar and vector fields are neglected. This leads to the assumption that the central charge of extremal black holes depends only to the gravitational field. In [25], the authors considered the assumption for theories that their actions contain gravity, scalar fields and a multiple of $U(1)$ vector fields and two topological terms. The assumption also considered in [20] for the Kerr-Newmann-(A)dS black hole in the Einstein-Maxwell theory with cosmological constant, where the authors showed that there is no contribution to the central charge of dual CFT from $U(1)$ gauge field in the Einstein-Maxwell theory with cosmological constant.

The other interesting fact is that there are more than one dual CFT for some charged rotating black holes. For example, there is one class of CFT that is associated to the rotation of black hole and the second class of CFT which is associated to the charge of black hole for the four-dimensional Kerr-Newman black holes. These two CFTs are called CFTs in $J$ and $Q$ pictures, respectively [26, 27, 28].

The angular momentum of the Kerr-Newman black hole is related to the rotational symmetry of black hole in $\phi$ direction. In the same way, the charge of the black hole is in correspondence to the gauge symmetry of the black hole. One can interpret the gauge symmetry of the black hole as the rotational symmetry of the uplifted black hole in the fifth-direction $\chi$. In fact, we call the two new CFTs correspond to two rotational symmetries of the uplifted black

[1]This presentation is entirely based on published papers in Journal of High Energy Physics (JHEP) 0908:045, 2009 by A.M. Ghezelbash and Classical and Quantum Gravity 30 (2013) 135005 by A.M. Ghezelbash and H.M. Siahaan.

hole in five dimensions as $\phi'$ and $\chi'$ pictures, respectively [28]. The general picture is a picture that can be derived by using the modular group $SL(2, \mathbb{Z})$ for a torus made by $(\phi, \chi)$.

We also can mention about the class of rotating charged Kerr-Sen black holes in four dimensions [29] which are the exact solutions of the low energy limit of heterotic string theory in four dimensions. The Kerr-Sen black hole solutions include three non-gravitational fields that do not contribute to the central charges of the dual chiral CFT [30]. In an interesting paper [31], the authors show that the Kerr-Sen black hole does not possess any well defined $Q$ picture, in contrast to the Kerr-Newman black hole.

We also must note that the hidden conformal symmetry generators can be extended by including a deformation parameter in the radial equation of motion, for the Kerr-Sen black. The extended deformed hidden conformal symmetries provide the hidden conformal symmetry for the charged Gibbons-Maeda-Garfinkle-Horowitz-Strominger black holes , in the limit of zero rotation [32, 33]. The hidden conformal symmetry to the four-dimensional spacetimes with rotational parameter and NUT twist has been condidered in [34]. For other related works, see [35, 36].

In this presentation which entirely is based on author's published papers [30], [31], we consider the class of extremal Kerr-Sen black hole as a class of solutions in the low energy limit of heterotic string theory and show that the antisymmetric tensor field does not contribute to the central charge of the dual CFT. In fact, we verify the validity domain of the suggestion in [25] to include the non-gravitational antisymmetric tensor fields. The Kerr-Sen black hole has been studied in regard to its hidden symmetries, null geodesics, photon capture and its singularities in [37] and [38] and also in black hole lensing in the strong deflection limit [39] and the massive complex scalar field [40]. We also investigate the existence of the extended deformed hidden conformal symmetry for the Kerr-Newman black holes in different pictures.

## 17.2 The Kerr-Sen Black Hole and the entropy in dual CFT

We recall that the effective action of heterotic string theory in four dimensions is given by

$$S = - \int d^4x \sqrt{-\det G} e^{-\Phi} (-R + \frac{1}{12} H^2 - G^{\mu\nu} \partial_\mu \Phi \partial_\nu \Phi + \frac{1}{8} F^2) \qquad (17.1)$$

where $H^2 = H_{\mu\nu\rho} H^{\mu\nu\rho}$ and $F^2 = F_{\mu\nu} F^{\mu\nu}$. $G_{\mu\nu}$ and $\Phi$ are the metric and the dilaton field respectively, $F_{\mu\nu} = \partial_\mu A_\nu - \partial_\nu A_\mu$ is the field strength for the gauge field $A_\mu$ associated with a $U(1)$ subgroup of $E_8 \times E_8$ and

$$H_{\mu\nu\rho} = \partial_\mu B_{\nu\rho} + \partial_\nu B_{\rho\mu} + \partial_\rho B_{\mu\nu} - \frac{1}{4} (A_\mu F_{\nu\rho} + A_\nu F_{\rho\mu} + A_\rho F_{\mu\nu}) \qquad (17.2)$$

where the last three terms are the gauge Chern-Simons terms. The Kerr-Sen black hole is given by [29]

$$
\begin{aligned}
ds^2 &= -\frac{r^2 + a^2 \cos^2(\theta) - 2mr}{r^2 + a^2 \cos^2(\theta) + 2mr \sinh^2(\alpha/2)} dt^2 \\
&+ \frac{r^2 + a^2 \cos^2(\theta) + 2mr \sinh^2(\alpha/2)}{r^2 + a^2 - 2mr} dr^2 \\
&+ (r^2 + a^2 \cos^2(\theta) + 2mr \sinh^2(\alpha/2)) d\theta^2 \\
&- \frac{4mra \cosh^2(\alpha/2) \sin^2(\theta)}{r^2 + a^2 \cos^2(\theta) + 2mr \sinh^2(\alpha/2)} dt d\phi \\
&+ \{(r^2 + a^2)(r^2 + a^2 \cos^2(\theta)) + 2mra^2 \sin^2(\theta) \\
&+ 4mr(r^2 + a^2) \sinh^2(\alpha/2) + 4m^2 r^2 \sinh^4(\alpha/2)\} \\
&\times \frac{\sin^2(\theta)}{r^2 + a^2 \cos^2(\theta) + 2mr \sinh^2(\alpha/2)} d\phi^2 \quad (17.3)
\end{aligned}
$$

and the dilaton and gauge field components are

$$
\Phi = -\ln \frac{r^2 + a^2 \cos^2(\theta) + 2mr \sinh^2(\alpha/2)}{r^2 + a^2 \cos^2(\theta)} \quad (17.4)
$$

$$
A_t = \frac{2mr \sinh(\alpha)}{r^2 + a^2 \cos^2(\theta) + 2mr \sinh^2(\alpha/2)} \quad (17.5)
$$

$$
A_\phi = \frac{-2mra \sinh(\alpha) \sin^2(\theta)}{r^2 + a^2 \cos^2(\theta) + 2mr \sinh^2(\alpha/2)} \quad (17.6)
$$

and the only non-vanishing component of antisymmetric tensor field is

$$
B_{t\phi} = \frac{2mra \sinh^2(\alpha/2) \sin^2(\theta)}{r^2 + a^2 \cos^2(\theta) + 2mr \sinh^2(\alpha/2)}. \quad (17.7)
$$

The black hole solution (17.3) has mass $M = m \cosh^2(\alpha/2)$, charge $Q = \frac{m}{\sqrt{2}} \sinh \alpha$ and angular momentum $J = ma \cosh^2(\alpha/2)$. We rewrite the metric as [30]

$$
\begin{aligned}
ds^2 &= -(1 - \frac{2m\tilde{r}}{\rho^2}) d\tilde{t}^2 + \rho^2 (\frac{d\tilde{r}^2}{\Delta} + d\theta^2) \\
&- \frac{4m\tilde{r}a}{\rho^2} \sin^2 \theta d\tilde{t} d\tilde{\phi} + \{\tilde{r}^2 + a^2 + \frac{2m\tilde{r}a^2 \sin^2 \theta}{\rho^2}\} \sin^2 \theta d\tilde{\phi}^2 \quad (17.8)
\end{aligned}
$$

where $\rho^2 = \tilde{r}^2 + a^2 \cos^2 \theta$ and $\Delta = \tilde{r}^2 - 2M\tilde{r} + a^2$. The parameter $\varrho$ is related to $m = M - \frac{Q^2}{2M}$ and $\alpha$ in (17.3) by $\varrho = 2m \sinh^2(\alpha/2) = Q^2/M$. The dilaton

, gauge field and the antisymmetric tensor field are given by

$$\Phi = -\ln\frac{\tilde{r}(\tilde{r}+\varrho)+a^2\cos^2\theta}{\tilde{r}^2+a^2\cos^2\theta} \tag{17.9}$$

$$A_{\tilde{t}} = \frac{2\sqrt{2}\tilde{r}Q}{\rho^2} \tag{17.10}$$

$$A_{\tilde{\phi}} = \frac{-2\sqrt{2}\tilde{r}Qa\sin^2\theta}{\rho^2} \tag{17.11}$$

$$B_{\tilde{t}\tilde{\phi}} = \frac{\tilde{r}\varrho a\sin^2\theta}{\rho^2}. \tag{17.12}$$

The event horizon of black hole (17.8) is

$$r_H = M + \sqrt{M^2-a^2} \tag{17.13}$$

where to avoid any naked singularity, we should impose

$$|J| \le M^2 - \frac{1}{2}Q^2. \tag{17.14}$$

For the black hole (17.8), the corresponding angular velocity at horizon, Hawking temperature and entropy are

$$\Omega_H = \frac{J}{2M(M^2+\sqrt{M^4-J^2})} \tag{17.15}$$

$$T_H = \frac{\sqrt{(M^4-4J^2}}{4\pi M(M^2+\sqrt{M^4-J^2})} \tag{17.16}$$

$$S = 2\pi M(M+\sqrt{M^2-\frac{J^2}{M^2}}). \tag{17.17}$$

We notice in the extremal limit where $J = M^2 - \frac{1}{2}Q^2$, the angular velocity and Hawking temperature reduce to $\frac{1}{2M}$ and 0, respectively and the entropy reduces simply to $S = 2\pi J$; independent of the mass of black hole.

The near-horizon limit of the extremal black hole (17.8) is given by [30]

$$ds^2 = \{M^2(1+\cos^2\theta)\}\{\frac{-dt^2+dy^2}{y^2}+d\theta^2+$$

$$+ \frac{4\sin^2\theta}{((1+\cos^2\theta))^2}(d\phi+\frac{dt}{y})^2\} \tag{17.18}$$

which is definitely not asymptotically flat. We also find that he near-horizon dilaton field is given by [30]

$$\Phi = \ln\frac{(2M^2-Q^2)(1+\cos^2\theta)}{Q^2\sin^2\theta+2M^2(1+\cos^2\theta)} \tag{17.19}$$

and the near-horizon $U(1)$ field strength is given by [30]

$$F = \frac{2\sqrt{2}Q(2M^2 - Q^2)\sin^2\theta}{y^2(Q^2\sin^2\theta + 2M^2(1 + \cos^2\theta))}dy \wedge dt$$
$$- \frac{8\sqrt{2}M^2(2M^2 - Q^2)\sin(2\theta)Q}{y(Q^2\sin^2\theta + 2M^2(1 + \cos^2\theta))^2}(yd\theta \wedge d\phi + d\theta \wedge dt). \quad (17.20)$$

We find that near-horizon gauge field is given by

$$A = -\frac{2\sqrt{2}Q(2M^2 - Q^2)\sin^2\theta}{(Q^2\sin^2\theta + 2M^2(1 + \cos^2\theta))}(d\phi + \frac{dt}{y}) \quad (17.21)$$

where the three-form field strength $H_{\mu\nu\sigma}$ reduces to [30]

$$H = \{\mathcal{H}\frac{dy}{y^2} - \frac{1}{y}\mathcal{H}'d\theta\} \wedge dt \wedge d\phi \quad (17.22)$$

where

$$\mathcal{H}(\theta) = \frac{2(2M^2 - Q^2)^2Q^2\sin^4\theta}{\{Q^2\sin^2\theta + 2M^2(1 + \cos^2\theta)\}^2}. \quad (17.23)$$

We can introduce the new antisymmetric tensor field $\mathcal{B}$ by

$$\mathcal{B} = -\frac{\mathcal{H}(\theta)}{y}dt \wedge d\phi \quad (17.24)$$

such that

$$H = d\mathcal{B}. \quad (17.25)$$

In the case of vanishing $\varrho$, the metric becomes the near-horizon geometry of the Kerr solution, as in [1].

We recall that asymptotic symmetry group of a spacetime is the group of allowed symmetries that obey the boundary conditions. So, the definition of the charge associated with a symmetry depends both on the action as well as the boundary conditions. Here, we compute the charges associated with asymptotic symmetry group of Kerr-Sen solution by considering all possible contributions from the different fields in the action (17.1). There are four contributions to the associated charge of asymptotic symmetry group of Kerr-Sen solution, from gravitational tensor, dilaton, $U(1)$ gauge field and antisymmetric tensor field $\mathcal{B}_{\mu\nu}$. We have [30]

$$Q_{\zeta,\Lambda,\Psi} = \frac{1}{8\pi}\int_{\partial\Sigma}(k_\zeta^g[h;g] + k_\zeta^\Phi[h,\phi;g,\Phi] + k_{\zeta,\Lambda}^A[h,a;g,A] + k_{\zeta,\Psi}^{\mathcal{B}}[h,b;g,\mathcal{B}]) \quad (17.26)$$

where $h, a, b$ and $\phi$ mean the infinitesimal variations of $g, A, \mathcal{B}$ and $\Phi$ fields, respectively, and $\partial\Sigma$ is the boundary of a spatial slice. We should note, thanks to equation (17.25), there is no contribution to the charge (17.26) from Chern-Simons terms. The gravitational and dilaton contribution two-forms $k_\zeta^g[h;g]$ and $k_\zeta^\Phi[h,\phi;g,\Phi]$ are given by [41, 42, 43]

$$k_\zeta^g[h;g] = -\delta\mathbf{Q}_\zeta^g + \mathbf{Q}_{\delta\zeta}^g + i_\zeta\Theta[h] - \mathbf{E}_\mathcal{L}[\mathcal{L}_\zeta g, h] \quad (17.27)$$
$$k_\zeta^\Phi[h,\phi;g,\Phi] = -i_\zeta\Theta_\Phi \quad (17.28)$$

where $\Theta_\Phi = *(\phi d\Phi)$, $\Theta[h] = *\{(D^\beta h_{\alpha\beta} - g^{\mu\nu}D_\alpha h_{\mu\nu})dx^\alpha\}$ and

$$\mathbf{E}_{\mathcal{L}}[\mathcal{L}_\zeta g, h] = *\{\frac{1}{2}h_{\alpha\gamma}(D^\gamma \zeta_\beta + D_\beta \zeta^\alpha)dx^\alpha \wedge dx^\beta\} \tag{17.29}$$

and $\mathbf{Q}_\zeta^g$ is the Koumar two-form

$$\mathbf{Q}_\zeta^g = \frac{1}{2} * (D_\mu \xi_\nu - D_\nu \xi_\mu)dx^\mu \wedge dx^\nu. \tag{17.30}$$

The last two terms in equation (17.26) are contributions of one-form gauge field $A$ and two-form $\mathcal{B}$ field to the charge. In general for a $\hat{p}$-form $P$ with the associated $(\hat{p}+1)$-form field strength $R$, the contribution is given by [43]

$$k_{\zeta,\Pi}^P[h, p; g, P] = -\delta\mathbf{Q}_{\zeta,\Pi}^P + \mathbf{Q}_{\delta\zeta,\delta\Pi}^P - i_\zeta\Theta^\mathbf{P} - \mathbf{E}_{\mathcal{L}}^P[\mathcal{L}_\zeta P + d\Pi, p] \tag{17.31}$$

where

$$\Theta^\mathbf{P} = p \wedge *R \tag{17.32}$$

$$\mathbf{E}_{\mathcal{L}}^P[\mathcal{L}_\zeta P + d\Pi, p] = *\{\frac{1}{2(\hat{p}-1)!}p_{\mu\rho_1\cdots\rho_{\hat{p}-1}}(\mathcal{L}_\zeta P + d\Pi)_\nu^{\rho_1\cdots\rho_{\hat{p}-1}}dx^\mu \wedge dx^\nu\} \tag{17.33}$$

and the two-form $\mathbf{Q}_{\zeta,\Pi}^P$ is

$$\mathbf{Q}_{\zeta,\Pi}^P = (i_\zeta P + \Pi) \wedge *R. \tag{17.34}$$

We find the contribution of antisymmetric tensor field as

$$\begin{aligned}
k_{\zeta,\Psi}^\mathcal{B}[h, b; g, \mathcal{B}] &= \frac{1}{12}\{\zeta^\lambda(\epsilon_{\mu\nu\rho\beta}b_{\lambda\alpha} + \epsilon_{\mu\nu\rho\alpha}b_{\beta\lambda} + \epsilon_{\mu\nu\rho\lambda}b_{\alpha\beta})H^{\mu\nu\rho}\}dx^\alpha \wedge dx^\beta \\
&\quad - \frac{1}{6}\epsilon_{\mu\nu\rho\sigma}\{\frac{1}{2}\xi^\lambda b_{\nu\lambda}H^{\mu\nu\rho} + (\frac{1}{2}\xi^\lambda \mathcal{B}_{\nu\lambda} + \Psi_\nu) \\
&\quad \times (\delta H^{\mu\nu\rho} + \frac{1}{2}hH^{\mu\nu\rho})\}dx^\nu \wedge dx^\sigma \\
&\quad + \frac{1}{8}\epsilon^{\mu\nu}{}_{\rho\sigma}b_\mu^\alpha(\mathcal{L}_\zeta\mathcal{B} + d\Psi)_{\nu\alpha}dx^\rho \wedge dx^\sigma. \tag{17.35}
\end{aligned}$$

We consider the same boundary condition in [1] for the near-horizon metric and consider the boundary conditions for the gauge field as

$$a_\mu \sim \mathcal{O}(r, 1/r^2, 1, 1/r) \tag{17.36}$$

and for the dilaton as

$$\phi \sim \mathcal{O}(1) \tag{17.37}$$

and for the antisymmetric tensor field, as

$$b_{\mu\nu} \sim \mathcal{O}\begin{pmatrix} 0 & 1/r^2 & 1/r & 1 \\ & 0 & 1/r^2 & 1/r \\ & & 0 & 1/r \\ & & & 0 \end{pmatrix}. \tag{17.38}$$

The near-horizon metric has a class of commuting diffeomorphisms, labeled by $n = 0, \pm 1, \pm 2, \cdots$

$$\zeta_n = -e^{-in\varphi}(\partial_\varphi + inr\partial_r). \qquad (17.39)$$

This diffeomorphism generates a Virasoro algebra without any central charge

$$[\zeta_m, \zeta_n] = -i(m - n)\zeta_{m+n}. \qquad (17.40)$$

The charge (17.26) generates the symmetry $(\zeta, \Lambda, \Psi)_n$ and the algebra of the asymptotic symmetric group is given by the Dirac bracket algebra of these charges

$$
\begin{aligned}
\{Q_{\zeta,\Lambda,\Psi}, Q_{\tilde\zeta,\tilde\Lambda,\tilde\Psi}\}_{D.B.} &= (\delta_{\tilde\zeta} + \delta_{\tilde\Lambda} + \delta_{\tilde\Psi})Q_{\zeta,\Lambda,\Psi} \\
&= \frac{1}{8\pi} \int_{\partial\Sigma} (k_\zeta^g[\mathcal{L}_{\tilde\zeta}g; g] + k_\zeta^\Phi[\mathcal{L}_{\tilde\zeta}g, \mathcal{L}_{\tilde\zeta}\Phi; g, \Phi] \\
&+ k_{\zeta,\Lambda}^A[\mathcal{L}_{\tilde\zeta}g, \mathcal{L}_{\tilde\zeta}A + d\tilde\Lambda; g, A] \\
&+ k_{\zeta,\Psi}^B[\mathcal{L}_{\tilde\zeta}g, \mathcal{L}_{\tilde\zeta}B + d\tilde\Psi; g, B]). \qquad (17.41)
\end{aligned}
$$

Taking the background geometry $\hat{g}$ and fields $\hat\Phi$, $\hat{A}$ and $\hat{B}$ by (17.18), (17.19), (17.21) and (17.24), we obtain

$$
\begin{aligned}
\{Q_{\zeta,\Lambda,\Psi}, Q_{\tilde\zeta,\tilde\Lambda,\tilde\Psi}\}_{D.B.} &= Q_{[(\zeta,\Lambda,\Psi),(\tilde\zeta,\tilde\Lambda,\tilde\Psi)]} \\
&+ \frac{1}{8\pi} \int_{\partial\Sigma} (k_\zeta^{\hat{g}}[\mathcal{L}_{\tilde\zeta}\hat{g}; \hat{g}] + k_\zeta^\Phi[\mathcal{L}_{\tilde\zeta}\hat{g}, \mathcal{L}_{\tilde\zeta}\hat\Phi; \hat{g}, \hat\Phi] \\
&+ k_{\zeta,\Lambda}^A[\mathcal{L}_{\tilde\zeta}\hat{g}, \mathcal{L}_{\tilde\zeta}\hat{A} + d\tilde\Lambda; \hat{g}, \hat{A}] \\
&+ k_{\zeta,\Psi}^B[\mathcal{L}_{\tilde\zeta}\hat{g}, \mathcal{L}_{\tilde\zeta}\hat{B} + d\tilde\Psi; \hat{g}, \hat{B}]). \qquad (17.42)
\end{aligned}
$$

A straightforward and lengthy calculation shows that the algebra of the asymptotic symmetry group is a Viraso algebra generated by $(\zeta, \Lambda, \Psi)_n$ with the central charge [30]

$$c = c_g + c_\Phi + c_A + c_B. \qquad (17.43)$$

The four contributions to the central charge are generated by the last four central terms in (17.42), respectively. Moreover, we find that the chosen boundary conditions for the metric tensor (as the same boundary condition in [1]), dilaton (given by (17.37)), gauge field (given by (17.36)) and antisymmetric tensor field (given by (17.38)), keep all the conserved charges as well as the central charges completely finite. Explicitly, we find that

$$c_g = 12J \qquad (17.44)$$

and the other central charges vanish. These results explicitly show that the non-gravitational fields (including the antisymmetric tensor field) do not contribute to the central charge of the dual CFT. Replacing the Dirac brackets by commutators yields a quantum Virasoro algebra with the central charge

$$c = 12J \qquad (17.45)$$

for the dual chiral CFT corresponding to Kerr-Sen black hole (17.8). To find the entropy of dual chiral CFT, we need to find Frolov-Thorne temperature [44]. A straightforward calculation shows

$$T_{FT} = \frac{1}{2\pi}.$$
(17.46)

Finally, we obtain the microscopic entropy in dual chiral CFT by using the Cardy relation, as

$$S = \frac{\pi^2}{3} c T_{FT} = 2\pi J.$$
(17.47)

This microscopic result for the entropy is exactly the same as macroscopic entropy of black hole (17.17) in the extremal limit.

Although Kerr-Sen black hole in the limit of $a \to 0$ reduces to Gibbons-Maeda-Garfinkle-Horowitz-Strominger charged black hole, but Kerr/CFT correspondence fails for Gibbons-Maeda-Garfinkle-Horowitz-Strominger black hole. This is quite reasonable since in derivation of microscopic entropy, we implicitly assumed that the angular velocity of the horizon is not zero.

## 17.3 The hidden conformal symmetries for Kerr-Newman black holes

The Kerr-Newman metric is the exact solution to Einstein-Maxwell equations that describes a spacetime outside of an electrically charged rotating massive object. The metric of Kerr-Newman spacetime is given by [26, 28]

$$
\begin{aligned}
ds^2 &= -\frac{\Delta - a^2 \sin^2\theta}{\varrho} \left[ dt + \frac{(2Mr - Q^2)a\sin^2\theta}{\Delta - a^2\sin^2\theta} d\phi \right]^2 \\
&+ \varrho\frac{dr^2}{\Delta} + \varrho d\theta^2 + \frac{\varrho\Delta\sin^2\theta}{\Delta - a^2\sin^2\theta} d\phi^2
\end{aligned}
$$
(17.48)

where

$$\varrho = r^2 + a^2\cos^2\theta$$
(17.49)

$$\Delta = r^2 - 2Mr + a^2 + Q^2.$$
(17.50)

In the limit of $a = 0$, the metric (17.48) reduces to the Reissner-Nordstrom solution, and in the limits of $a = 0$, $Q = 0$, the metric (17.48) reduces to Schwarzschild spacetime. The components of gauge field are given by

$$A = -\frac{Qr}{\varrho}\left(dt - a\sin^2\theta d\phi\right).$$
(17.51)

We note that the inner and outer horizons, $r_-$ and $r_+$ are given by

$$r_{\pm} = M \pm \sqrt{M^2 - a^2 - Q^2}.$$
(17.52)

For an extremal Kerr-Newman black holes, $M^2 = a^2 + Q^2$ which provides $r_+ = r_- = M$. The physical quantities of black hole such as Bekenstein-Hawking entropy, Hawking temperature, angular velocity and the electric potential at the horizon are given by

$$S_{BH} = \pi(r_+^2 + a^2) \qquad (17.53)$$

$$T_H = \frac{r_+ - r_-}{4\pi(r_+^2 + a^2)} \qquad (17.54)$$

$$\Omega_H = \frac{a}{r_+^2 + a^2} \qquad (17.55)$$

$$\Phi_H = \frac{Qr_+}{r_+^2 + a^2} \qquad (17.56)$$

respectively.

According to the results of paper [31], the radial equation for the scalar field has two poles on outer horizon $r_+$ and inner horizon $r_-$, where the Kerr-Newman metric function $\Delta$ (17.50) vanishes. Moreover, it is found that far from the extremality, the coordinate $r$ is far enough from $r_-$, so, one can drop the linear and quadratic terms in frequency in the radial equation [31, 33]. In fact, one can deform the radial equation near the inner horizon $r_-$ by deformation parameter $\kappa$ as

$$\partial_r(\Delta\partial_r R(r)) + \left[ \frac{[(2Mr_+ - Q^2)\omega - am - Qr_+e]^2}{(r - r_+)(r_+ - r_-)} \right.$$
$$\left. - \frac{[(2M\kappa r_+ - Q^2)\omega - am - Q\kappa r_+e]^2}{(r - r_-)(r_+ - r_-)} - l(l+1) \right] R(r) = 0$$

$$(17.57)$$

where $\kappa$ satisfies $(2M^2\kappa - Q^2)am\omega << 2\sqrt{M^2 - a^2 - Q^2}(r - r_-)$ as well as $(2M^2\kappa - Q^2)^2\omega^2 << 2\sqrt{M^2 - a^2 - Q^2}(r - r_-)$ [31]. In other words, one can consider the deformation of the radial equation near the inner horizon only, given by (17.57), as the radial equation describes the dynamics of the test field outside of the null spacelike singularity. The other interesting feature of deformation of the inner horizon is that it doesn't change the location of other singularities of the radial equation (17.57) that are located on the outer horizon and far infinity. We consider the deformed equation (17.57) in the $J$ picture

$$\partial_r(\Delta\partial_r R(r))$$
$$+ \left[ \frac{[(2Mr_+ - Q^2)\omega - am]^2}{(r - r_+)(r_+ - r_-)} - \frac{[(2M\kappa r_+ - Q^2)\omega - am]^2}{(r - r_-)(r_+ - r_-)} \right] R(r)$$
$$= l(l+1)R(r). \qquad (17.58)$$

We consider the following vector fields

$$L_\pm = e^{\pm \rho t \pm \sigma \phi} \left( \mp \sqrt{\Delta} \partial_r + \frac{C_1 - \gamma r}{\sqrt{\Delta}} \partial_t + \frac{C_2 - \delta r}{\sqrt{\Delta}} \partial_\phi \right) \qquad (17.59)$$

$$L_0 = \gamma \partial_t + \delta \partial_\phi. \qquad (17.60)$$

These vector fields make the $sl(2, \mathbb{R})$ algebra which is given by $[L_\pm, L_0] = \pm L_\pm$ and $[L_+, L_-] = 2L_0$ [31, 33]. We also require the quadratic Casimir operator of $sl(2, \mathbb{R})$ represents the deformed radial equation (17.58). Hence we find

$$L_0^2 - \frac{1}{2}(L_+L_- + L_-L_+) = \partial_r(\Delta \partial_r) + \frac{((2Mr_+ - Q^2)\omega - am)^2}{(r - r_+)(r_+ - r_-)}$$
$$- \frac{((2M\kappa r_+ - Q^2)\omega - am)^2}{(r - r_-)(r_+ - r_-)} \qquad (17.61)$$

where the generators $L_\pm$ and $L_0$ have the following automorphism,

$$L_\pm \to -L_\pm \quad , \quad L_0 \to L_0. \qquad (17.62)$$

The $sl(2, \mathbb{R})$ algebra and so the quadratic Casimir operator is invariant under (17.62).

We find the following two equations for the coefficients of $\partial_r$ and $\partial_r^2$ in (17.61)

$$\rho C_1 + \sigma C_2 + M = 0 \qquad (17.63)$$

and

$$1 + \rho \gamma + \sigma \delta = 0. \qquad (17.64)$$

Moreover, the coefficients of $\partial_\phi^2$ and $\partial_t^2$ in (17.61) give two other equations as [31]

$$-\delta^2 (r - r_+)(r - r_-) + C_2^2 - 2C_2\delta r + \delta^2 r^2 = a^2 \qquad (17.65)$$

and

$$C_1^2 - \gamma^2 (r - r_+)(r - r_-) - 2C_1\gamma r + \gamma^2 r^2$$
$$= \frac{(2Mr_+ - Q^2)^2}{(r_+ - r_-)} ((r - r_-) - \kappa^2 (r - r_+))$$
$$- \frac{4MQ^2 r_+}{(r_+ - r_-)} (r - r_- - \kappa (r - r_+)) + Q^4. \qquad (17.66)$$

Finally, we get the following equation which is the coefficient of $\partial_\phi \partial_t$ in (17.61)

$$- C_2 C_1 + \delta r C_1 - \delta r^2 \gamma + \gamma (r - r_+)(r - r_-)\delta + C_2 \gamma r$$
$$= -\frac{2Mr_+ a}{(r_+ - r_-)} ((r - r_-) - \kappa (r - r_+)) + 2aQ^2. \qquad (17.67)$$

From equation (17.65), we find two classes of solutions,

$$\delta_a^J = \frac{2a}{r_+ - r_-}, \qquad C_{2a}^J = \frac{a(r_+ + r_-)}{r_+ - r_-} \tag{17.68}$$

$$\delta_b^J = 0, \qquad C_{2b}^J = a. \tag{17.69}$$

Substituting (17.68) and (17.69) into equations (17.66) and (17.67), we find $C_1$ and $\gamma$ that are given by

$$\gamma_a^J = \frac{2Mr_+(\kappa + 1) - 2Q^2}{r_+ - r_-}, \qquad C_{1a}^J = \frac{2Mr_+(\kappa r_+ + r_-)}{r_+ - r_-} - Q^2\left(\frac{r_+ + r_-}{r_+ - r_-}\right) \tag{17.70}$$

$$\gamma_b^J = \frac{2Mr_+(\kappa - 1)}{r_+ - r_-}, \qquad C_{1b}^J = \frac{2Mr_+(\kappa r_+ - r_-)}{r_+ - r_-} - Q^2. \tag{17.71}$$

Solving (17.63) and (17.64) for $\sigma$ and $\rho$ gives all the conformal generators (17.59) and (17.60). We note that multiplying all the coefficients in table 1 with $-1$ also are solutions to equations (17.63), (17.64), (17.65), (17.66) and (17.67). However these solutions correspond to invariance of the Casmir operator $L_0^2 - \frac{1}{2}(L_+L_- + L_-L_+)$ by renaming the vector fields as $L_0 \to -L_0$ , $L_\pm \to -L_\mp$. We also note that in the limit of $Q = 0$, the vector fields in the $J$ picture reduce correctly to the generators of deformed conformal symmetry for the Kerr black holes [33].

So, finally we get the explicit expressions for the deformed conformal generators in branch a

$$L_\pm^a = e^{\mp 2\pi T_R \phi}\left[\mp\sqrt{\Delta}\partial_r - \frac{1}{2\pi T_H}\frac{r - M}{\sqrt{\Delta}}(\Omega_H\partial_\phi + \partial_t)\right.$$
$$\left. + \frac{1}{2\pi\Omega_H(T_L + T_R)}\frac{r - r_+}{\sqrt{\Delta}}\partial_t\right] \tag{17.72}$$

$$L_0^a = \frac{1}{2\pi T_H}(\Omega_H\partial_\phi + \partial_t) - \frac{1}{2\pi\Omega_H(T_L + T_R)}\partial_t \tag{17.73}$$

and branch b

$$L_\pm^b = e^{\pm 2\pi\Omega(T_L + T_R)t \mp 2\pi T_L \phi}\left[\mp\sqrt{\Delta}\partial_r + \frac{2Mr_+ - Q^2}{\sqrt{\Delta}}(\Omega\partial_\phi + \partial_t)\right.$$
$$\left. + \frac{1}{2\pi\Omega_H(T_L + T_R)}\frac{r - r_+}{\sqrt{\Delta}}\partial_t\right] \tag{17.74}$$

$$L_0^b = -\frac{1}{2\pi\Omega_H(T_L + T_R)}\partial_t \tag{17.75}$$

where $T_H$ and $\Omega_H$ are defined in (17.54) and (17.55) and the left and right moving CFT temperatures are given by $T_R = \frac{r_+ - r_-}{4\pi a}$, $T_L = \frac{T_R(1 + \kappa)}{1 - \kappa} - \frac{Q^2 T_R}{Mr_+(1 - \kappa)}$. One can verify that by taking the left and right central charges as

$$c_R = c_L = \frac{6aMr_+(1 - \kappa)}{\sqrt{M^2 - a^2 - Q^2}} \tag{17.76}$$

we find the exact Bekenstein-Hawking entropy for Kerr-Newman black holes (17.53), if we use the Cardy formula

$$S_{Cardy} = \frac{\pi^2}{3}(c_R T_R + c_L T_L). \tag{17.77}$$

We note that for the special case of deformation parameter given by $\kappa = r_+/r_-$, we find the generators of hidden conformal symmetry for the Kerr-Newman black holes [26]. In fact, for $\kappa = r_+/r_-$, the deformed generators $L_\pm^a$ and $L_0^a$ (up to automorphisms (17.62)) reduce to conformal generators $H_\pm$ and $H_0$ in [26] according to $L_k^a = -iH_k$ where $k = +, -, 0$. The generators in branch b for $\kappa = r_+/r_-$ reduce to the other copy of conformal generators $\bar{H}_k$ in [26] by the mapping $L_k^b = i\bar{H}_k$. The left and right temperatures (17.3) as well as central charge (17.76) reduce to the corresponding results in [26] after setting $\kappa = r_-/r_+$.

An interesting open question is to derive the deformed central charges by using either ASG or stretched horizon techniques.

## 17.4 Concluding Remarks

In this article, which entirely is based on published papers [30] and [31], we considered the class of extremal Kerr-Sen black holes in the low energy limit of heterotic string theory. We found the near-horizon metric of the black hole, as well as the near-horizon limits of the other non-gravitational fields of the theory by taking the near-horizon procedure. We found that the contribution of the Chern-Simons terms to the three-form field strength of the theory is very crucial. In fact, the contribution of the antisymmetric tensor field to the three-form field strength in near-horizon limit, is divergent. Moreover, the contribution of the Chern-Simons terms to the three-form field strength in near-horizon limit, also is divergent. However, these two divergences cancel out exactly when we consider both contributions to the three-form field stength. We found an important result that states near the horizon (which has the topology of warped AdS$_3$), the three-form field strength depends explicitly on a new antisymmetric tensor field, and not to the Maxwell gauge filed. By choosing the proper boundary conditions for the gravitational field, dilaton, gauge field and the antisymmetric tensor field, we found the diffeomorphisms that generate Virasoro algebra without any central charge. The generator of diffeomorphisms which is a conserved charge, can be used to construct an algebra under Dirac brackets. This algebra is the same as diffeomorphism algebra but just with some extra central terms. These central terms, in general contribute to the the central charge of the Virasoro algebra. We showed that the only non-zero contribution to the central charge of the dual conformmal field theory comes from gravitational field. So, we extended the validity of conjecture (that the central charges from the non-gravitational fields vanish) to theories that include the antisymmetric tensor fields. The central charge together with Frolov-Thorne temperature enable us to find the microscopic entropy of the extremal Kerr-Sen black hole in dual chiral CFT. The microscopic entropy is exactly the same as macroscopic Bekenstein-Hawking entropy of the extremal black hole. Our work provides further supportive evidence in favor of a Kerr/CFT correspondence in the low energy limit of heterotic string theory that contains three non-gravitational fields. Moreover, we find an extended family of hidden conformal symmetry for the Kerr-Newman black holes that are characterized by deformation parameter $\kappa$. The deformation of the inner horizon of the black hole in the radial equation of scalar field is justified by the fact that the back-reaction of the scalar field on the black hole geometry makes the inner horizon replaced by a null curvature spacelike singulatity. The deformed hidden conformal generators constructed explicitly in different conformal pictures for the Kerr-Newman black holes. The deformed hidden conformal symmetry in the $J$ picture reduces to the hidden conformal symmetry of the Kerr-Newman where the deformation parameter $\kappa$ is set $r_-/r_+$.

# Acknowledgments

This proceeding presentation and also participation in the Minkowski conference were supported by the Natural Sciences and Engineering Research Council of Canada.

# References

[1] M. Guica, T. Hartman, W. Song and A. Strominger, [arXiv:0809.4266].

[2] S.N. Solodukhin, *Phys. Lett.* **B454** (1999) 213.

[3] S. Carlip, *Phys. Rev. Lett.* **82** (1999) 2828.

[4] G. Barnich and F. Brandt, *Nucl. Phys.* **B633** (2002) 3.

[5] G. Barnich and G. Compere, *J. Math. Phys.* **49** (2008) 042901.

[6] M. Park, *Nucl. Phys.* **B634** (2002) 339.

[7] G. Kang, J. Koga and M. Park, *Phys. Rev.* **D70** (2004) 024005.

[8] E. Witten, *Adv. Theor. Math. Phys.* **2** (1998) 253.

[9] J. Maldacena, *Adv. Theor. Math. Phys.* **2** (1998) 231.

[10] K. Sfetsos and K. Skenderis, *Nucl. Phys.* **B517** (1998) 179.

[11] D.Z. Freedman, S.D. Mathur, A. Matusis and L. Rastelli, *Nucl. Phys.* **B546** (1999) 96.

[12] S.S. Gubser, I.R. Klebanov and A.M. Polyakov, *Phys. Lett.* **B428** (1998) 105.

[13] A.M. Ghezelbash, K. Kaviani, S. Parvizi and A.H. Fatollahi, *Phys. Lett.* **B435** (1998) 291.

[14] H. Lü, J. Mei and C.N. Pope, *JHEP* **0904** (2009) 054.

[15] H. Isono, T.-S. Tai and W.-Y. Wen, [arXiv:0812.4440].

[16] T. Azeyanagi, N. Ogawa and S. Terashima, [arXiv:0812.4883].

[17] J.-J. Peng and S.-Q. Wu, *Phys. Lett.* **B673** (2009) 216.

[18] S.M. Carroll, M.C. Johnson and L. Randall, [arXiv:0901.0931].

[19] T. Azeyanagi, N. Ogawa and S. Terashima, *JHEP* **0904** (2009) 061.

[20] T. Hartman, K. Murata, T. Nishioka and A. Strominger, *JHEP* **0904** (2009) 019.

[21] Y. Nakayama, *Phys. Lett.* **B673** (2009) 272.

[22] A.M. Ghezelbash, *Mod. Phys. Lett.* **A27** (2012) 1250046.

[23] A.M. Ghezelbash, *Gen. Rel. Grav.* **48** (2016) 102.

[24] A.M. Ghezelbash and H.M. Siahaan, *Phys. Rev.* **D89** (2014) 024017.

[25] G. Compere, K. Murata and T. Nishioka, *JHEP* **0905** (2009) 077.

[26] C.-M. Chen, Y.-M. Huang, J.-R. Sun, M.-F. Wu and S.-J. Zou, *Phys. Rev.* **D 82**, 066004 (2010).

[27] B. Chen and J.-j. Zhang, *Nucl. Phys.* **B 856**, 449 (2012).

[28] B. Chen and J.-j. Zhang, *JHEP* **1108**, 114 (2011).

[29] A. Sen, *Phys. Rev. Lett.* **69**, 1006 (1992).

[30] A. M. Ghezelbash, *JHEP* **0908**, 045 (2009).

[31] A. M. Ghezelbash and H. M. Siahaan, *Class. Quantum Grav.* **30** (2013) 135005.

[32] G. W. Gibbons and K.-i. Maeda, Nucl. Phys. **B 298**, 741 (1988); D. Garfinkle, G. T. Horowitz and A. Strominger, *Phys. Rev.* **D 43**, 3140 (1991) [Erratum-ibid. **D 45**, 3888 (1992)].

[33] D. A. Lowe and A. Skanata, *J. Phys.* **A 45**, 475401 (2012).

[34] A.M. Ghezelbash, V. Kamali and M.R. Setare, *Phys. Rev.* **D82** (2010) 124051.

[35] D. Gates, D. Kapec, A. Lupsasca, Y. Shi and A Strominger, [arXiv:1809.09092].

[36] A. Pathak, A.P. Porfyriadis, A. Strominger and O. Varela, *JHEP* **2017**, 90 (2017).

[37] K. Hioki, U. Miyamoto, *Phys. Rev.* **D78** (2008) 044007.

[38] A. Burinskii and G. Magli, *Annals Israel Phys. Soc.* **13** (1997) 296.

[39] G.N. Gyulchev and S.S. Yazadjiev, *Phys. Rev.* **D75** (2007) 023006.

[40] S.Q. Wu and X. Cai, *J. Math. Phys.* **44** (2003) 1084.

[41] G. Barnich and G. Compere, *Phys. Rev. Lett* **95** (2005) 031302.

[42] M. Banados, G. Barnich, G. Compere and A. Gomberoff, *Phys. Rev.* **D73** (2006) 044006.

[43] G. Compere, [arXiv:0708.3153].

[44] V.P. Frolov and K.S. Thorne, *Phys. Rev.* **D39** (1989) 2125.

# 18    IS THE FLAT SPACETIME RELATED TO

## A KINEMATICAL STRUCTURE?

MOHAMMED SANDUK

**Abstract**    The three wave hypothesis (TWH) has been proposed and developed during the last century. It is based on the de Broglie wave hypothesis and covariant æther. In 2007, the author found that the TWH might be attributed to a kinematical classical system of two rolling circles. In 2012, based on the model of two rolling circles, the author has shown that the position vector of a point in this model can be transformed to a complex vector under the effect of partial observation. In this project, we present this concept of transformation as a lab observation. Accordingly, the velocity equation of a point is transformed to an equation analogise the relativistic quantum mechanics equation (Dirac equation). The cause behind this transformation is the problem of partial observation. Then, the flat spacetime may look like an approximation for the partially observed structure. Then, these analogies could suggest that both of the quantum mechanics and the special relativity are emergent, and both of them are unified, and of the same origin.

## 18.1   Introduction

Snyder introduced quantized spacetime in 1947. As an example, the commutator of position momentum ( $[x, p] = i\hbar$ ) is represented as [1]:

$$[x, p_x] = i\hbar \left[ 1 + \left( \frac{a}{\hbar} \right)^2 p_x^2 \right], \tag{18.1}$$

where $a$ is a natural unit of length. The standard commutator is restored when $a \to 0$. Snyder equation is based on a linear combination assumption. Within this frame, the Einstein Hamiltonian of a lattice of Planck length ( $\lambda_P$ ) has

C. Duston, M. Holman (Eds), *Spacetime Physics 1907 - 2017. Selected peer-reviewed papers presented at the First Hermann Minkowski Meeting on the Foundations of Spacetime Physics, 15-18 May 2017, Albena, Bulgaria* (Minkowski Institute Press, Montreal 2019). ISBN 978-1-927763-48-3 (softcover), ISBN 978-1-927763-49-0 (ebook).

308

been represented as [2]:

$$E^2 = c^2p^2 + m^2c^4 + \alpha \left(\frac{c}{\hbar}\right)^2 \lambda_P^2 p^4, \qquad (18.2)$$

where $\alpha$ is a dimensionless constant. Then with linearization, Glinka modified the Dirac equation to include $\lambda_P$ [2].

However, Planck length (scale) and Planck time are constants and are based on the universal physical constants, according to Planck units system. In other words, they are not related to a physical theory [3], but they are introduced via dimensional analysis technique.

In the above two equations, the ordinary flat spacetime continuum arises when constant units of length ($\lambda \equiv a$ or $\lambda_P \dots$) equal zero ($\lambda_p \to 0$). This $\lambda_p$ works as a boundary between two different worlds.

The Loop Quantum Gravity (LQG) is the most modern theory for gravity. LQG is based on the geometrical formulation of general relativity, and then a quantization process [4]. Due to the quantization, the spacetime is granular. The size of this structure is the Planck length [5].

Interestingly, during the fifties of the last century, Dirac proposed a possibility of existing of a quantum-mechanical æther, which does not violate Lorentz symmetry [6, 7]. Das and Vagenas argued that this proposed *quantum æther* "can provide an economical way of having an invariant quantum gravity or Planck scale" [8].

### 18.1.1 The æther and a kinematic model

Based on the de Broglie's particle-wave duality [9-12], and the assumption of the existence of a covariant æther [13–16], Three Wave Hypothesis (TWH) was developed by Horodecki [17–20] during the eighties of the last century.

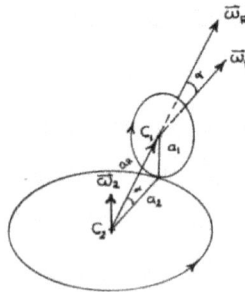

**Figure 18.1:** The bevel gear (de Broglie wave- covariant æther) [21].

In 2007, the author considered the TWH in angular form and in a single representation of waves [21, 22]. This consideration exhibits similarities with a system of two perpendicular circles (Figure 1). This form looks like a classical kinematical bevel gear model. This rolling circles system is totally far from the quantum mechanics, and can not be observed experimentally. Thus, we called it a hidden structure (unobservable). The wave function is an abstract model in the quantum mechanics, and it is non-physical, hence unobservable.

This real kinematical model which is related originally to the de Broglie wave- covariant æther, and may stand behind the concept of complex form in the quantum mechanics.

## 18.2 The kinematical model and the complex function

The relationship between the two rolling circles system and the complex function has no relation with quantum mechanics and has been considered in 2012 [23]. In considering a rolling circles system in a plane (Figure 2), the position vector of a point $(P)$ on the circle of centre $C_1$ is:

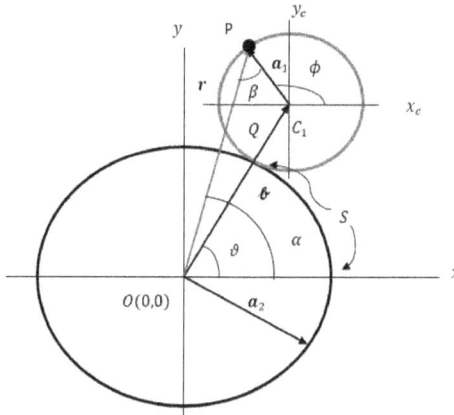

Figure 18.2: Rolling circles model in a plane.

$$r = b \left\{ cos\left(\vartheta - \phi + \beta\right) \pm \sqrt{-sin^2\left(\vartheta - \phi + \beta\right) + \left(\frac{a_1}{b}\right)^2} \right\}. \tag{18.3}$$

The symbols are shown in Figure 2. Let us call the large circle a guiding circle and the small a guided circle. The angles can be regarded as angular displacements, due to the motion of the point, or $k_2 \cdot s = \vartheta$ , $a_2 = \frac{1}{k_2}$ , $\omega_1 t = \phi$ , and $\omega_\beta t = \beta$ . The position vector is in relative to the origin O.

The velocity of the point in relative to the origin O is:

$$\frac{\partial r(r,t,X)}{\partial t} = \frac{\partial(a_2 + b\sqrt{X})}{\partial t}\left\{\cos(k_2 \cdot s - \omega_1 t + \omega_\beta t)\right.$$

$$\pm\sqrt{-sin^2(k_2 \cdot s - \omega_1 t + \omega_\beta t) + X}\bigg\}$$

$$+ (a_2 + b\sqrt{X})\left\{-(-\omega_1 + \omega_\beta)\sin(k_2 \cdot s - \omega_1 t + \omega_\beta t) \pm \left(\tfrac{1}{2}\right) \times\right.$$

$$\times \frac{(-2)(-\omega_1 + \omega_\beta)\sin(k_2 \cdot s - \omega_1 t + \omega_\beta t)\cos(k_2 \cdot s - \omega_1 t + \omega_\beta t) + \frac{\partial X}{\partial t}}{\sqrt{-\sin^2(k_2 \cdot s - \omega_1 t + \omega_\beta t) + X}}\bigg\}$$

$$(18.4)$$

where

$$X \equiv \left(\frac{|a_1|}{|b|}\right)^2 = \left(\frac{a_1}{b}\right)^2. \tag{18.5}$$

## 18.2.1 Rotations unite vectors

There are two angular rotations $\omega_1$ and $\omega_2$ in the system. The unit vectors of these rotations are $\hat{e}_\phi$ and $\hat{e}_\vartheta$, respectively. The signs $\pm$ in Eqs. )3, and 4( are related to the two possibilities of rotation of the point, or they are related to $\hat{e}_\phi$.

The dot product of $\hat{e}_\vartheta$ with any perpendicular unit vector let $\hat{e}_\varphi$ is

$$\hat{e}_\vartheta \cdot \hat{e}_\varphi + \hat{e}_\vartheta \cdot \hat{e}_\varphi = 0. \tag{18.6}$$

The same for:

$$\hat{e}_\vartheta \cdot \hat{e}_\vartheta + \hat{e}_\varphi \cdot \hat{e}_\varphi = 2. \tag{18.7}$$

The square of the unit vectors

$$\hat{e}_\vartheta \cdot \hat{e}_\vartheta = \hat{e}_\phi \cdot \hat{e}_\phi = 1. \tag{18.8}$$

The $\hat{e}_\phi$ and $\hat{e}_\vartheta$ are non-commutative

$$\hat{e}_\vartheta \times \hat{e}_\phi + \hat{e}_\phi \times \hat{e}_\vartheta = 0. \tag{18.9}$$

## 18.2.2 The partial observation

Physics cannot work without observation (as an example of the observation tools is the light). In macroscopic physics, the studied object is obvious and well distinguished. That is related mainly to the fact of concept of high resolution of the using light or the Rayleigh criterion (spatial resolution). Thus, there is no problem attributed to that observation effect in macroscopic physics.

In microscopic world where the dimensions are comparable with wavelength of the using light, it is expected to arise a problem of inability of distinguishing. The transformation from the real system of two rolling circles to a complex form has been found via an assumption of *partial observation* concept [23]. The partial observation is based on the Rayleigh criterion (spatial resolution). For a lab observer, the system looks like an object, note Figure 3. The observation with aid of monochromatic light is affected under the following resolution conditions:

$$a_1 << d_\lambda << a_2, \quad \text{and} \quad \omega_1 >> \omega_\lambda >> \omega_2,$$

where $d_\lambda$ and $omega_\lambda$ are the spatial and angular frequency resolutions. This leads to regard

$$a_1 \to 0, \omega_2 \to 0, \quad \text{and} \quad X = 0. \tag{18.10}$$

In this case, the angle $\beta$ cannot be recognised either. The observer cannot distinguish the system.

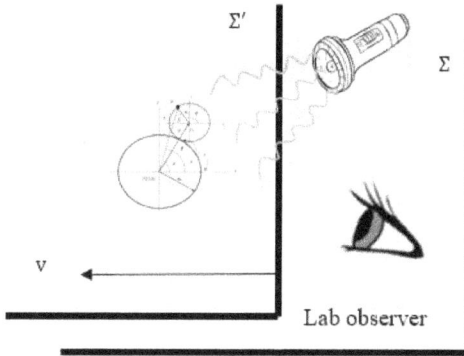

**Figure 18.3:** The lab observer frame of reference.

The substitution of Eq. (10) into Eq. (3) leads to this transformation

$$\mathcal{Z}(s,t,0) = a_{2m} \exp \pm i \left( k_{2m} \cdot s - \omega_{1m} t \right). \tag{18.11}$$

The subscript $\mathcal{Z}$ indicates resolved (measured) values. The real position vector (**r**) is transformed to the complex vector $\mathcal{Z}$ . It is an equation of real vector deformation.

The equation of velocity (Eq. (4)) is transformed as well (Appendix) to

$$i\frac{\partial \mathcal{Z}}{\partial t} = (-ivA \cdot \nabla + B\omega_{1m}) \mathcal{Z} \tag{18.12}$$

where $\mathbf{A} = \mp i\hat{e}_\theta$ and $B = \pm 1$. Equation (12) describes the evolution process of the function $\mathcal{Z}$ . This equation is for a motion in a complex plane. It is a complex velocity equation.

The lab observer cannot recognise the rotations of the system, and mathematically $A$ and $B$ are not the normal unit vector. Thus, the properties of the rotation unit vectors mentioned above will not be considered by the lab observer. Using the $A$ and $B$ instead of the $\hat{\mathbf{e}}_\vartheta$ and $\hat{\mathbf{e}}_\phi$ (in the above Eqs. (6, 7, 8, and 9)), one can find that:

$$(\pm i\,\hat{\mathbf{e}}_\vartheta) \cdot (\pm i\,\hat{\mathbf{e}}_\varphi) + (\pm i\,\hat{\mathbf{e}}_\vartheta) \cdot (\pm i\,\hat{\mathbf{e}}_\varphi) = 0. \tag{18.13}$$

The same for

$$(\pm i\,\hat{\mathbf{e}}_\vartheta) \cdot (\pm i\,\hat{\mathbf{e}}_\vartheta) + (\pm i\,\hat{\mathbf{e}}_\varphi) \cdot (\pm i\,\hat{\mathbf{e}}_\varphi) = 2. \tag{18.14}$$

The square of $\mathbf{B}$ is

$$\mathbf{B}^2 = 1, \tag{18.15}$$

and for $\mathbf{A}$

$$\mathbf{A}^2 = \mathbf{A} \cdot \mathbf{A} = 1. \tag{18.16}$$

Accordingly,

$$\mathbf{A}^2 = \mathbf{B}^2 = 1. \tag{18.17}$$

Finally, since $\mathbf{B}$ is not a vector, then in dealing with the $\mathbf{A}$ and $\mathbf{B}$ as in Eq. 9, one can say:

$$\mathbf{A} \times \mathbf{B} + \mathbf{B} \times \mathbf{A} = 0. \tag{18.18}$$

But there is a problem, where $A$ and $B$ are not normal unit vectors. This mathematical problem can be solved by using Dirac coefficient techniques.

With aid of these properties, the second derivative is

$$i^2 \frac{\partial^2 \mathcal{Z}}{\partial t^2} = \left(-v^2 \nabla^2 + \omega_{1m}^2\right) \mathcal{Z}. \tag{18.19}$$

## 18.3   The observed system

The lab observer can not recognise the two rolling circles system. He/she deals with the abstract world. This situation is similar to the case of dealing with the wave function in quantum mechanics. Therefore, some of the quantum mechanics techniques (postulates) are used to achieve physical information. Hence, the lab observer is going to deal with a probabilistic nature. The observation process acts a separable filter, as in Figure 4.

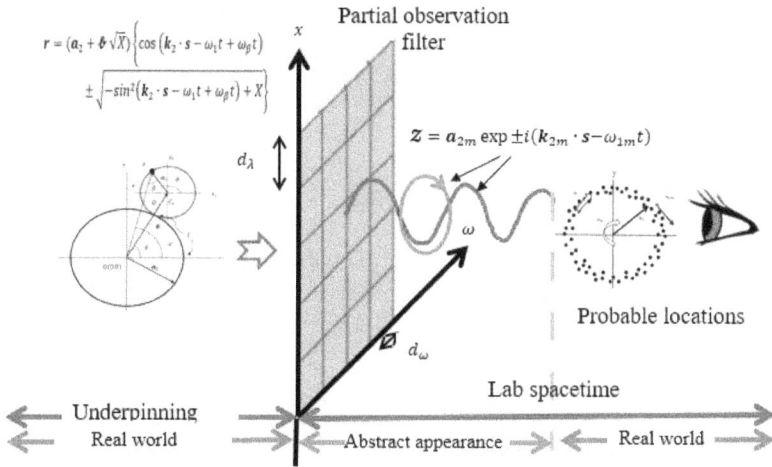

**Figure 18.4:** The system and the lab observation

The system is collapsed to a guided point, and this may be interpreted as a duality of point and probability wave. The system (Figure 2) is reduced to a point (represented by $\omega_1$ ) and its guide is a statistical wave (represented by $a_2$ ), and both features are on an equal footing.

The probable locations of the point are distributed in a circular form of an average radius $a_{2m}$ ( $\langle r \rangle = a_2$) as shown in Figure 5. The $\omega_{1m}$ is attributed to the point, or it is a characteristic quantity of the point. The observable velocity of the point is $v_p$ as shown in Figure 5.

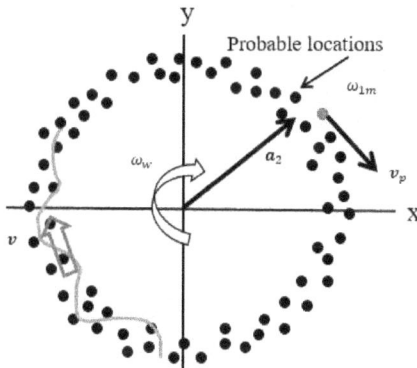

**Figure 18.5:** The observable system, the probable locations of point (in real space).

The velocity of the probability wave ($v$) for the lab observer is (Figure 5):

$$v = \frac{\partial a_{2m'}}{\partial t} = a_{2m'} \omega_{w'}.  \tag{18.20}$$

The quantities with the prime symbol are the observed quantities by lab observer due to the moving frame of reference (Figure 3). This velocity $(v)$ appears in Eqs. (12, 19), and Eq. (3A) which is related to the change of the touch point $(Q)$ location (Figure 3).

## 18.4   Lab transformations

The equivalent pictures of the guiding probability wave and the guided point can be formulated mathematically with aid of the point's parameters $(v_p$ & $\omega_{1m})$, which are corresponded the probability wave's parameters( $v$ & $a_{2m'}$ ) as:

$$v_p \omega_{1m'} \equiv \frac{v^2}{a_{2m'}}. \tag{18.21}$$

In Eq. (21), the left-hand side is related to the guided point form and the right-hand side is the guiding probability wave. With the aid of Eq. (20), Eq. (21) can be reformulated as a ratio:

$$\frac{v}{v_p} = \frac{a_{2m'} \omega_{w'}}{v_p} = \frac{\omega_{1m'}}{\omega_{w'}} = \mu_L. \tag{18.22}$$

Eq. (22) may serve as a lab system (guided point-guiding wave system) ratio $(\mu_L)$. From Eq. (21) one can say that the velocity of the point is

$$v_p = \frac{v^2}{\omega_{1m'}\, a_{2m'}}, \tag{18.23}$$

or the guiding distance (Figure 5) of the point is

$$a_{2m'} = \frac{v^2}{\omega_{1m'}\, v_p}, \tag{18.24}$$

and one can find as well

$$v^2 = v_p \omega_{1m'} a_{2m'} = v_p v_{ph}, \tag{18.25}$$

where $v_{ph}$ is phase velocity and is related to the phase of Eq. (11). We have to mention here that, the phase velocity can be calculated, but it does not refer to any real wave.

With aid of operator postulate, one can say that (Eq.(19)):

$$\omega^2 = v^2 \nabla^2 - \omega_{1m}^2 = \omega_{1m'}^2. \tag{18.26}$$

$\omega \equiv \omega_{1m'}$ is due to the lab observer can not see only one angular frequency that is related to the system.

For the lab observer, The relationship between the $\omega_{1m}$ and $\omega_{1m'}$ can be obtained from Eq. (26) with the aide of Eqs. (24, and 5A):

$$\frac{\omega_{1m}}{\sqrt{1 - \frac{v_p^2}{v^2}}} = \omega_{1m'}. \tag{18.27}$$

Eq. (27) can be formulated as

$$\omega_{1m'}^2 v^2 = \omega_{1m'}^2 v_p^2 + \omega_{1m}^2 v^2. \tag{18.28}$$

Substituting Eq. (27) in Eq. (24), shows that:

$$a_{2m}\sqrt{1 - \frac{v_p^2}{v^2}} = a_{2m'}, \tag{18.29}$$

where

$$a_{2m} \equiv \frac{v^2}{\omega_{1m}\, v_p}. \tag{18.30}$$

Substituting Eq. (27) in Eq. (21) shows that

$$\frac{\omega_w}{\sqrt{1 - \frac{v_p^2}{v^2}}} = \omega_{w'}, \tag{18.31}$$

where

$$\omega_w \equiv \frac{\omega_{1m}\, v_p}{v}. \tag{18.32}$$

According to Eqs. (20), (31)

$$v = a_{2m'}\omega_{w'} = a_{2m}\sqrt{1 - \frac{v_p^2}{v^2}}\,\frac{\omega_w}{\sqrt{1 - \frac{v_p^2}{v^2}}} = a_{2m}\omega_w. \tag{18.33}$$

Thus, the velocity $v$ is invariant. From Eq. (31) the $\omega_{1m'}$ is a real quantity, then $v_p < v < v_{ph}$

# 18.5  Conclusions and remarks

• *Comparisons*

The above sections demonstrate the lab observer's equations for the observed system. Table 1 shows two sets of equations, the conventional equations of quantum mechanics and special relativity, and the analogical model forms that we have obtained. The quantum mechanics equations are presented without $\hbar$ to show kinematical forms.

| Conventional definition | Conventional equations of quantum mechanics and special relativity | Analogical model forms | Analogy definition |
|---|---|---|---|
| Dirac wave function | $\psi_D = u_D \exp i(k \cdot x - \omega t)$ | $\mathcal{Z} = a_{2m} \exp \pm i(k_{2m} \cdot s - \omega_{1m} t)$ | Z-complex vector |
| Dirac equation | $i\dfrac{\partial \psi}{\partial t} = \left(-ic\boldsymbol{\alpha} \cdot \nabla + \beta\omega\right)\psi$ | $i\dfrac{\partial \mathcal{Z}}{\partial t} = \left(-ivA \cdot \nabla + B\omega_{1m}\right)\mathcal{Z}$ | Complex velocity equation |
| Klein-Gordon equation | $\dfrac{\partial^2 \psi}{\partial t^2} = \left[c^2\nabla^2 - \omega^2\right]\psi$ | $\dfrac{\partial^2 \mathcal{Z}}{\partial t^2} = \left[v^2\nabla^2 - \omega_{1m}^2\right]\mathcal{Z}$ | Complex acceleration equation |
| Light speed | $c < v$ <br> $c$ is constant | $v < v_p$ <br> $v$ is constant | Wave speed |
| Lorentz factor | $\dfrac{1}{\sqrt{1 - \dfrac{v^2}{c^2}}}$ | $\dfrac{1}{\sqrt{1 - \dfrac{v_p^2}{v^2}}}$ | Lab transformation |
| Four-vector $\times \left(\dfrac{\hbar}{c^2}\right)^2$ | $\omega^2 c^2 = \omega^2 v^2 + \omega_\circ^2 c^2$ | $\omega_{1m}^2 v^2 = \omega_{1m}^2 v_p^2 + \omega_{1m}^2 v^2$ | Lab space |

**Table 1.** Comparisons of the equations of conventional quantum mechanics and special relativity with the analogical model forms

- *Is spacetime an emergent phenomenon?*

The similarity between the special relativity equations and the equations of the analogise model as shown in Table 1 may lead to say that the spacetime is related to the same underpinning that of the quantum mechanics. Then, the spacetime is not just emergent but is related to the same origin of the quantum mechanics. The rolling circles model under the partial observation may show unification of the special relativity with quantum mechanics.

The spacetime continuum may be attributed to the effect of the partial observation on the kinematical system of two rolling circles. It looks as the ordinary Lorentzian spacetime arises as $X = 0$ , or $a_1 = 0$ . This condition may lead to the spacetime continuum and quantum realm.

The derived complex forms of the model are relative to the origin of the system (O). As we found that, these forms are similar to the relativistic forms, which are in relative to the inertial frame of reference. Then, does the lab observer regard the origin of the system (O) as the inertial frame of reference. If this is the case, then the complex vector is in relative to the inertial frame of reference.

- *The origin of curved space*

In his words, Einstein mentions the necessity of *æther*, "We may say that according to the general theory of relativity space is endowed with physical qualities; in this sense, therefore, there exists an æther. According to the general theory of relativity space without æther is unthinkable" [24]. In addition to that, he regards the gravity not as a force but a manifestation of the local geometry of spacetime.

Within the approximation $X = 0$, one can get flat spacetime form, which leads to a singularity as well. The space of underpinning is curvature space. If the above system is related to the *æther* then, is the gravity related to that deep underpinning curved space? The answer is out of the scope of this project and needs a lot of work.

- *The minimum length*

The present work is not based on an assumption of minimum length (Planck's scale), but it has been shown the existence of a small structure with small length and angular frequency ( $a_1$ and $\omega_2$ ). These small quantities are unobservable, not just because they are small, but due to the problem of partial observation of the system. In other words, there is a problem of system recognition.

**Acknowledgments**

The author is indebted to an anonymous referee for his remarks and report, and the organisers of "The First Hermann Minkowski Meeting on the Foundations of Spacetime Physics" for the given opportunity to present this work.

**Appendix**

Under partial observation conditions ($? = 0$ and $\omega_\beta t = 0$ ) Eq. (4) becomes:

$$\frac{\partial r\,(r,t,0)}{\partial t} = \frac{\partial a_{2m}}{\partial t}\left\{\cos\left(k_{2m}\cdot s - \omega_{1m}t\right) \pm\sqrt{-\sin^2\left(k_{2m}\cdot s - \omega_{1m}t\right)}\right\}$$
$$+ a_{2m}\left\{\omega_{1m}\sin\left(k_{2m}\cdot s - \omega_{1m}t\right)\right. \tag{1A}$$
$$\left.\pm\frac{\omega_{1m}\sin\left(k_{2m}\cdot s - \omega_{1m}t\right)\cos\left(k_{2m}\cdot s - \omega_{1m}t\right)}{i\sin\left(k_{2m}\cdot s - \omega_{1m}t\right)}\right\}.$$

The vector differentiation is

$$\frac{\partial a_{2m}}{\partial t} = a_{2m}\omega\hat{\mathbf{e}}_\vartheta = v\hat{\mathbf{e}}_\vartheta. \tag{2A}$$

Then, Eq. (1A) becomes:

$$i\frac{\partial r\,(r,t,0)}{\partial t} = \left(ive_\vartheta\cdot k_{2m}\right)a_{2om}\left\{\cos\left(k_{2m}\cdot s - \omega_{1m}t\right) \pm i\sin\left(k_{2m}\cdot s - \omega_{1m}t\right)\right\}$$
$$+ a_{2m}\omega_{1m}\left\{i\sin\left(k_{2m}\cdot s - \omega_{1m}t\right) \pm\cos\left(k_{2m}\cdot s - \omega_{1m}t\right)\right\}. \tag{3A}$$

In exponential form, Eq. (3A) becomes:

$$i\frac{\partial \mathbf{r}\,(r,t,0)}{\partial t} = (iv\hat{\mathbf{e}}_\vartheta \cdot \mathbf{k}_{2m})\,a_{2m}\exp i \pm (\mathbf{k}_{2m} \cdot \mathbf{s} - \omega_{1m}t)$$

$$+ \mathbf{B}\omega_{1m}\mathbf{a}_{2m}\exp i \pm (\mathbf{k}_{2m} \cdot \mathbf{s} - \omega_{1m}t)\,. \tag{4A}$$

Regarding that

$$i\,v\,\hat{\mathbf{e}}_\vartheta \cdot \mathbf{k}_{2m} = i\,v\,\hat{\mathbf{e}}_\vartheta \cdot \nabla = -i\,v\,(-\hat{\mathbf{e}}_\vartheta) \cdot \nabla. \tag{5A}$$

Then let

$$\pm i\nabla = \frac{1}{a_{2m}} = \mathbf{k}_{2m}, \quad \text{(operator)}$$

$$\mathbf{B} = \pm 1, \quad \text{and} \quad \mathbf{A} = \mp i\,\hat{\mathbf{e}}_\vartheta, \tag{6A}$$

and with the aid of Eqs. (5A) then Eq. (4A) becomes

$$i\frac{\partial \mathcal{Z}}{\partial t} = (-i\,v\,\mathbf{A} \cdot \nabla + B\omega_{1m})\,\mathcal{Z}. \tag{7A}$$

### References

1. Snyder, H.S., Quantized Space-Time, Phys. Rev. 71 1947, 38–41 (1947).

2. Glinka, L.A., CP violation, massive neutrinos, and its chiral condensate: new results from Snyder noncommutative geometry, Apeiron 17 (4), pp. 223–242 (2010).

3. Meschini, D., Planck-scale physics: facts and beliefs, arXiv:gr-qc/ 0601097v1 23 (2006).

4. Rovelli, C., and Vidotto, F, Covariant Loop Quantum Gravity: An Elementary Introduction to Quantum Gravity and Spinfoam Theory (Cambridge Monographs on Mathematical Physics) Cambridge University Press (13 Nov. 2014)

5. Rovelli, C. and Smolin, L., Discreteness of area and volume in quantum gravity, Nuclear Physics B, **442**, 593–619 (1995).

6. Dirac, P. A. M., Is there an Æther?, Nature **168**, 906–907 (1951).

7. Dirac P. A. M., Quantum mechanics and the æther, Sci. Mon., **78**, 142–146 (1954).

8. Saurya Das and Elias C. Vagenas, Quantum æther and an invariant Planck scale, A Letters Journal Exploring the Frontiers of Physics, 96 (2011) 50005

9. De Broglie, L., "13 remarques sur divers sujets de physique théorique," Ann. Fond. L. de Broglie 1 (2), 116 (1976).

10. De Broglie, L., Quanta de lumière, diffraction et interférences, C.R. Acad. Sei. 177, 548 (1923).

11. De Broglie, L., Les quanta, la théorie cinétique des gaz et le principe de Fermat, C.R.A.S. 177, 630 (1923).

12. De Broglie, L., Reeherehes sttr la théorie des quanta, Doct. thesis, Univ. Paris (1924) [English translation: J.W. Haslett, Am. J. Phys. 40 (1972) 1315].

13. Einstein, A., Äther und Relativitäits-theorie, Springer, Berlin, (1920).

14. Einstein, A., Über den Äther, Sehweiz. Natarforseh. C-esellseh. Verhandl. 85 (1924) 85.

15. Einstein, A., Raum-, Feld- und Ätherproblem in der Physik, 2nd World power Conf. Berlin, Transactions, 19, pp. 1–5; Raum, Äther und Feld in der Physik, Forum Philosophicum, VoL 1., pp. 173–180 (1930).

16. Einstein, A., Relativity and the problem of space, in: Ideas and opinions, Crown, New York, pp. 360–377 (1945).

17. Horodecki, R., De Broglie wave and its dual wave, Phys. Lett. 87A, pp. 95–97 (1981).

18. Horodecki, R., Dual wave equation, Phys. Lett. 91A, 269–271 (1982).

19. Horodecki, R., The extended wave-particle duality, Phys. Lett. 96A, pp. 175–178 (1983).

20. Horodecki, R., Superluminal singular dual wave, Lett. Novo Cimento 38, 509–511 (1983).

21. Sanduk, M. I., Does the Three Wave Hypothesis Imply a Hidden Structure? Apeiron, 14, No.2, pp. 113–125 (2007).

22. Sanduk, M. I., Three Wave Hypothesis, Gear Model and the Rest Mass, arXiv:0904.0790 [physics.gen-ph] (2009).

23. Sanduk, M., A kinematic model for a partially resolved dynamical system in a Euclidean plane, Journal of Mathematical Modelling and Application, 1, No.6, pp. 40–51. ISSN: 2178–2423 (2012).

24. Einstein, A., "Ether and the Theory of relativity" (1920), republished in *Sidelights on Relativity* (The lecture was published by Methuen, London, 1922).

# 19 THE ROLE OF TIME IN REPARAMETRIZATION-INVARIANT SYSTEMS

V. G. GUEORGUIEV

**Abstract**  The relativistic particle Lagrangian is used to justify the importance of Reparametrization-Invariant Systems and in particular the first-order homogeneous Lagrangians in the velocities. The usual gravitational interaction term along with the observational fact of finite propagational speed is used to justify the Minkowski spacetime physical reality. Our justification implies only one time-like coordinate in addition to the spatial coordinates along which particles propagate with a finite speed. By using the freedom of choosing time-like parametrization for a process, it is argued that the corresponding causal structure results in the observed common Arrow of Time and nonnegative masses for the physical particles. The meaning of the time parameter $\lambda$ is further investigated within the framework of Reparametrization-Invariant Systems. Such systems are studied from the point of view of the Lagrangian and extended Hamiltonian formalism. The corresponding extended Hamiltonian $\boldsymbol{H}$ defines the classical phase space-time of the system via the Hamiltonian constraint $\boldsymbol{H} = 0$ and guarantees that the Classical Hamiltonian $H$ corresponds to $p_0$ – the energy of the particle when the parametrization $\lambda = t$ is chosen. A connection has been demonstrated between the positivity of the energy $E > 0$ and the normalizability of the wave function by using the extended Hamiltonian that is relevant for the proper-time parametrization. It is demonstrated that the choice of the extended Hamiltonian $\boldsymbol{H}$ is closely related to the meaning of the process parameter $\lambda$. The two familiar roles that $\lambda$ can take upon – the coordinate time $t$ and the proper–time $\tau$ are illustrated using the simplest one-dimensional reparametrization invariant systems.

C. Duston, M. Holman (Eds), *Spacetime Physics 1907 - 2017. Selected peer-reviewed papers presented at the First Hermann Minkowski Meeting on the Foundations of Spacetime Physics, 15-18 May 2017, Albena, Bulgaria* (Minkowski Institute Press, Montreal 2019). ISBN 978-1-927763-48-3 (softcover), ISBN 978-1-927763-49-0 (ebook).

# 19.1   Introduction

It has taken thousands of years for natural philosophers and thinkers to arrive at and accept Newton's first principle: an object maintains its state, of rest or constant velocity propagation through space, unless a force acts on it. At first sight, such principle seems to be untrue due to our everyday experience which shows that for an object to maintain its constant velocity an external influence is needed. The accumulation of knowledge and technological progress have made it possible for Newton to find the framework and formulate the three main principles that are now the cornerstone of Newtonian Mechanics.

In *Newtonian Mechanics, time* is a parameter that all observers connected via Galilean transformations will find to be the same - as long as they use the same identical clocks to keep a record of their time. The Galilean transformations are reflecting the symmetry under which the lows of the Newtonian Mechanics are form-invariant [1, 2]. The spatial coordinates of the processes studied may have different values for different inertial observers, but these observers can compare their observations and would find an agreement upon utilization of the Galilean transformations. In this sense, the time coordinate is disconnected/disjoint from the configuration space $M$, which is used to label the states of the system/process, however, it is essential for the definition of the velocity vectors in the cotangent space $TM$.

In *Special Relativity (SR), time* becomes related to the observer and the Lorentz transformations intertwine space and time together in a Minkowski space-time [1, 3, 4]. This way the time duration of a process could be measured by different observers to be different even if they use identical laboratory clocks. However, all observers can identify a time duration related to an observer that is at rest with respect to the process coordinate frame (co-moving frame). This is the *proper-time* duration of a process. Then all observers that are connected by Lorentz transformations will arrive at the same value for the proper-time duration of a process. Special Relativity unifies the time coordinate with the spatial coordinates of an observer to a space–time – the configuration space of the coordinates of events. This way, from the point of view of an observer, the space-time is divided into three important subsets: the time-like paths, space-like curves, and light-like paths or equivalently into a past and future cones inside the light-cone defined by the light-like paths connected to the observer, and the space-like exterior of the rest of space-time. Lorentz transformations preserve the local light-cone at any point in the space-time and thus the causal structure of the time-like paths describing a physical process.

General Relativity (GR) goes even further by allowing comparison between observers related by any coordinate transformations, as long as there is an equivalent local observer who's space-time is of Minkowski type. This means that time-records associated with identical clocks that undergo arbitrary physically acceptable motion/process can be compared successfully - that is, the observers will reach a mutually acceptable agreement on what is going on when studying a causal process. In this framework, a larger class of observers, beyond those in Newtonian and Special Relativity frameworks, can connect their laboratory time duration of a process to the proper-time duration measured by an observer in a co-moving frame along the time-like process. The essential ingredient of GR is the invariance of the proper-time interval $d\tau$ and the proper-length interval $dl$; this is achieved by the notion of *parallel transport* that preserves the magnitude of a vector upon its transport to nearby points in the configuration space-time. The symmetry transformations of the space-time associated with this larger class of observers are the largest possible set – the *diffeomorphisms of the space-time*. A theory that has such symmetry is called *covariant theory*. All modern successful theories in physics are build to be explicitly covariant [5, 6].

Considering the above view of describing physical reality, and in particular, that any physically acceptable observer can use their own coordinate time as parametrization for a physical process then it seems reasonable to impose **the principle of re-parametrization invariance** along with the *principle of the covariant formulation* when constructing models of the physical processes [6, 7, 8]. This means that along with the laboratory coordinates that label the events in the local space-time of an observer, who is an arbitrary and therefore can choose the coordinates in any way suitable for the description of a natural phenomenon within the means of the laboratory apparatus. The observer should also be free to choose an arbitrary parametrization of the process as long as it is useful for the process considered. As long as the formulation of the model is covariant then there would be a suitable diffeomorphism transformation between any two physical observers that will allow them to reach agreement on the conclusions drawn from the data. Thus the process is independent of the observer's coordinate frame. The re-parametrization invariance of the process then means that the process is also independent, not only on the coordinate frame of the observer, but it is also independent on the particular choice of process parametrization selected by the observer who is studying the process . Formulating a covariant theory is well known in various sub-fields of physics, but if we embrace *the principle of re-parametrization invariance* then there are two important questions to be addressed:

1. How do we construct such models?

2. What are the meaning and the roles of the arbitrarily chosen time-parameter $\lambda$ for a particular process?

The first question, *"How do we construct re-parametrization invariant models"* has been already discussed, in general terms, by the author in a previous publications [7, 6] along with further relevant discussions of the possible relations to other modern theories and models [6, 8, 9]. The important line of reasoning is that *fiber bundles* provide the mathematical framework for classical mechanics, field theory, and even quantum mechanics when viewed as a classical field theory. Parallel transport, covariant differentiation, and *gauge symmetry* are very *important structures associated with fiber bundles* [7, 4]. When asking: "What structures are important to physics?", we should also ask: "Why one fiber bundle should be more 'physical' than another?", "Why does the 'physical' base manifold seems to be a four-dimensional Minkowski manifold?" [7, 6, 10, 11, 12], and "*How should one construct an action integral for a given fiber bundle?*" [6, 13, 14, 15, 16, 17]. Starting with the tangent or cotangent bundle seems natural because these bundles are related to the notion of classical point-like objects. Since we accrue and test our knowledge via experiments that involve classical apparatus, the physically accessible fields should be generated by matter and should couple to matter as well. Therefore, the *matter Lagrangian should contain the interaction fields, not their derivatives*, with which classical matter interacts [18, 7, 9]. The important point here is that probing and understanding physical reality goes through a classical interface that shapes our thoughts as classical causality chains. Therefore, understanding the essential mathematical constructions in classical mechanics and classical field theory is important, even though quantum mechanics and quantum field theory are regarded as more fundamental than their classical counterparts. Two approaches, the *Hamiltonian* and the *Lagrangian framework*, are very useful in physics [13, 2, 19, 14, 20]. In general, there is a transformation that relates these two approaches. For a reparametrization-invariant theory [20, 21, 22, 23], however, there are *problems in changing from Lagrangian to the Hamiltonian approach* [2, 20, 19, 24, 25].

In this paper, we illustrate these problems and their resolution for simplest one-dimensional reparametrization-invariant systems relevant to the physical reality.

We start the discussion by first reviewing the main points from [7, 6, 8] as pertained to point particles: Section 19.2 has a sub-section 19.2.1 on the Lagrangian for a relativistic particle as an example of a reparametrization-invariant system, followed by a sub-section 19.2.2 where the general properties of homogeneous Lagrangians in the velocities are stated, the section concludes with a list of pros and cons of the first-order homogeneous Lagrangians. The next Section 19.3 is revisiting our argument why a space-time with a maximum speed of propagation through space, when modeled via first-order homogeneous Lagrangian based on a metric tensor, should be locally a Minkowski space-time with common Arrow of Time and non-negative mass for the particles. The meaning of the process parameter $\lambda$ within the extended Hamiltonian framework is discussed in 19.4 using the simplest possible one-dimensional reparametrization invariant systems. Our conclusions and discussions are given in Section 19.5.

## 19.2  Justifying the Reparametrization Invariance

### 19.2.1  Relativistic Particle Lagrangian

From everyday experience, we know that localized particles move with a finite speed in a three-dimensional space. However, in an extended-configuration space (space-time), when the time is added as a coordinate ($x^0 = ct$), massive particles move along a space-time trajectory such that $u \cdot u = 1$. Here, $u^\mu$ are the coordinates of a general 4-velocity vector $v^\mu = dx^\mu/d\lambda$ but with a special choice of the parametrization parameter $\lambda$; that is, $u^\mu = dx^\mu/d\tau$. While $\lambda$ is an arbitrary parametrization, $\tau$ is a special choice of parametrization that is invariant with respect to any coordinate transformations between reasonable physical observers, it is the *proper–time* ($\tau$) mathematically defined via a metric tensor $g_{\mu\nu}$ ($d\tau^2 = g_{\mu\nu}dx^\mu dx^\nu$). In particular when the metric tensor takes the form of the Minkowski metric ($g_{\mu\nu} = \eta_{\mu\nu}$) then one can talk about local Lorentz equivalent observers. In this case, the *action for a massive relativistic particle* has a nice geometrical meaning: the "time distance" along the particle trajectory [4]:

$$S_1 = \int d\lambda L_1(x,v) = \int d\lambda \sqrt{g_{\mu\nu}v^\mu v^\nu},  \tag{19.1}$$

$$\sqrt{g_{\mu\nu}v^\mu v^\nu} \to \sqrt{g_{\mu\nu}u^\mu u^\nu} = 1 \Rightarrow S_1 \to \int d\tau.$$

However, for massless particles, such as photons, the 4-velocity is a null vector ($g_{\mu\nu}v^\mu v^\nu = 0$). Thus, proper time is not well defined and furthermore, one has to use a different Lagrangian to avoid problems due to division by zero when evaluating the final Euler-Lagrange equations. The appropriate 'good' action is then [4]:

$$S_2 = \int L_2(x,v)d\lambda = \int g_{\mu\nu}v^\mu v^\nu d\lambda.  \tag{19.2}$$

For a massive particle, the Euler-Lagrange equations obtained from $S_1$ and $S_2$ are equivalent, we will discuss this equivalence in more details later. In the above discussion, we considered an arbitrary parametrization $\lambda$ and the proper-time parametrization $\tau$ for a massive particle. The physical meaning of the proper-time $\tau$ is usually considered to be the passing of clock time of a co-moving observer. Another important parametrization is the coordinate time $t$ corresponding to the clock time of an arbitrary physical observer that is studying the motion of the massive particle. Contemporary physics models are expected to be invariant with respect to coordinate transformations between physical observers. This is achieved by constructing Lagrangians as a scalar object from various vector and tensor quantities that correspond to the measurements of an arbitrary observer. In mathematical terms, this is a diffeomorphism invariance. Thus, the physics content of a process under study is the same and therefore independent of the coordinate system of an observer. Clearly, diffeomorphism invariance is an important symmetry that reflects the expectation that observing a process should not affect the process itself. Thus, various observers should find a way to understand each-others measurements in a consistent way as long as they pertain to the same process under the study.

In the example above, the process under study is the motion of a massive particle. In this respect, the process corresponds to a one-dimensional manifold which is a curve in a higher dimensional space-time. It seems that the relationships of the points along the curve, in particular, the ordering of the points and their relative measures, should be something about the curve (the process under study). Thus, various observers should be able to find a consistent way to understand the curve and its properties. Therefore, the description of the curve should be independent of the choices an observer can make in order to describe the curve. In particular, the choice of parametrization of the curve should be irrelevant to the understanding of the corresponding process. Thus, a *reparametrization invariant formulation* would be the corresponding symmetry that the description should obey. While a model build on $L_1$ above do obey such symmetry, its quadratic version based on $L_2$ does not seem to obey it even-though the corresponding Euler-Lagrange equations are equivalent. Even more, the Euler-Lagrange equation does obey parametrization-rescaling symmetry that is easily seen when the Lagrangian is a homogenous function of the velocity (see below). A way to resolve this puzzle is to recognize that $L_2$ can be viewed as a reparametrization invariant Lagrangian in a particular fixed gauge [26]:

$$S'_2 = \int g_{\mu\nu} v^\mu v^\nu e^{-1} d\lambda. \tag{19.3}$$

Here $e$ is an auxiliary field that makes the action $S'_2$ reparametrization invariant by choosing $e \to e(d\lambda/d\tilde{\lambda})$ when $\lambda \to \tilde{\lambda}(\lambda)$. Since now $S'_2$ is reparametrization invariant then one can choose a gauge parametrization such that $e = 1$ and thus arriving at $S_2$ but under proper-time parametrization $\lambda = \tau$ where $g_{\alpha\beta} u^\alpha u^\beta = 1$.

The above could be demonstrated on a more complicated Lagrangian as a specific choice of parametrization such that $g_{\alpha\beta}(x) v^\alpha v^\beta$ is constant. Indeed, if one starts with the re-parametrization invariant Lagrangian $L = qA_\alpha v^\alpha - m\sqrt{g_{\alpha\beta}(x)v^\alpha v^\beta}$ and defines **proper time** gauge $\tau$ such that: $d\tau = \sqrt{g_{\alpha\beta}dx^\alpha dx^\beta} \Rightarrow \sqrt{g_{\alpha\beta}u^\alpha u^\beta} = 1$, then one can effectively consider $L = qA_\alpha u^\alpha - (m - \chi)\sqrt{g_{\alpha\beta}u^\alpha u^\beta} - \chi$ as our model Lagrangian. Here $\chi$ is a Lagrange multiplier to enforce $\sqrt{g_{\alpha\beta}u^\alpha u^\beta} = 1$ that breaks the reparametrization invariance explicitly. Then one can write it as $L = qA_\alpha u^\alpha - (m - \chi)\frac{g_{\alpha\beta}u^\alpha u^\beta}{\sqrt{g_{\alpha\beta}u^\alpha u^\beta}} - \chi$ and using $\sqrt{g_{\alpha\beta}u^\alpha u^\beta} = 1$ one arrives at $L = qA_\alpha u^\alpha - (m - \chi)g_{\alpha\beta}u^\alpha u^\beta - \chi$. One can deduce a specific value for $\chi$ ($\chi = m/2$) by requiring that $L = qA_\alpha u^\alpha - m\sqrt{g_{\alpha\beta}(x)u^\alpha u^\beta}$ and $L = qA_\alpha u^\alpha - (m - \chi)g_{\alpha\beta}u^\alpha u^\beta - \chi$ produce the same Euler-Lagrange equations under the constraint $\sqrt{g_{\alpha\beta}u^\alpha u^\beta} = 1$. Then, by dropping the overall constant term, this finally results in the familiar equivalent Lagrangian: $L = qA_\alpha u^\alpha - \frac{m}{2}g_{\alpha\beta}u^\alpha u^\beta$ where $\tau$ has the usual meaning of proper-time parametrization such that $\sqrt{g_{\alpha\beta}u^\alpha u^\beta} = 1$ but it is imposed after deriving all the equation from the Lagrangian under consideration.

The equivalence between $S_1$ and $S_2$ is very robust. Since $L_2$ is a homogeneous function of order 2 with respect to the four-velocity $\vec{v}$, the corresponding Hamiltonian function $(H = v\partial L/\partial v - L)$ is exactly equal to $L$ $(H(x,v) = L(x,v))$. Thus, $L_2$ is conserved, and so is the length of $\vec{v}$ and therefore $L_1$ is conserved as well. Any homogeneous Lagrangian in $\vec{v}$ of order $n \neq 1$ is conserved because $H = (n-1)L$. When $dL/d\tau = 0$, then one can show that the Euler-Lagrange equations for $L$ and $\tilde{L} = f(L)$ are equivalent under certain minor restrictions on $f$. To see this, consider the Euler-Lagrange equation for $L$:

$$\frac{d}{d\lambda}\left(\frac{\partial L}{\partial v^i}\right) - \frac{\partial L}{\partial x^i} = 0,$$

and compare it with the Euler-Lagrange equation for $\tilde{L} = f(L)$ that can be written as:

$$\left(\frac{f''}{f'}\frac{dL}{d\lambda}\right)\left(\frac{\partial L}{\partial v^i}\right) + \frac{d}{d\lambda}\left(\frac{\partial L}{\partial v^i}\right) - \left(\frac{\partial L}{\partial x^i}\right) = 0$$

Clearly, these equations will be equivalent if the Lagrangians $L$ and $\tilde{L} = f(L)$ are constants of motion; that is, $dL/d\lambda = 0$, and $f'$ and $f''$ are well behaved. This is an interesting type of *equivalence that applies to homogeneous Lagrangians* $(L(\beta v) = \beta^n L(v))$. It is different from *the usual equivalence* $L \to \tilde{L} = L + d\Lambda/d\tau$ or the more *general equivalence* discussed in ref. [27, 17, 28, 19]. Any solution of the Euler-Lagrange equation for $\tilde{L} = L^\alpha$ would conserve $L = L_1$ since $\tilde{H} = (\alpha - 1)L^\alpha$ when $\alpha \neq 1$, while for $\alpha = 1$ it can be enforced as a choice of parametrization. All these solutions are solutions of the Euler-Lagrange equation for $L$ as well; thus $L^\alpha \subset L$. In general, conservation of $L_1$ is not guaranteed since $L_1 \to L_1 + d\Lambda/d\lambda$ is also a homogeneous Lagrangian of order one equivalent to $L_1$. This suggests that there may be a choice of $\lambda$, a "gauge fixing", such that $L_1 + d\Lambda/d\lambda$ is conserved even if $L_1$ is not. The above discussion applies to any homogeneous Lagrangian.

## 19.2.2 Homogeneous Lagrangians of First Order

Suppose we don't know anything about classical physics, which is mainly concerned with trajectories of point particles in some space $M$, but we are told we can derive it from a variational principle if we use the right action integral $S = \int L d\tau$. By following the above example we wonder: "should the smallest 'distance' be the guiding principle?" when constructing $L$. If yes, "How should it be defined for other field theories?" It seems that a reparametrization-invariant theory can provide us with a *metric-like structure* [24, 25, 29, 16], and thus a possible *link between field models and geometric models* [30].

In the example of the relativistic particle, the Lagrangian and the trajectory parameterization have a geometrical meaning. In general, however, parameterization of a trajectory is quite arbitrary for any observer. If there is such thing as the smallest time interval that sets a space-time scale, then this would imply a *discrete space–time structure* since there may not be any events in the smallest time interval. The Planck scale is often considered to be such a special scale [31, 32]. Leaving aside hints for quantum space-time from loop quantum gravity and other theories, we ask: "Should there be any preferred trajectory parameterization in a smooth 4-dimensional space-time?" and "Aren't we free to choose the standard of distance (time, using natural units $c = 1$)?" If so, then *we should have a smooth continuous manifold and our theory should not depend on the choice of parameterization.*

If we examine the Euler-Lagrange equations carefully:

$$\frac{d}{d\tau}\left(\frac{\partial L}{\partial v^{\alpha}}\right) = \frac{\partial L}{\partial x^{\alpha}}, \tag{19.4}$$

we notice that any homogeneous Lagrangian of order $n$ ($L(x, \alpha\vec{v}) = \alpha^{n}L(x, \vec{v})$) provides a *reparametrization invariance of the equations* under the rescaling transformations of the parametrization $\tau \to \tau/\alpha, \vec{v} \to \alpha\vec{v}$. Next, note that the action $S$ involves an integration that is a natural structure for orientable manifolds ($M$) with an $n$-form of the volume. Since a trajectory is a one-dimensional object, then what we are looking at is an embedding $\phi : \mathbb{R}^{1} \to M$. This means that we push forward the tangential space $\phi_{*} : T(\mathbb{R}^{1}) = \mathbb{R}^{1} \to T(M)$, and pull back the cotangent space $\phi^{*} : T(\mathbb{R}^{1}) = \mathbb{R}^{1} \leftarrow T^{*}(M)$. Thus a 1-form $\omega$ on $M$ that is in $T^{*}(M)$ ($\omega = A_{\mu}(x)\,dx^{\mu}$) will be pulled back on $\mathbb{R}^{1}$ ($\phi^{*}(\omega)$) and there it should be proportional to the volume form on $\mathbb{R}^{1}$ ($\phi^{*}(\omega) = A_{\mu}(x)\,(dx^{\mu}/d\tau)d\tau \sim d\tau$), allowing us to integrate $\int \phi^{*}(\omega)$:

$$\int \phi^{*}(\omega) = \int L d\tau = \int A_{\mu}(x)\,v^{\mu}d\tau.$$

Therefore, by selecting a 1-form $\omega = A_{\mu}(x)\,dx^{\mu}$ on $M$ and using $L = A_{\mu}(x)\,v^{\mu}$ we are actually solving for the embedding $\phi : \mathbb{R}^{1} \to M$ using a chart on $M$ with coordinates $x : M \to \mathbb{R}^{n}$. The Lagrangian obtained this way is homogeneous of first-order in $v$ with a very simple dynamics. The corresponding Euler-Lagrange equation is $F_{\nu\mu}v^{\mu} = 0$ where $F$ is a 2-form ($F = dA$); in electrodynamics, this is the Faraday's tensor. If we relax the assumption that $L$ is a pulled back 1-form and assume that it is just a *homogeneous Lagrangian of order one*, then we find a reparametrization-invariant theory that may have an interesting dynamics. The above mathematical reasoning can be viewed as *justification for the known classical forces of electromagnetism and gravitation* and perhaps even of *new classical fields beyond electromagnetism and gravitation* [7, 6, 9].

### 19.2.3 Pros and Cons of Homogeneous Lagrangians of First Order

Although most of the features listed below are more or less self-evident, it is important to compile a list of properties of the first-order homogeneous Lagrangians in the velocity $\vec{v}$.

Some of the good properties of a theory with a first-order homogeneous Lagrangian are:

(1) First of all, the action $S = \int L(x, \frac{dx}{d\tau})d\tau$ is a reparametrization invariant.

(2) For any Lagrangian $L(x, v = \frac{dx}{dt})$ one can construct a *reparametrization-invariant Lagrangian by enlarging the configuration space* $\{x\}$ to an extended configuration space - the space-time $\{ct, x\}$ [2, 20]. However, it is an open question whether there is a full equivalence of the corresponding Euler-Lagrange equations.

(3) Parameterization-independent path-integral quantization could be possible since the action $S$ is reparametrization invariant.

(4) The reparametrization invariance may help in dealing with singularities [23].

(5) It is easily generalized to $D$–dimensional extended objects ($p$–branes /$d$–branes) [7, 6].

The list of trouble-making properties in a theory with a first-order homogeneous Lagrangian includes:

(1) There are constraints among the Euler-Lagrange equations [2], since $\det \left( \frac{\partial^2 L}{\partial v^\alpha \partial v^\beta} \right) = 0$.

(2) It follows that the Legendre transformation $(T(M) \leftrightarrow T^*(M))$, which exchanges velocity and momentum coordinates $(x, v) \leftrightarrow (x, p)$, is problematic [19].

(3) There is a problem with the canonical quantization approach since the Hamiltonian function is identically ZERO ($H \equiv 0$).

The pro (2) and the con (3) above are of key importance. The procedure that can be utilized as mentioned in pro (2) above is very simple: $L(x, \frac{dx}{dt}) \rightarrow \dot{t} L(x, \frac{\dot{x}}{\dot{t}})$ where the dotted notation is a derivative with respect to the parametrization $\lambda$, that is, $\dot{t} = \frac{dt}{d\lambda}$ and $\dot{x} = \frac{dx}{d\lambda}$. This means that every Lagrangian based theory can be reformulated in a reparametrization invariant form. This is a different symmetry from the diffeomorphism invariance of the theory, which is still satisfied by construction. However, the parameter $\lambda$ does not have to be the typical physical time parameterization of a process like its own process time - the proper time $\tau$, nor the coordinate time $t$ for the observer that is studying the process. In this sense, $\lambda$ could be truly arbitrary and thus demonstrating the existence of a larger class of theories that do satisfy the principle of re-parametrization invariance as envisioned in the introduction. The problem with these larger class of theories is in the con (3). Which makes the standard treatment quite difficult and unusual due to the presence of constraints among the equations of motion con (1) above. In this paper we are mostly concerned with physical processes that are associated with one-dimensional manifolds and their reparametrization. However, as the pro (5) is pointing, the formulation is relevant to 2-dimensional sub-manifolds, which is the domain of string theory, and extends to high-dimensional sub-manifolds with reparametrization invariant Lagrangians such as the Nambu-Gotto Lagrangians [9]. The reparametrization invariance, which is a diffeomorphism of the submanifold corresponding to a physical process, is a far-reaching idea and it is different from the coordinate diffeomorphisms of the target space. However, it is beyond the scope of this paper to go into string-theory, p-branes, and gravity which represent sub-manifolds with dimension bigger than one.

*Constraints among the equations of motion* are not an insurmountable problem since there are procedures for quantizing such theories [33, 34, 35, 36, 37]. For example, instead of using $H \equiv 0$ one can use some of the constraint equations available, or a conserved quantity, as Hamiltonian for the quantization procedure. Changing coordinates $(x, v) \leftrightarrow (x, p)$ seems to be difficult, but it may be resolved in some special cases by using the assumption that a gauge $\lambda$ has been chosen so that $L \to L + \frac{d\Lambda}{d\tau} = \breve{L} = const$. We would not discuss the above-mentioned quantization troubles since they are outside of the scope of this paper. A new approach that resolves $H \equiv 0$ and naturally leads to a Dirac-like equation is under investigation, for some preliminary details see ref. [6]. Here we focus on understanding the meaning and role of the general parameter $\lambda$ by extending the Hamiltonian framework to an extended phase-space (phase-space-time) with a *covariant extended Poisson Bracket* $[\![,]\!]$, which is consistent with the *Canonical Quantization* process, along with an *extended Hamiltonian* $H$ that defines the *extended phase-space-time* via $H \equiv 0$. By following the Canonical Quantization formalism $(H \to \hat{H})$ the Hilbert space of the quantum system can be defined via the corresponding extended Hamiltonian $\hat{H}$ as the linear space of states $\Psi$ that satisfy $\hat{H}\Psi = 0$. As a byproduct, we can use this formulation to justify the Schrödinger's equation. A similar approach to the Schrödinger's equation has been discussed in Ref. [20]. Furthermore, we do not concern ourselves with the questions about the algebra of observables nor with issues of unboundedness of important physical operators or alternative approaches to the quantization framework that use exponentiated versions of self-adjoint operators instead. Such issues are outside of the scope of the current paper. Finally, before we discuss the extended Hamiltonian framework, we would like to use the structure of the first–order homogenous Lagrangians to mathematically justify few other important features of the physical reality that we often take for granted which are closely related to the Minkowski's work.

# 19.3 One-Time Physics, Causality, Arrow of time, and the Maximum Speed of Propagation

In our everyday life, most of us take time for granted, but there are people who are questioning its *actual existence* or consider *models with more than one time-like coordinate* [38, 39, 40, 41, 42, 43, 44, 45, 46, 47, 48, 49, 50, 51, 52, 1, 10, 53, 54, 11]. Since we are trying to understand the meaning of an arbitrary time-like parameter $\lambda$ within the framework of reparametrization invariant systems, it seems important to think about the possible number of time-parameters. Here, we briefly argue that a one-time-physics, in case of massive point particles, is essential to assure *causality via finite propagational speed* through space of these massive point particles [6]. Then the *common arrow of time*, which is often viewed as related to the *increasing entropy* as commanded by second law of thermodynamics, becomes instead a consequence of the *positivity of the rest mass*.

## 19.3.1 One-Time Physics, Maximum Speed of Propagation, and Space-Time Metric Signature

Why the space-time seems to be one time plus three spatial dimensions have been discussed by using arguments *a la* Wigner [10, 11]. However, these arguments are deducing that the space-time is 1+3 because only this signature is consistent with particles with finite spin. In our opinion, one should turn this argument backwards claiming that one should observe only particles with finite spin because the signature is 1+3. Here we present an argument that only one-time physics is consistent with a finite spatial propagational speed [6]. The local *Lorentz symmetry* implies the existence of a local observer with Minkowski like coordinate frame.

Our main assumptions are: (I) a gravity-like term $\sqrt{g(\vec{v}, \vec{v})}$ is always present in the matter Lagrangian and (II) the corresponding matter Lagrangian is a real-valued function. Therefore, physical processes like propagation of a particle must be related to positive-valued term $g(\vec{v}, \vec{v}) \geq 0$. Here $\vec{v}$ is the rate of change of the space-time coordinates with respect to some arbitrarily chosen parameter $\lambda$ that describes the evolution of the process (propagation of a particle). That is, $v^\alpha = dx^\alpha/d\lambda$. By speed we mean the magnitude of the spatial velocity with respect to a laboratory time $t = x^0$ coordinate $v^\alpha_{space} = dx^\alpha/dx^0$.

The use of a covariant formulation allows one to select a local coordinate system so that the metric is diagonal $(+, +, .., +, -, ...-)$. If we denote the $(+)$

coordinates as time coordinates and the (−) as spatial coordinates, then there are three essential cases:

(1) No time coordinates. Thus $g(\vec{v}, \vec{v}) = -\sum_{\alpha} (v^{\alpha})^2 < 0$, which contradicts $(g(\vec{v}, \vec{v}) \geq 0)$.

(2) Two or more time coordinates.
Thus $g(\vec{v}, \vec{v}) = (v^0)^2 + (v^1)^2 - \sum_{\alpha=2}^{n} (v^{\alpha})^2 \Rightarrow 1 + w^2 \geq \vec{v}^2_{space}$.

(3) Only one time coordinate. Thus $g(\vec{v}, \vec{v}) = (v^0)^2 - \sum_{\alpha=1}^{n} (v^{\alpha})^2 \Rightarrow 1 \geq \vec{v}^2_{space}$.

Clearly, for two or more time coordinates, we don't have bounded from above speed of propagation ($\vec{v}^2_{space} = (dl/dt)^2$). For example, when the coordinate time ($t$) is chosen so that $t = x^0 \Rightarrow v^0 = 1$ then along another time-like coordinate $x^1$ the speed will be $w^2 = (dx^1/dt)^2$ – which could be anything in magnitude. **Only the space-time with only one time** accounts for a finite spatial speed of propagation - therefore, a causal structure. When going from one point of space to another, it takes time and thus there is a *natural causal structure* [55, 54]. The details of the causal structure will depend on the interactions that can take place when two objects are at the same point in the space-time since there is a natural future-and-past cone in such space-time.

## 19.3.2 Causality, the Common Arrow of Time, and the Non-Negativity of the Mass

In the previous section, we deduced that the space–time metric has to reflect that there is only one coordinate time and the rest of the coordinates should be spatial which is a requirement for finite spatial speed of propagation that induces causality. If nature is really reparametrization-invariant, then any observer studying a process can use its own time coordinate $t$ or any other suitable time-parameter $\lambda$, to label an unfolding process. However, when comparing to other observers who study the same process, it will be more advantageous to use a proper–time parametrization $\tau$ which is usually related to an observer who is following/moving along with the process (particle propagation in its co-moving frame). To be able to study a process using any laboratory time-coordinate $t$ and to deduce the process proper–time parametrization for the purpose of comparing to other arbitrary observers **would then imply a reparametrization-invariant system.**

The process has to be related to a massive system because actual observers are also massive and cannot move as fast as light or other massless particles due to the previously deduced $1 + n$ signature of the metric. As long as there is a term $m(v)\sqrt{g(\vec{v}, \vec{v})}$ in the Lagrangian $L$ then the relationship $m(0)d\tau = m(v)\sqrt{g(\vec{v}, \vec{v})}d\lambda$ could be used to define proper–time $\tau$. Here the mass $m(v)$ has to be a homogeneous function of order zero with respect to the velocity *v*. If one considers the obvious choice $\lambda = t$ for any particular laboratory observer then this would imply that time is going forward for the observer as well as for the process - as long as $m(v) \neq 0$ for any physical value of $v$. This would be valid for any two observers that can study each other's motion as well. Therefore, all observers, which can study at least one common process in nature and the other observers studying the same process as well, will find a common arrow of time. Using the freedom of parametrization an observer may decide to use $\lambda = \tau$ for a process and then this gives us the relation of the rest mass to moving mass $m(0) = m(u)\sqrt{g(\vec{u}, \vec{u})}$ as deduced in the theory of Special Relativity for $u^\alpha = dx^\alpha/d\tau$ [4].

All processes and observers should have the same sign of their mass, if not then we can envision a non-interacting pair that has opposite sign of their rest masses $(m'(0) = -m''(0))$ moving in the same way (with $\vec{v}$ the same) with respect to us; we expect that they will have the same proper–time; however, as a pair their proper-time would not be accessible to us since $m'(0) + m''(0) = 0$. Given that all masses have to be of the same sign and the relationship between mass and energy, we conclude that $m(v) > 0$. Thus, the non-negativity of the masses of particles and the positivity of the mass of physical observers. We will see later that the positivity of the energy is related to the positivity of the norm of the corresponding quantum system in its proper-time quantization within the extended Hamiltonian formalism 19.4.2.

In the light of the above discussion, the Common Arrow of Time is a result of the *positivity of the mass of physical observers* and has nothing to do with the entropy of a closed system and the second law of thermodynamics [56, 57, 58, 59, 60, 45, 61].

## 19.4 The Meaning of $\lambda$ and the Role of the Hamiltonian Constraint

In this section, we discuss the meaning of the time (evolution) parameter $\lambda$ as related to the choice of expressing the *Hamiltonian constraint* of a reparametrization-invariant system based on first-order homogeneous Lagrangians in the velocities. Up to our best knowledge, the general *functional form of the first-order homogeneous Lagrangians* in $n$-dimensional space-time is not fully understood yet [6, 24, 17].

### 19.4.1 The Picture from Lagrangian Mechanics' Point of View

For the simplest possible case of *only one space-time coordinate*, we have an explicit unique form for the Lagrangian based on the Euler's equation for

334

homogeneous functions of first-order in the velocity:

$$v\frac{\partial L(q,v)}{\partial v} = L(q,v) \Rightarrow L(q,v) = \phi(q)v \tag{19.5}$$

The action $\mathcal{A}$ will take a very simple form:

$$\mathcal{A} = \int L(q,v)d\lambda = \int \phi(q)vd\lambda = \int \phi(q)dq \tag{19.6}$$

The Euler-Lagrange equations are now:

$$\frac{dp}{d\lambda} = \frac{\partial L}{\partial q}, \ p := \frac{\partial L}{\partial v} = \phi(q) \tag{19.7}$$

The Hamiltonian function is then:

$$H = pv - L \equiv 0 \tag{19.8}$$

At this point, we have two choices for the meaning of the coordinate $q$. It could be a spatial coordinate or a time coordinate. We are especially interested in the case of time-like coordinate, so we will focus on this case and the space-like case will be discussed elsewhere [62]. In what follows, $q$ will be set to be the laboratory time coordinate $t$ and the rate of its change $dt/d\lambda$ will be denoted with $u$ but we will not use $E$ nor $p_0$ for $p$ which in this case carry the correct meaning of $p$:

$$L_1(t,u) = \phi(t)u \Rightarrow u := \frac{dt}{d\lambda}, \ p := \frac{\partial L_1}{\partial u} = \phi(t), \tag{19.9}$$

$$\frac{dp}{d\lambda} = \frac{\partial L_1}{\partial t} \Rightarrow \frac{d\phi(t)}{d\lambda} = u\frac{\partial\phi(t)}{\partial t} \tag{19.10}$$

The corresponding Hamiltonian function is then:

$$H = pu - L \equiv 0 \tag{19.11}$$

but we cannot say anything about the rate with which $u$ is changing. The action $\mathcal{A}$ will take the value $\Delta$ for the overall observed motion:

$$\mathcal{A} = \int L(t,u)d\lambda = \int u\phi(t)d\lambda = \int \phi(t)dt = \Delta \tag{19.12}$$

Since the model is reparametrization invariant, we can define a quantity that different observers can deduce from observations and compare - this is the proper–time parametrization $\tau$:

$$d\tau = \phi(t)dt \tag{19.13}$$

In this parametrization the action $\mathcal{A}$ will take simpler form:

$$\mathcal{A} = \int L(t,u)d\lambda = \int u\phi(t)d\lambda = \int \phi(t)dt = \int d\tau = \Delta\tau \qquad (19.14)$$

and different observers will be able to compare different phases of the process and deduce overall scale factor that will allow agreement of the observational data.

Furthermore, for the equivalent Lagrangians $L_{(n)} = (L_1)^n$ we have explicit time dependence. Thus, the corresponding Hamiltonian functions will not be integrals of the motion. For example, $H_{(2)} = L_1^2 = \phi(t)^2 u^2$. However, the proper-time parametrization will make $L_1 = 1$ or by requiring $\frac{dL_1}{d\lambda} = 0$ we will arrive at:

$$\frac{dL_1}{d\lambda} = 0 \Rightarrow \frac{du}{d\lambda} = -u^2 \frac{d\ln\phi(t)}{dt} \qquad (19.15)$$

which has the general solutions $\lambda_0 d\lambda = \phi(t)dt$ as discussed elsewhere [62].

However, if $\phi(t) = \phi_0$ is a constant then $u = \zeta$ is a constant; therefore, the rate of change of $t$ and $\lambda$ are proportional. This means that we can choose the unit of the process time $\lambda$ to be the same as the coordinate time $t$ which makes $u = 1$. Therefore, the action integral will give us:

$$\mathcal{A} = \int L(t,u)d\lambda = \int \phi(t)dt = \Delta\tau \to \phi_0 \int dt = \phi_0 \Delta t = \phi_0 \zeta \Delta\lambda \qquad (19.16)$$

Alternatively, if we start with $\lambda = t$, then we have $u = 1$, $L = \phi(t)$, and $p$ should be assumed to be $\phi(t)$ since that was the case for all other choices of parameterization. Then we can consider the proper time $\tau$ as a new choice of parametrization to study the system. In the proper-time parametrization $L = 1$ which is explicitly a constant. It seems that for massive particles/systems one can always expect $L$ to be non-zero and thus in the proper-time parametrization to be set to 1. Since $p = p_0 = \phi(t)$ should be related to the energy of the system, then a process can be considered classical with conserved energy if a quantity (energy) can be associated to the process and it is independent of the observational time interval $\Delta t$ via:

$$E = E(\Delta t) := \langle\phi\rangle_{\Delta t} = \frac{1}{\Delta t}\int_0^{\Delta t} \phi(t)dt = \frac{1}{\Delta t}\int_0^{\Delta\tau} d\tau = \frac{\Delta\tau}{\Delta t} \qquad (19.17)$$

However, if one studies natural processes at shorter and shorter time scales then one may encounter systems where the proper-time is poorly defined due to fluctuations of $\phi(t)$ and the above formula is not applicable because of fluctuations at very small time scales. The observation of such scale $\delta$ can signal the onset of quantum phenomenon.

336

## 19.4.2 The Picture from Hamiltonian Mechanics Point of View

Consider now the same system but from the Hamiltonian point of view using the extended Poisson bracket [62]. The main relationships in the Lagrangian formulation based on the Lagrangian $L_1(t, u) = \phi(t)u$ are $u := \frac{dt}{d\lambda}$, $p := \frac{\partial L_1}{\partial u} = \phi(t)$, and $H = pu - L \equiv 0$. If one considers the choice of parametrization $\lambda$ to be the laboratory time coordinate $t$ then we have $u = 1$, $L_1(t, u) = \phi(t)$, and $H = p - L \equiv 0$ which is consistent with the general expression $p := \frac{\partial L_1}{\partial u} = \phi(t)$ that holds for general choice of parameterizations. Thus the general constraint would be to make sure that:

$$p_0 = \phi(t) \tag{19.18}$$

Here we use explicitly the sub-index zero to emphasize that this is to be related to the energy momentum of a system.

**Hamiltonian Constraint for $\lambda$ in Coordinate-Time Role ($\lambda = t$)**

The above expression immediately suggests an extended Hamiltonian in the spirit of $H \to \boldsymbol{H} = H - p_0$ that will have the form:

$$\boldsymbol{H} = \phi(t) - p_0 \tag{19.19}$$

Now an interesting question is: How the phase-space coordinates evolve and what is the meaning of $\lambda$ for such choice of the extended Hamiltonian? To answer this question we look at the evolution equation for the function $t$ and for $p_0$:

$$\frac{dt}{d\lambda} = [\![t, \boldsymbol{H}]\!] \Rightarrow \frac{dt}{d\lambda} = [\![t, (-p_0)]\!] = 1 \tag{19.20}$$

Thus this immediately tells us that the choice of $\lambda$ is actually the laboratory time coordinate. Now we have to confirm the consistency by looking at the evolution of $p_0$:

$$\frac{dp_0}{d\lambda} = [\![p_0, \boldsymbol{H}]\!] \Rightarrow \frac{dp_0}{d\lambda} = [\![p_0, \phi(t)]\!] = \frac{\partial \phi(t)}{\partial t} \tag{19.21}$$

Thus, the choice of $\boldsymbol{H} = \phi(t) - p_0 \equiv 0$ corresponds to $\lambda = t$ indeed.

**Hamiltonian Constraint for $\lambda$ in the Proper-Time Role ($\lambda = \tau$)**

The constraint in the equation (19.18) has many possible realizations. Another possibility is:

$$\boldsymbol{H} = 1 - \frac{p_0}{\phi(t)} \tag{19.22}$$

What is the meaning of $\lambda$ for this form of $\boldsymbol{H}$ ? Again we look at the evolution of $t$ and $p_0$:

$$\frac{dt}{d\lambda} = [\![t, \boldsymbol{H}]\!] \Rightarrow \frac{dt}{d\lambda} = \frac{1}{\phi(t)} [\![t, (-p_0)]\!] = \frac{1}{\phi(t)} \qquad (19.23)$$

Thus, this is the proper time parametrization choice since $d\lambda = \phi(t)dt = d\tau$. Again we check the consistency by looking at $p_0$ :

$$\frac{dp_0}{d\lambda} = [\![p_0, \boldsymbol{H}]\!] \Rightarrow \frac{dp_0}{d\lambda} = -p_0[\![p_0, \frac{1}{\phi(t)}]\!] = -p_0\frac{\partial \left(\phi(t)\right)^{-1}}{\partial t}$$
$$= -p_0 \left(\frac{-1}{\phi(t)^2}\right) \frac{\partial \phi(t)}{\partial t} \qquad (19.24)$$

Since we have to keep $p_0 = \phi(t)$ this finally gives the same expression as in the laboratory coordinate time because $d\lambda = \phi(t)dt = d\tau$:

$$\frac{dp_0}{d\tau} = \frac{1}{\phi(t)} \frac{\partial \phi(t)}{\partial t} \Rightarrow \frac{1}{\phi(t)} \frac{dp_0}{dt} = \frac{1}{\phi(t)} \frac{\partial \phi(t)}{\partial t} \qquad (19.25)$$

## The Quantum Mechanics Picture and the Positivity of the Energy

If we apply the Canonical Quantization formalism to the extended Hamiltonian framework [62] with the $\boldsymbol{H}$ for the time coordinate parametrization $\lambda = t$, we obtain the standard Schrödinger equation:

$$\boldsymbol{H} = \phi(t) - p_0 \rightarrow \hat{\boldsymbol{H}}\psi(t) = 0 \Rightarrow i\hbar\frac{\partial \psi}{\partial t} = \phi(t)\psi \qquad (19.26)$$

where the wave function solutions $\psi(t)$ are given by:

$$\psi(t) = \frac{1}{\mathcal{N}} \exp\left[-\frac{i}{\hbar}\left(\overset{\Delta t < \delta}{\int\limits_{0}} \phi(t)dt + p_0 \overset{\Delta t \gg \delta}{\int\limits_{0}} dt\right)\right] \qquad (19.27)$$

For coordinate-time interval of the process $\Delta t \gg \delta$ such that the energy $p_0 = E$ is conserved, as discussed earlier in the Lagrangian formulation of this system (19.17), this is the familiar plane wave with normalization factor $\mathcal{N} = 1$. However, for fluctuation of $\phi(t)$ at short time scale $\Delta t \lesssim \delta$ that does not show energy conservation for the process then the wave function is related to the integral of $\phi(t)$ and the normalization $\mathcal{N}$ may now depend on the size of $\delta$ and the structure of the relevant Hilbert space. Generally, the *inner product* in the space of solutions that turns it into a Hilbert space could be tricky, but a running average may be useful. However, for the plane waves one can use the standard inner product where $\Delta$ would be a sufficiently long observational window for the process such that energy is conserved and thus slight variations in the window time duration $\Delta$ are producing consistent results:

$$\langle \Psi | \Phi \rangle_\Delta = \frac{1}{\Delta} \int \Psi^* \Phi dt$$

In the case of the proper-time parametrization $\lambda = \tau$ one is facing the question of ordering of the operators that can be resolved by the requirement of Hermiticity (self-adjointness) of the extended Hamiltonian with respect to the usual QM rules:

$$H = 1 - \frac{1}{2} \left( \frac{1}{\phi(t)} p_0 + p_0 \frac{1}{\phi(t)} \right) \tag{19.28}$$

The corresponding Schrödinger like equation now will have an additional term:

$$\psi(t) - \frac{1}{2} \left( \frac{1}{\phi(t)} \hat{p}_0 \psi(t) + \hat{p}_0 \frac{\psi(t)}{\phi(t)} \right) = 0 \tag{19.29}$$

$$\psi(t) - \frac{i\hbar}{2} \frac{1}{\phi(t)} \left( \frac{\partial \psi}{\partial t} + \phi(t) \frac{\partial}{\partial t} \left( \frac{\psi(t)}{\phi(t)} \right) \right) = 0 \tag{19.30}$$

$$i\hbar \frac{\partial \psi}{\partial t} = \left[ \phi(t) + \frac{i\hbar}{2} \left( \frac{\partial \ln \phi(t)}{\partial t} \right) \right] \psi(t) \tag{19.31}$$

Therefore, the amplitude of the original plane wave will be modulated now by an additional factor $\varrho(t)$ satisfying:

$$\frac{\partial \varrho(t)}{\partial t} = \frac{1}{2} \left( \frac{\partial \ln \phi(t)}{\partial t} \right) \varrho(t) \tag{19.32}$$

This factor will not disappear for $\Delta t \gg \delta$ when the energy $p_0 = E$ is conserved. It will have the form $\varrho(t) = \sqrt{\phi(t)}$ and now the wave function will be:

$$\psi(t) = \frac{1}{\mathcal{N}} \sqrt{\phi(t)} \exp \left[ -\frac{i}{\hbar} \left( \int_0^{t<\delta} \phi(t) dt + p_0 \int_0^{t\gg\delta} dt \right) \right] \tag{19.33}$$

This modifies the wave function normalization to $\mathcal{N} = \sqrt{p_0}$ for processes when energy conservation is observed.

The result is very interesting since **the positivity of the norm now requires positivity of the energy** $E = p_0 > 0$ since $\phi(t) \to p_0$. In the rest frame this should correspond to the rest mass of the particle.

$$\|\psi\|^2 = \langle \psi | \psi \rangle_\Delta = \frac{1}{\mathcal{N}^2 \Delta} \int_0^{\Delta \gg \delta} \phi(t) dt \xrightarrow[\Delta \to \infty]{} \frac{p_0}{\mathcal{N}^2} > 0.$$

By looking at the mathematical expression (19.33) one see that the complex conjugated wave function $\psi(t)^*$ should be viewed as the wave function for the time reversal process of the original process.

**The Notion of Time Reversal**

In the discussion above, we have shown that the meaning of the process time parameterization $\lambda$ is intimately related to the choice of the Hamiltonian constraint $\boldsymbol{H}$ as expressed in the laboratory. Changing $\boldsymbol{H}$ to its negative $\boldsymbol{H} \rightarrow -\boldsymbol{H}$ does not change the phase space determined by the Hamiltonian constraint $\boldsymbol{H} = 0$, but changes the choice of parametrization $\lambda$ to $\xi$ that are now time reversal to each other $d\xi = -d\lambda$. We can see this by comparing the evolution equations of the coordinate time $t$:

$$\frac{dt}{d\lambda} = [\![t, \boldsymbol{H}]\!] \rightarrow \frac{dt}{d\xi} = [\![t, (-\boldsymbol{H})]\!] = -\frac{dt}{d\lambda} \qquad (19.34)$$

Thus, if we consider

$$\boldsymbol{H} = p_0 - \phi(t) \qquad (19.35)$$

in our example above we will have:

$$\frac{dt}{d\xi} = [\![t, \boldsymbol{H}]\!] \rightarrow \frac{dt}{d\xi} = [\![t, p_0 - \phi(t)]\!] = -1 \qquad (19.36)$$

from where we deduce that $d\xi = -dt$. If we observed that the energy $E = p_0$ didn't change during the process then this will correspond to a time reversal process. For example, if there are two "identical" clocks one in the laboratory and the other outside and we observe and compare the time from both. Then we can conclude that one clock is running backwards. This way, it will be possible for models based on reparametrization invariance formalism to have *time reversal as a symmetry* along with the common arrow of time due to the positivity of the energy (the rest mass of the observers).

## 19.4.3 The Meaning of $\lambda$ and $\boldsymbol{H}$

Within a larger framework one can consider the phase-space momentum co-ordinates $p_i$ to be the generators of forward motion along the corresponding coordinates $q_i$, while the time and energy coordinate stand out in that $p_0$ will correspond to backward coordinate time transformation [62]. In a similar way, the extended Hamiltonian $\boldsymbol{H}$ defines the evolution of a system's observables $f$ along a process parametrized by $\lambda$. In the observer's coordinate frame $\boldsymbol{H}$ defines the relevant phase space via $\boldsymbol{H} = 0$ along with equations that tell the observer how the process will unfold from one stage (state), determined by a point in the phase–space, to the nearby stage (state) - another point in the phase space. This is different from the Lagrangian formulation where the configuration space $M$ and its co-tangential space $TM$ that contains the coordinates and the velocities have to be "predetermined" in a way that has nothing to do with the Lagrangian $L$. The Lagrangian, however, tell us how the process should be embedded in the tangential space $TM$ by using the Euler-Lagrange equations of motion expressed in a specific laboratory coordinate frame. The phase-space, in this case, is determined by the initial conditions and it is expected to be a sub-manifold of $TM$ upon the evolution using the Euler-Lagrange equations. In the laboratory coordinate frame, the choice $\lambda = t$ is the natural first choice for the process parametrization. However, upon investigation of the system in the Lagrangian formulation one may arrive at the notion of a proper–time $\tau$ that may be a more useful choice of parametrization of a process that should be detached from the choice of a laboratory coordinate frame in the sense that this is the unique laboratory frame where all the special velocities are zero and the time–speed $u$ is 1. That is, in an arbitrary laboratory frame the various momenta are determined from $p_\mu = \partial L/\partial v^\mu$ and evolve according to Euler-Lagrange equations $dp_\mu/d\lambda = \partial L/\partial x^\mu$, but there is the unique co-moving frame where $v^0 = dx^0/d\lambda = dt/d\lambda = 1$ and $v^i = dq^i/d\lambda = 0$. Then, for homogeneous Lagrangians of first-order the phase space should be determined by an additional requirement such as parallel transport that conserves the norm of the vectors $(dl/d\lambda = 0)$.

In the extended Hamiltonian framework, the phase-space is determined from $\boldsymbol{H} = 0$ and the evolution of the coordinates and momenta are governed by the evolution equation via the extended Poisson bracket $df/d\lambda = [\![f, \boldsymbol{H}]\!]$ the specific choice of $\boldsymbol{H}$ then tells us the details about the coordinate frame where the observer is studying the process. The reparametrization-invariance was explicit in the Lagrangian framework due to the use of first-order homogeneous Lagrangians in the velocities. In the extended Hamiltonian formulation this is somehow encoded in the extended Hamiltonian $\boldsymbol{H}$ and the structure of the phase–space determined from $\boldsymbol{H} = 0$. To understand how the extended Hamiltonian should change when we change the choice of parametrization we can consider the extended Poisson bracket evolution equation for two different parameterizations that are related by $\lambda(\xi)$:

$$\frac{df}{d\lambda} = [\![f, \boldsymbol{H}_\lambda]\!] \rightarrow \left(\frac{d\xi}{d\lambda}\right)\frac{df}{d\xi} = [\![f, \boldsymbol{H}_\lambda]\!] \rightarrow \left(\frac{d\xi}{d\lambda}\right)[\![f, \boldsymbol{H}_\xi]\!] = [\![f, \boldsymbol{H}_\lambda]\!]$$

This can be satisfied if $d\xi \boldsymbol{H}_\xi = d\lambda \boldsymbol{H}_\lambda + d\lambda I$ where $I$ is an integral of the process $([\![I, \boldsymbol{H}]\!] = 0)$ such that $I = 0$ over the phase space determined by $\boldsymbol{H} = 0$. To illustrate this we consider $\boldsymbol{H}_t = \phi(t) - p_0$ and $\boldsymbol{H}_\tau = 1 - \frac{p_0}{\phi(t)}$. From the specific expressions of these two hamiltonians, we can see that $\boldsymbol{H}_t = \phi(t)\boldsymbol{H}_\tau$ thus $d\tau\boldsymbol{H}_\tau = dt\boldsymbol{H}_t$ and therefore $d\tau = \phi(t)dt$ which is the usual definition of the relationship of the proper-time to the coordinate time as we have seen in the discussion above.

If we apply this framework to a moving particle with constant velocity $v$ along the spatial coordinate $q$ we have $q = vt + q(0)$ where $t$ is the new parametrization. Therefore, $\boldsymbol{H}_q = p_1 - p_1(0)$ should be related to $\boldsymbol{H}_t = v\boldsymbol{H}_q + vI = v(p_1 - p_1(0)) + vI$. The question now is what is the integral of motion $I$? To find it, we should realize that the configuration space now is two-dimensional $(t, q)$ and the phase–space then will also include $(p_0, p_1)$. Therefore, $I$ can be determined from the evolution of the equation for the coordinate $t$ and from the requirement that $I$ is an integral of motion:

$$\frac{df}{dt} = [\![f, \boldsymbol{H}_t]\!] \rightarrow \frac{dt}{dt} = [\![t, \boldsymbol{H}_t]\!] = [\![t, vI]\!] = 1 \Rightarrow I = -\frac{p_0}{v}$$

This way the corresponding general expression for $\boldsymbol{H}_t$ becomes:

$$\boldsymbol{H}_t = v(p_1 - p_1(0)) - (p_0 - p_0(0)) \rightarrow \vec{v}.\vec{p} - p_0 + E$$

Although this is the physically more relevant system to study due to its possibility to include at least one spatial coordinate and the necessary one-time-coordinate within a Minkowski space-time, it is beyond the scope of the paper which is to analyze the simplest reparametrization-invariant one–dimensional systems for physically relevant consequences and to understand the meaning of the reparametrization parameter $\lambda$.

Based on the examples and the discussions above, one can conclude that the role of the reparametrization parameter $\lambda$ is of a placeholder parameter that is to be clarified after a specific choice of the expression for $\boldsymbol{H}$. However, the usual dynamic time-like meaning of $\lambda$ is often associated with the expression for $\boldsymbol{H}$ that defines the hole phase-space or Hilbert space of the system either via $\boldsymbol{H} = \boldsymbol{0}$ or via the expression for $\hat{\boldsymbol{H}}\psi = \boldsymbol{0}$.

## 19.5   Conclusions and Discussions

Following our main motivation on the importance of reparametrization-invariant models, we have studied the meaning and the roles of the parameter $\lambda$ for the simplest reparametrization-invariant system in one-dimension. From the examples studied we conclude that the proper-time is uniquely identified as the parameterization where the corresponding Lagrangian becomes a constant of motion with its value equal to 1. In the corresponding extended Hamiltonian formulation, the corresponding extended Hamiltonian $\boldsymbol{H}$ is easily identifiable in the coordinate $t$ parametrization. While we have shown and confirmed the corresponding expression for the extended Hamiltonian $\boldsymbol{H}$ in the proper-time parametrization $\tau$, it is not clear how to identify the functional form of $\boldsymbol{H}$ in more general $n$-dimensional systems. However, the connection between the explicit form of the extended Hamiltonian $\boldsymbol{H}$ and the meaning of the parameter $\lambda$ has been illustrated clearly. The quantum mechanical equivalent of such systems has been studied and in the coordinate $t$ parametrization, the usual plane wave has been recovered. An interesting result has emerged from the study of the system using the extended Hamiltonian $\boldsymbol{H}$ for the proper-time parametrization. The wave function now is modulated by the field $\phi(t)$ and in the limit of energy conservation on the macroscopic scale, the energy is forced to be positive in order to have a normalizable wave function. This implies the positivity of the rest mass. Models based on reparametrization-invariance are likely to have *time reversal as a symmetry* along with the *common arrow of time* due to the positivity of the rest mass of the particles. The next step in our study on the reparametrization-invariant models is to follow the above procedures and to apply them to the next simplest case - the 2-dimensional system with one time and one spatial coordinate.

# References

[1] E. Anderson, *The Problem of Time: Quantum Mechanics Versus General Relativity*, Fundamental Theories of Physics, Springer International Publishing, 2017.

[2] H. Goldstein, *Classical Mechanics*, Addison-Wesley series in physics, Addison-Wesley Publishing Company, 1980.

[3] D. Mattingly, Living Reviews in Relativity **8**, 5 (2005).

[4] W. Pauli, *Theory of Relativity*, Pergamon Press, New York, 1958.

[5] S. Gryb, Classical and Quantum Gravity **27**, 215018 (2010).

[6] V. G. Gueorguiev, *Matter, Fields, and Reparametrization-Invariant Systems*, volume IV, pages 168–177, Coral Press, Sofia, Bulgaria, 2003.

[7] V. G. Gueorguiev, *The Relativistic Particle and its d-brane Cousins*, pages 148–158, St. Kliment Ohridski University Press, Sofia, 2003.

[8] V. G. Gueorguiev, *Aspects of Diffeomorphism Invariant Theory of Extended Objects*, pages 234–242, World Scientific, 2004.

[9] V. G. Gueorguiev, ArXiv Mathematical Physics e-prints (2005).

[10] M. N. Borštnik and H. B. Nielsen, Physics Letters B **486**, 314 (2000).

[11] H. van Dam and Y. J. Ng, Physics Letters B **520**, 159 (2001).

[12] C. Saçlıoğlu, Classical and Quantum Gravity **18**, 3287 (2001).

[13] C. W. Kilmister, *Lagrangian Dynamics: an Introduction for Students*, Plenum Press, New York, 1967.

[14] J. F. Cariñena, L. A. Ibort, G. Marmo, and A. Stern, Phys. Rep. **263**, 153 (1995).

[15] R. Leighton, M. Sands, and R. Feynman, *Feynman Lectures On Physics*, Addison-Wesley, Massachusetts, 1965.

[16] E. Gerjuoy, A. R. P. Rau, and L. Spruch, Reviews of Modern Physics **55**, 725 (1983).

[17] M. Rivas, ArXiv Physics e-prints (2001) [physics/0106023].

[18] P. A. M. Dirac, Proceedings of the Royal Society of London Series A **246**, 333 (1958).

[19] X. Gràcia and J. M. Pons, Journal of Physics A Mathematical General **34**, 3047 (2001).

[20] A. Deriglazov, *Classical mechanics: Hamiltonian and Lagrangian formalism; 2nd ed.*, Springer, Cham, 2016.

[21] I. D. Lawrie and R. J. Epp, Phys. Rev. D **53**, 7336 (1996).

[22] S. Gryb and K. Thébault, Classical and Quantum Gravity **33**, 065004 (2016).

[23] H. Kleinert, Physics Letters B **224**, 313 (1989).

[24] H. Rund, *The Hamilton-Jacobi theory in the calculus of variations: its role in mathematics and physics*, New university mathematics series, Van Nostrand, Huntington, N.Y., 1966.

[25] C. Lanczos, *The Variational Principles of Mechanics*, Dover Books On Physics, Dover Publications, 1970.

[26] M. B. Green, J. H. Schwarz, and E. Witten, *Superstring theory. Volume 1 - Introduction*, Cambridge University Press, 1987.

[27] S. Hojman and H. Harleston, Journal of Mathematical Physics **22**, 1414 (1981).

[28] L. M. Baker and D. B. Fairlie, Nuclear Physics B **596**, 348 (2001).

[29] L. D. Landau and E. M. Lifshitz, *The classical theory of fields*, Oxford: Pergamon Press, 4th rev. engl. ed. edition, 1975.

[30] R. Rucker, *Geometry, Relativity and the Fourth Dimension*, Dover Books on Mathematics, Dover Publications, New York., 2012.

[31] J. Magueijo and L. Smolin, Physical Review Letters **88**, 190403 (2002).

[32] R. Ragazzoni, M. Turatto, and W. Gaessler, ApJ **587**, L1 (2003).

[33] P. A. M. Dirac, Proceedings of the Royal Society of London Series A **246**, 326 (1958).

[34] C. Teitelboim, Phys. Rev. D **25**, 3159 (1982).

[35] M. Henneaux and C. Teitelboim, *Quantization of Gauge Systems*, Princeton paperbacks, Princeton, USA: Univ. Pr., 1994.

[36] K. Sundermeyer, editor, *Constrained Dynamics*, volume 169 of *Lecture Notes in Physics*, Berlin Springer Verlag, 1982.

[37] M. de León, J. C. Marrero, D. M. de Diego, and M. Vaquero, Journal of Mathematical Physics **54**, 032902 (2013).

[38] D. N. Page and W. K. Wootters, Phys. Rev. D **27**, 2885 (1983).

[39] J. Géhéniau and I. Prigogine, Foundations of Physics **16**, 437 (1986).

[40] F. H. Gaioli and E. T. Garcia-Alvarez, General Relativity and Gravitation **26**, 1267 (1994).

[41] H.-T. Elze and O. Schipper, Phys. Rev. D **66**, 044020 (2002).

[42] H.-T. Elze, Quantum Mechanics Emerging from "Timeless" Classical Dynamics, in *Trends in General Relativity and Quantum Cosmology*, edited by C. V. Benton, page 79, 2006.

[43] A. Albrecht and A. Iglesias, Phys. Rev. D **77**, 063506 (2008).

[44] A. Bicego, ArXiv e-prints (2010).

[45] S. Viznyuk, ArXiv e-prints (2011).

[46] M. Bojowald, P. A. Höhn, and A. Tsobanjan, Classical and Quantum Gravity **28**, 035006 (2011).

[47] E. Prati, Generalized clocks in timeless canonical formalism, in *Journal of Physics Conference Series*, volume 306 of *Journal of Physics Conference Series*, page 012013, 2011.

[48] C. Wetterich, Foundations of Physics **42**, 1384 (2012).

[49] J. Barbour, M. Lostaglio, and F. Mercati, General Relativity and Gravitation **45**, 911 (2013).

[50] J. A. Vaccaro, Foundations of Physics **45**, 691 (2015).

[51] C. Marletto and V. Vedral, Phys. Rev. D **95**, 043510 (2017).

[52] A. Albrecht and A. Iglesias, Phys. Rev. D **77**, 063506 (2008).

[53] A. Elçi, Journal of Physics A Mathematical General **43**, 285302 (2010).

[54] R. Renner and S. Stupar, *Time in Physics*, Tutorials, Schools, and Workshops in the Mathematical Sciences, Springer International Publishing, 2017.

[55] J. D. Bekenstein, Phys. Rev. D **48**, 3641 (1993).

[56] S. M. Carroll and J. Chen, ArXiv High Energy Physics - Theory e-prints (2004).

[57] S. M. Carroll and J. Chen, General Relativity and Gravitation **37**, 1671 (2005).

[58] V. G. Gurzadyan, S. Sargsyan, and G. Yegorian, On the time arrows, and randomness in cosmological signals, in *European Physical Journal Web of Conferences*, volume 58 of *European Physical Journal Web of Conferences*, page 02005, 2013.

[59] D. N. Page, Nature **304**, 39 (1983).

[60] P. C. W. Davies, Nature **312**, 524 (1984).

[61] I. D. Lawrie, Phys. Rev. D **83**, 043503 (2011).

[62] V. G. Gueorguiev, The Role of Time in Reparametrization-Invariant Systems (math-ph-1903.02483), 2019.

www.ingramcontent.com/pod-product-compliance
Lightning Source LLC
Chambersburg PA
CBHW021918190326
41519CB00009B/823